Chiral Pollutants:
Distribution, Toxicity
and Analysis by
Chromatography
and Capillary
Electrophoresis

Chiral Pollutants: Distribution, Toxicity and Analysis by Chromatography and Capillary Electrophoresis

Imran Ali

National Institute of Hydrology, Roorkee, India

and

Hassan Y. Aboul-Enein

King Faisal Specialist Hospital, Riyadh, Saudi Arabia

John Wiley & Sons, Ltd

This publication is designed to provide accurate and authoritative information in regard to the subject
matter covered. It is sold on the understanding that the Publisher is not engaged in rendering
professional services. If professional advice or other expert assistance is required, the services of a
competent professional should be sought.

Other Wiley Editorial Offices

John Wiley & Sons Inc., 111 River Street, Hoboken, NJ 07030, USA

Jossey-Bass, 989 Market Street, San Francisco, CA 94103-1741, USA

Wiley-VCH Verlag GmbH, Boschstr. 12, D-69469 Weinheim, Germany

John Wiley & Sons Australia Ltd, 33 Park Road, Milton, Queensland 4064, Australia

John Wiley & Sons (Asia) Pte Ltd, 2 Clementi Loop #02-01, Jin Xing Distripark, Singapore 129809

John Wiley & Sons Canada Ltd, 22 Worcester Road, Etobicoke, Ontario, Canada M9W 1L1

Wiley also publishes its books in a variety of electronic formats. Some content that appears
in print may not be available in electronic books.

Library of Congress Cataloging-in-Publication Data

Aboul-Enein, Hassan Y.
  Chiral pollutants : distribution, toxicity, and analysis by
chromatography and capillary electrophoresis / Hassan Y. Aboul-Enein and
Imran Ali.
       p. cm.
Includes bibliographical references and index.
  ISBN 0-470-86780-9 (cloth : alk. paper)
  1. Environmental toxicology. 2. Enantiomers--Toxicology. 3.
Enantiomers--Separation. 4. Chromatographic analysis. 5. Capillary
electrophoresis. I. Ali, Imran. II. Title.
RA1226 .A26 2004
615.9′02--dc22
                                        2003024672

British Library Cataloguing in Publication Data

A catalogue record for this book is available from the British Library

ISBN 0-470-86780-9

Typeset in 11/13pt Times New Roman by Laserwords Private Limited, Chennai, India
Printed and bound in Great Britain by Antony Rowe Ltd, Chippenham, Wiltshire
This book is printed on acid-free paper responsibly manufactured from sustainable forestry
in which at least two trees are planted for each one used for paper production.

Dedication

Dedicated to the memories of my late parents:
Basheer Ahmed and Mehmudan Begum

Imran Ali

To my loving family:
Nagla, Youssef, Faisal and Basil
for their support and encouragement

Hassan Y. Aboul-Enein

Contents

Preface

One of the two enantiomers of a chiral pollutant may be more toxic than the other, and about 25 % of agrochemicals are chiral in nature – including pesticides, which are applied in agricultural and forestry activities in the form of their raccmates. The biological transformation of chiral pollutants can be stereoselective, such that the uptake, metabolism and excretion of the enantiomers may be very different. Therefore, the enantiomeric composition of chiral pollutants may be changed during these processes. The metabolites of chiral compounds are often chiral. Therefore, to predict the exact toxicities of pollutants, determination of the concentrations of both enantiomers is essential, and hence environmental scientists are eagerly seeking techniques for their analysis. Moreover, diverse groups of people – ranging from regulators to the materials industries, clinicians and nutritional experts, agriculturalists and environmentalists – are also now demanding data on the ratio of pollutant enantiomers, rather than their total concentrations.

Various approaches to chiral resolution have been developed for the analysis of pharmaceuticals and drugs but, unfortunately, few reports and monographs are available on the chiral separation of pollutants. Therefore, we have set out to write this book, which deals with the distribution, toxicities and art of analysis of chiral pollutants by gas chromatography and liquid chromatography; that is, by high performance liquid chromatography (HPLC), sub- and supercritical fluid chromatography (SFC), capillary electrochromatography (CEC) and thin layer chromatography (TLC). Additionally, a chapter has been included on the chiral analysis of pollutants by capillary electrophoresis. This book also describes the types, structures and properties of chiral stationary phases, and the applications and future scope of chiral resolution. Moreover, we have attempted to explain the optimization of chiral analysis, which will be helpful in the design of future experiments in this area. Attempts have also been made to explain chiral recognition mechanisms in detail. We very much hope that this book will be a useful source of information for scientists, researchers, academics and graduate students who are working in the field of the chiral analysis of pollutants.

Imran Ali
drimran_ali@yahoo.com
Roorkee, India

Hassan Y. Aboul-Enein
Riyadh, Saudi Arabia

Acknowledgements

I express my deep sense of gratitude and warmest felicitations to my wife Seema Imran, who has helped me and supported me while I have carried out this work. My lovely and sweet thanks are also offered to my dearest son, Al-Arsh Basheer Baichain, who has given me freshness and fragrance continuously during the completion of this difficult task. I would also like to acknowledge my other family members and relatives who have helped me, directly and indirectly, during this period.

I must also pay my sincere and respectful thanks to Professor Vinod K. Gupta, Department of Chemistry, Indian Institute of Technology, Roorkee, India, who helped me to complete this book. Moreover, his moral support, which I received continuously, has been the biggest help and the most memorable event in my life. Finally, the administration of the National Institute of Hydrology, Roorkee, India is also acknowledged for allowing me to write this book.

Imran Ali

I would like to express my thanks to the administration of the King Faisal Specialist Hospital and Research Centre for their support of this work. Special thanks are extended to Ms Jennifer Cossham and the editorial staff of John Wiley & Sons, Ltd, for their assistance in publishing this book. I am particularly grateful to my wife, Nagla El-Mogaddady, for her forbearance and support throughout the preparation of this book, and it is to her that I extend my deepest gratitude.

Hassan Y. Aboul-Enein

About the Book

This book describes the distribution and toxicity of, and analytical techniques for, environmental chiral pollutants. The techniques discussed are gas and liquid chromatography and capillary electrophoresis, the different liquid chromatographic approaches being high performance liquid chromatography (HPLC), sub- and supercritical fluid chromatography (SFC), capillary electrochromatography (CEC) and thin layer chromatography (TLC). This book is divided into ten chapters. The first chapter is an introduction to the principles of chirality. This is followed by Chapters 2–9, which discuss the distribution, toxicity, sample preparation and chiral resolution of environmental pollutants by chromatography and capillary electrophoresis, and include details of the distribution, toxicities, sample preparation and analysis of chiral pollutants. Moreover, optimization of the experimental parameters of chromatographic and capillary electrophoretic techniques is also discussed, and hence this book may be considered as an applied text in the area of chiral pollutant analysis. Discussions have also been included on the types, structures and properties of chiral stationary phases and their applications to the analysis of chiral pollutants. Chiral recognition mechanisms have also been considered, which may be useful in the design of future research in this field of study. The final chapter considers the regulatory framework with regard to chirality around the world, together with perspectives on the large-scale production of pure enantiomers and the impact of chirality on economic growth.

About the Authors

Dr Imran Ali obtained his M.Sc. (1986) and Ph.D. (1990) degrees from the Indian Institute of Technology, Roorkee, India. At present, he is working as a Scientist in the National Institute of Hydrology, Roorkee, India. His research areas of interest are the chiral analysis of biologically and environmentally active chiral compounds, and metal ion speciation using chromatographic and capillary electrophoresis techniques. He also has expertise in water quality and wastewater treatment methodologies. Dr Ali is the author or co-author of more than 70 journal articles, book and encyclopedia chapters, and of a book entitled *Chiral Separations by Liquid Chromatography and Related Technologies*, published by Marcel Dekker, Inc., in New York. Dr Ali has been awarded a 'Khosla Research Award – 1987' by The Indian Institute of Technology, Roorkee, India, for work on the chiral resolution of amino acids. He is a life member of the Indian Science Congress Association.

Professor Hassan Y. Aboul-Encin is a Principal Scientist and Head of the Pharmaceutical Analysis and Drug Development Laboratory at King Faisal Specialist Hospital and Research Centre, Riyadh, Saudi Arabia. He is the author or co-author of over 500 refereed journal articles, 30 book chapters and 270 conference presentations. He is the author of six books, including *Chiral Separation by Liquid Chromatography and Related Technologies* (Marcel Dekker, Inc.) and *The Impact of Stereochemistry on Drug Development and Use* (John Wiley & Sons, Ltd). He is a member of the

Editorial Board of several journals, including *Talanta, Chirality, Biomedical Chromatography, Analytical Letters, Talanta* and *Chromatographia*.

Professor Aboul-Enein is a member of the World Health Organization (WHO) advisory panel on international pharmacopeia and pharmaceutical preparations, and he is a Fellow of the Royal Society of Chemistry (UK). He received his B.Sc. degree (1964) in pharmacy and pharmaceutical chemistry from Cairo University, Cairo, Egypt, and his M.Sc. (1969) and Ph.D. (1971) degrees in pharmaceutical and medicinal chemistry from the University of Mississippi, Oxford, USA. Professor Aboul-Enein's current research interests are in the field of pharmaceutical and biomedical analysis and drug development, with a special emphasis on chiral chromatography, ion-selective electrodes and other separation techniques.

Chapter 1

Introduction

1.1 The Importance of the Environment

The growth, health and persistence of human beings and other organisms all depend on the quality of the environment. Therefore, conservation and protection of the environment are essential in the present-day industrialized and developing world. Unfortunately, pollution of the environment is one of the most pressing problems of our age. The problem of the environment has now reached a level that poses a potential threat not only to health but also to entire populations. The quality of our environment is deteriorating day by day, due to the continuous discharge of undesirable constituents. The main sources of the contamination are the geometric increase in the global population, industrialization, domestic and agricultural activities, atomic explosions, and other environmental and global changes. If they are not properly controlled, these activities and changes can destroy the quality of our environment. Broadly, the environment is divided into three parts: the atmosphere, including the air sphere around the Earth; the lithosphere, which consists of the Earth itself; and the hydrosphere – all the water bodies, including the oceans and the surface and ground water. The hydrosphere and atmosphere components of the environment are directly and readily available for contamination by pollutants. Therefore, the quality of these components of the environment is deteriorating continuously, which is a matter of great concern. Again, the notorious pollutants find their way easily through water bodies and reach various levels in the food chain. The

Chiral Pollutants.   I. Ali and H. Y. Aboul-Enein
© 2004 John Wiley & Sons, Ltd   ISBN: 0-470-86780-9

atmosphere is only being contaminated by some gases and volatile organic pollutants. Furthermore, the ground and surface water in many places are not suitable for drinking purposes due to the presence of aesthetic and toxic pollutants. The air quality of some metropolitan cities is not safe according to minimum health requirements. Many toxic gases and organic pollutants, including lethal pesticides, phenols, plasticizers and so on, have been reported in the air. Briefly, the quality of water, air and edible foodstuffs is not safe in some places, and this poses a threat to human beings and other animals. Therefore, the conservation and improvement of the environment is essential and urgent [1–3]. In view of this, environmental authorities are seeking data and information on pollution levels and improvement measures in order to control the contamination of the environment.

1.2 Environmental Pollutants

Any undesirable and toxic chemical, commodity, organism or other object present in the environment may be considered as an environmental pollutant. The pollutant may be present in the form of a solid, a liquid or a gas. Among these, the presence of toxic pollutants poses a serious threat to human beings and other useful organisms. In general, environmental pollutants may be categorized into chemical and biological classes. The chemical pollutants are organic and inorganic compounds, while the biological contaminants are toxic microbes. Among the various organic environmental pollutants, pesticides, phenols, plasticizers and polynuclear aromatic hydrocarbons are the most toxic, while the toxic inorganic pollutants consist of some metal ions and their complexes. These organic and inorganic pollutants are considered to be the most toxic as they are carcinogenic in nature [4–12]. Most of these environmental pollutants enter into the human body through water and other foodstuffs. Therefore, the monitoring of pollutants in water bodies is essential. Prior to supplying water for drinking, bathing, agriculture and other purposes, it is important to determine the concentrations of these pollutants, if they are present. Of course, analysis of the total concentrations of the toxic pollutants is required and essential. There are many reports available in the literature on the analysis of the organic and inorganic pollutants present in various water bodies and the atmosphere, but the data presented are not reliable. This is due to the fact that some of the organic pollutants are chiral, and the data do not distinguish which mirror images of certain pollutants are present and which are harmful [13–15]. Because of this, knowledge about chirality, chiral pollutants and their methods of analysis is essential for environmental and industrial chemists, and for scientists working in other analytical laboratories. In view of these facts, in the following sections, an attempts has been made to explain the meaning of chirality and chiral pollutants, along with the methods of analysis of chiral pollutants.

1.3 Chirality and its Occurrence

The term 'chirality' is derived from the Greek word *kheir*, meaning 'handedness' [13]. Any object that lacks the three elements of symmetry – that is, a plan of symmetry, a centre of symmetry and an axis of symmetry – exists in more than one form. These forms are nonsuperimposible mirror images of each other and are known as chiral objects (enantiomers) and this property termed optical activity. This optical activity results from the refraction of right and left circulatory polarized light to different extents by chiral molecules (pollutants). The source of the rotation, and hence also the optical rotatory dispersion, is birefringence; that is, the unequal slowing down of right (R) and left (L) circularly polarized light ($n_R \neq n_L$, where n is the refractive index) as the light passes through the sample. On the contrary, 'circular dichroism' is the consequence of the difference in absorption of right and left circularly polarized light (*cpl*) ($\varepsilon_R \neq \varepsilon_L$, where ε is the molar absorption coefficient) [16, 17]. The rotation of polarized light is measured by a polarimeter and the angle of rotation (α) measured is expressed as follows:

$$[\alpha]_D^t = \alpha_{obs}/lc$$

where $[\alpha]_D^t$ denotes the specific rotation determined at $t\,^{\circ}C$ and using the D-line of sodium light, and α_{obs} is the observed angle of rotation, l is the length of the solution medium, in decimetres, and c is the concentration of the chiral pollutant, in g ml^{-1}. The value of $[\alpha]_D^t$ may be positive or negative, depending on the direction of rotation of the angle.

In radiation, the electric field associated with the light waves oscillates in all directions perpendicular to the direction of propagation, but in plane polarized light the electric field only oscillates in one direction, which is achieved by passing ordinary radiation through a Nicol prism. The electric field (E) and the magnetic field vector (H) oscillate at right angles to one another. The circularly polarized light (*cpl*) is described by examining the movement of the electric field only. The linearly polarized light is represented mathematically and graphically as a combination of left and right coherent rotating beams of *cpl*. In an isotropic medium, the two components travel at the same velocity but in opposite directions. In

1894, Curie [18] showed that both the electrical and the magnetic fields have individual mirror planes of symmetry. These planes are eliminated in collinear combination of the two fields. Two enantiomorphous combinations are possible; one in which the electric and magnetic fields are parallel, and another in which the component fields are antiparallel. If both fields oscillate at the same frequency, the two chiral combinations represent right- and left-handed circularly polarized electromagnetic radiation. Drude [19] proposed that the interaction of a chiral molecule with an electromagnetic field gives rise to a helical charge displacement in the molecule, and hence that an oscillatory charge displacement has a right-handed helical form in one optically active isomer and a left-handed form in its antipode. The electric and magnetic dipole moments that develop in a chiral molecule are parallel for the right-handed helical charge displacement and antiparallel for the other antipode, which results in positive or negative circular dichroic light absorption. In 1935, Lowry [20] observed circular dichroism in quartz crystals.

From elementary particles to humans, chirality is found in a wide range of objects [21]. This observation leads to the interpretation that chirality plays a very important and essential role in the existence of the universe, which is still a mystery. There are several examples that indicate the presence of chirality in our environment. In the old kingdoms of Upper and Lower Egypt, many examples of burial chamber mural paintings depict significant events in our modern view of chirality [13]. Additionally, out of the 1168 galaxies listed in the *Carnegie Atlas of Galaxies*, 540 are chiral when coupled with the direction of their recession velocities [22]. The influence of chirality can be observed in plants and animals, where numerous examples of asymmetric structures can be observed. For example, the helical structures of plants and animals make them asymmetric in nature. Briefly, chirality exists almost everywhere in the universe and is associated with the origin of the Earth [23].

1.4 The Chemical Evolution of Chirality

The chemical evolution of chirality began in 1809 with the discovery of Haüy [24] who postulated, from crystal cleavage observations, that a crystal and each constituent space-filling molecule are images of each other in overall shape. In 1819, Mitscherlich [25] postulated a law of isomorphisms which describes the similarity of crystal shape to an equivalent stoichiometry in chemical composition. In 1822, Herschel [26] made the connection

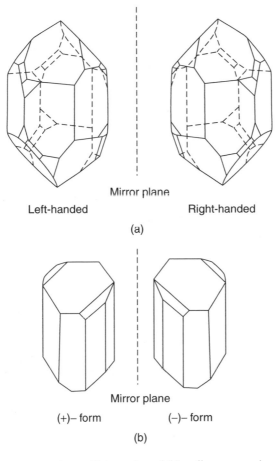

Left-handed   Mirror plane   Right-handed

(a)

(+)– form   Mirror plane   (–)– form

(b)

Figure 1.1 The stereostructures of (a) quartz and (b) sodium ammonium tartarate crystals.

between the morphological handedness of quartz crystals and the sign of the optical activity of the crystals. Herschel observed two types of quartz crystals, distinguished by the right- and left-handed screw sets of hemihedral facets, which reduce the crystal symmetry from hexagonal to trigonal. Furthermore, the author found that all of the crystals of the left-handed morphological set were levorotatory, while those of the right-handed set were dextrorotatory, producing opposition rotation of the plane polarized light; that is, clockwise and anticlockwise rotations (Figure 1.1). At that time, Herschel supposed that the morphological chirality of quartz crystals and their signs of rotation had a common molecular basis, but in 1824 Fresnel [27] determined that the optical activity in these crystals has a circular birefringence $n_L - n_R$, which is positive for dextrorotatory and negative for levorotatory media, where n_L and n_R are the refractive indices for left- and right-handed circularly polarized light, respectively. Furthermore, he proposed that such types of structures (molecules) have both left- or right-handed helical forms or arrangements. In 1844, Mitscherlich [28] reported that sodium ammonium salts of tartaric acid showed different activities with respect to *Penicillium glaucum*, but he could not explain this behaviour scientifically. Later on, in 1848, Pasteur [29] reported the differing destruction rates for *dextro* and *levo* ammonium tartarate by the mould *Penicillium glaucum*, by considering the work of Herschel and Fresnel. Pasteur repeated the crystallization of the sodium ammonium salt of tartaric acid and separated two types of crystals that had enantiomorphous crystal facets. Using Haüy's morphological principle, he inferred that the individual molecules of (+)- and (−)-tartaric acid were stereochemically dissymmetric in nature and related in the form of nonsuperimposible mirror images. In spite of these findings, these observations could not be explained properly at that time, due to the lack of extensive scientific knowledge on this issue. In 1874, Le Bel [30] and van't Hoff [31] independently proposed that the four valences of the carbon atom remain directed towards the vertices of an atom-centred tetrahedron. This finding led to the development of the theory of three-dimensional molecular structures of molecules, by which the phenomenon of chirality and Pasteur's discovery could be explained scientifically. Later on, the different biological properties of the enantiomers were explained as being due to their three-dimensional structures (configurations).

1.5 The Electronic Theory of Chirality

Enantiomers differ only in the spatial arrangement of the atom, and the electro-optical activity of the constituent atoms produces an optical molecular inequivalence by adding to the molecular optical activity of one enantiomer and substracting from that of the other. The electronic interaction discriminates between the binding energies of the corresponding electronic states – stationary or transitional – of the two mirror images of a chiral molecule. Therefore, an electro-energy increment is added to the binding energy of a given electronic state in one enantiomer and subtracted from its counterpart enantiomer, resulting in an electro-energy difference between the two enantiomers. Therefore, L-amino acids in a preferred conformation are more stable than D-amino acids. Similarly, D-aldotriose in a preferred conformation is more stable than its antipode.

1.6 The Importance of Chirality

As discussed above, chirality exists everywhere in the universe and hence it plays a vital part in some aspect of our lives. The consideration of chirality aspect is very important in the environment and some industries, particularly the pharmaceutical, agrochemical, food and beverages, and petrochemical industries. We have already discussed the importance of chirality in environmental pollutants, as the different enantiomers of the pollutants have different toxicities. In the pharmaceutical and drug industries, the existence of chirality became particularly important in the wake of the thalidomide tragedy in the 1960s.

Thalidomide was put on the market in the late 1950s as a sedative, in its racemic form. Even when applied in the therapeutic and harmless (+)-form, the *in vivo* interconversion into the harmful (−)-isomer was shown to be responsible for the disastrous malformations of embryos when thalidomide was administered to women during pregnancy [32–34]. In addition to creating a general awareness, the thalidomide tragedy resulted in stricter controls and reconsideration of the approval guidelines for newly developed drugs. To protect patients from unwanted and harmful enantiomers [35] and side effects, the possibility of different actions of the individual enantiomers with regard to pharmacology and toxicology had to be taken into account. In spite of the fact that the optical isomers of a racemic drug can exhibit different pharmacological activities in living systems [35–41], the bioactive synthetic compounds, most of which are chiral drugs, are administered as racemates [42].

A similar situation also pertains in the agrochemical industry, as many pesticides and other agrochemicals are chiral in nature. In 1981, Spencer [43] reported that out of 550 pesticides, 98 % were synthetic in nature with 17 % chiral molecules and less than 8 % have been marketed as single isomers. Lewis *et al.* [44] also reported that 25 % of pesticides are chiral in nature. Recently, Vetter [45] reviewed the enantioselective fate of chiral chlorinated hydrocarbons and their metabolites in environmental samples. Not all of the aspects of chirality in the agrochemical industry have been fully explored yet, but investigations are under way. Therefore, knowledge about chirality in the agricultural industry is also very important.

Chirality is also important in the food and beverage industries, as many food products contain several chiral substances. Control of the fermentation process and of storage affects when a single isomer is converted into a racemic mixture as time proceeds. The *S*-enantiomer of asparagine is bitter, while the *R*-antipode is sweet. Amino acids, which are essential for animal growth, are all chiral except for glycine. The chemical products used to produce flavours and fragrances are highly dependent on enantioseparation for their properties. The terpenes carvone and limonene are other chiral molecules that are used in the food and beverage industries.

Many hydrocarbons are by-products of the petroleum industry and some of them are chiral in nature. The most important chiral hydrocarbons are halogenated, polyalkane and so on. These hydrocarbons are also used as precursors for many synthetic formulations and products, in which chiral products are produced during the synthesis processes. The different oilfields have been produced from various materials at different times, and hence different ratios of stereoisomers can be found in samples from adjacent fields. Briefly, chirality is of considerable importance in the petrochemical industry.

1.7 Nomenclature for Chiral Pollutants

The nomenclature for the enantiomers can be explained as follows. Initially, the optical isomers were distinguished using (+) and (−) signs or *d* (*dextro*) and *l* (*levo*), indicating the direction in which the enantiomers rotate the plane of the polarized light. (+) or *d* stands for a rotation to the right (clockwise), whereas (−) or *l* indicates a rotation to the left (anticlockwise). The main drawback of such an assignment is that one cannot derive the number of chiral centres from it. This is possible when applying the *R/S* notation, which describes the absolute configuration (the spatial arrangement of the substituents) around the asymmetric carbon atom of the pollutant (the molecule). This assignment is based on the Cahn–Ingold–Prelog (CIP) convention [46]. It has almost replaced the older D/L notation, which correlates the configuration of a molecule with the configuration of D/L-glyceraldehyde according to the Fischer convention. Today, the latter nomenclature is predominantly restricted to amino acids and carbohydrates [47]. The assignment of *R* or *S* according to CIP follows the sequence rule; that is, the order of priority of the substituents on the centre of chirality. It can be determined on the basis of the decrease in the atomic number of the atoms directly bonded to the centre of chirality. In the case in which two or more of these atoms are identical, the next bonded atoms have to be considered, and then eventually the third bonded atoms and so on. In so doing, the branch containing the atoms with the highest atomic numbers attains the highest priority. If double or triple bonds connect the atoms, their weight is higher than that of two or three singly bonded atoms [47]. When isotopes of atoms are involved, the order of priority can be determined by putting in order the decline in their mass numbers. It is very interesting to observe that in closely related structures the nomenclature of the absolute configuration may change, whereas the spatial arrangement of the substituents is maintained. Other consequences of chirality are concerned with the metabolic processes. Several transformations, such as prochiral to chiral, chiral to chiral, chiral to diastereoisomer, chiral to nonchiral and chiral inversion, can occur [35, 48].

In general, the phenomenon of chirality exists in organic pollutants (at the molecular level). However, it is also found in some inorganic pollutants. In some pollutants, the carbon atoms remain attached to four different atoms or groups. This arrangement makes the whole pollutant (molecule) asymmetric in structure. This type of pollutant differs in three-dimensional configurations and exists into two forms, which are mirror images of each other. No matter what symmetry operation is applied to this sort of pollutant, one will never be able to superimpose the two mirror images upon each other. These mirror images are called optical isomers (since they have the capacity to rotate the plane polarized light), or stereoisomers, enantiomers, enantiomorphs, antipodes or chiral molecules. The phenomenon of the existence of these different enantiomers is called stereoisomerism or chirality. A 50 : 50 ratio of the enantiomers is called a racemic mixture, and does not rotate plane polarized light. The absence of rotation of plane polarized light is due to the equal and opposite rotation of the two enantiomers (50 : 50) and hence this phenomenon is called external compensation. Some enantiomers contain a plane of symmetry and hence are unable to rotate plane polarized light. This type of enantiomer are called the *meso-* form. The optical inactivity of the *meso-* form is due to the opposite signs of the rotation of its two halves, and hence the phenomenon of optical inactivity is called internal compensation. In addition to the central chirality, axial chirality can occur in allenes and cumulenes. In the former class, the substituents do not necessarily have to be different, since the second double bond causes the loss of the C_3 rotational symmetry element. In the latter class, only the members with an odd number of cumulated carbon atoms are potentially chiral, whereas an even number of carbon atoms results in *E*-/*Z*-isomerism (geometric isomerism) [27]. Another type of axial chirality is represented by atropisomers, which possess conformational chirality. As long as the *ortho*-substituents in tetra-substituted biaryls are large enough, the rotation around a C–C single bond will be hindered and will prevent the two forms from interconverting. Finally, there exists a planar chirality,

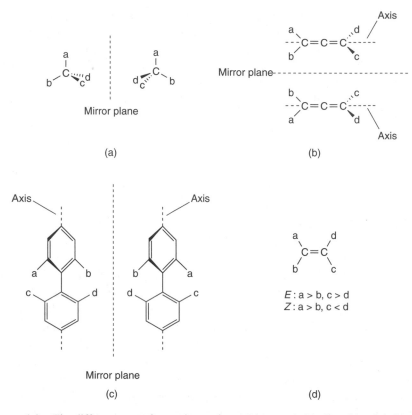

Figure 1.2 The different type of stereoisomerisms: (a) central chirality, (b) axial chirality, (c) atropisomerism and (d) E-/Z-isomerism (geometric).

which arises from the arrangement of atoms or groups of atoms relative to a stereogenic plane. However, this form of chirality is rather rare [32]. Helicity is a special form of chirality and often occurs in macromolecules such as biopolymers, proteins and polysaccharides [47]. A helix is always chiral due to its right-handed (clockwise) or left-handed (anticlockwise) arrangement. In the case in which a stereoisomer has more than one stereogenic centre, the number n of theoretically possible enantiomers can be derived from the formula 2^n. The phenomenon of stereoisomerism in different types of molecules (pollutants) is shown in Figure 1.2.

1.8 Chirality in Environmental Pollutants

It has recently been observed that one of the two enantiomers of a chiral pollutant/xenobiotic may be more toxic then the other [49]. This is important information for the environmental chemist when performing an environmental analysis. The biological transformation of chiral pollutants can be stereoselective, and so the uptake, metabolism and excretion of enantiomers may thus be very different [49, 50]. Therefore, the enantiomeric composition of chiral pollutants may be changed by these processes. Metabolites of chiral compounds are often chiral. Thus, to obtain information on the toxicity and biotransformation of chiral pollutants, it is necessary to explain chirality in environmental pollutants.

As discussed earlier, the number of the enantiomers of a pollutant depends on the number of chiral centres present in the pollutant. A simple example of this type of chirality is presented in Figure 1.3(a), which shows the enantiomers of [1,1,1-trichloro-2-(*o*-chlorophenyl)-2-(*p*-chlorophenyl)] (DDT) with one stereogenic centre. There are several pollutants that contain two stereogenic centres and exist as four stereoisomers. The four stereoisomers make up the two pairs of enantiomers. It is important to mention here that the stereoisomers, which are not mirror images of each other, are called diastereoisomers and, unlike enantiomers, they have different physical and chemical properties. Examples of these types of pollutant include the 1,3,4,6,7-hexahydro-4,6,6,7,8,8-hexamethylcyclopenta[g]-2-benzopyran-1-one (HHCB) metabolite of galaxolide, bis(2,3-dichloro-1-propyl) ether (*meso*- form) and so on (Figure 1.3(b)).

Some pollutants do not contain a chiral centre but are still chiral due to their overall chiral structures. The best examples of this type of chirality are found in biphenyls and cyclic pollutants. Biphenyls that contain four large groups in *ortho*- positions cannot freely rotate about the central single bond because of steric hindrance. In such pollutants, the two ring systems are oriented in perpendicular planes, or in any plane between angles 0 and 90°. The example of chirality in polychlorinated biphenyls (PCBs) is shown in Figure 1.3(c). It is not necessary that all four groups are responsible for the existence of chirality. Three or even two large groups, if placed properly, can also hinder free rotation, as required for the existence of chirality. Basically, in PCB the existence of chirality is controlled by the energy of free rotation of the central bond. The groups led by Schurig [51], König [52, 53] and Harju [54] have carried out detailed studies on the free rotation of PCBs. Interested readers should consult these publications. The chirality in cyclic pollutants can be explained easily by citing the example of hexachlorocyclohexane (HCH) pesticide. It is very interesting to note that out of eight isomers of HCH, only α-HCH is chiral. The other isomers of HCH are achiral due to the presence of some symmetry elements (a centre

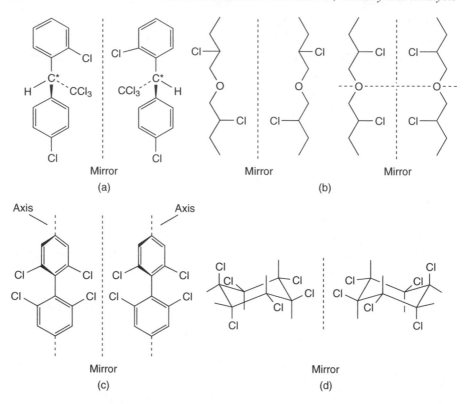

Figure 1.3 The enantiomers of some chiral pollutants: (a) DDT, (b) bis(2,3-dichloro-1-propyl)ether, (c) PCB and (d) α-HCH.

of symmetry, an axis of symmetry or a plane of symmetry). The structure of α-HCH is shown again in Figure 1.3(d).

1.9 Chirality and its Consequences in the Environment

Many xenobiotics and pollutants are chiral in nature and the two enantiomers of these pollutants may have different toxicities [13]. Additionally, the degradation of some chiral pollutants is stereospecific in the environment, and the degradation of some achiral pollutants may result in chiral toxic metabolites. Moreover, it has also been reported that enantiomers may react at different rates with achiral molecules in the presence of a chiral catalyst [13]. It is also obvious that most of the identities and the structures in nature are chiral and, therefore, that there is a greater chance that the environmental pollutants will react at different rates. Therefore, to predict the exact toxicities of pollutants, determination of the concentrations of both of the enantiomers is not just required but essential. In 1991, Kallenborn *et al.* [55] reported the enantioselective metabolism of α-hexachlorocyclohexane in the organs of eider duck, while in the same year Faller *et al.* [56] reported the degradation of α-hexachlorocyclohexane enantioselectively by marine bacteria. Therefore, environmental chemists are also looking for the optimum technique with which to determine the chiral ratio of xenobiotics in the environment. Furthermore, diverse groups of people, ranging from regulators to the materials industries, clinicians and nutritional experts, agriculturalists and environmentalists, are now demanding data on the ratio of the enantiomers, rather than the total concentrations of the racemic pollutants.

1.10 The Enantiomeric Ratio and Fractions of Chiral Pollutants

About 25 % of agrochemicals – including pesticides – are chiral in nature and, therefore, many of these chiral agrochemicals are applied in agricultural and forestry activities in the form of their racemates. The task of the environmental chemist involves the study of the conversion of enantiomers by biological processes and of their compositions. Several terms have been used to describe the extent of deviations from the racemic compositions. The most commonly used terms are the enantiomeric ratio (ER) and the enantiomeric fraction (EF) [57]. ERs are defined by the ratio of the enantiomers, which is directly obtained by integration of the analysis methods. If the directions of the rotation of plane polarized light by the enantiomers are known, the ER is formed by the quotient of the dextro- and levorotatory enantiomers (ER_+). In the case of a lack of standards for the enantiomers and no information on the directions of the plane polarized light, the ER is defined as the quotient of the first and the second eluted enantiomer ($ER_{1/2}$) in any chromatographic method, and is defined as follows:

$$ER = C_+/C_- \quad \text{or} \quad C_1/C_2 \tag{1.1}$$

where C is the concentration of the levo- $(-)$, dextro- $(+)$, first (1) or second (2) eluting enantiomer in the sample. A value for the absence of the second enantiomer ($C_2 = 0$) is mathematically not defined and has to be presented as the limit towards infinity. Enantiomeric ratios extend from infinity (only C_+ or C_1) to the limit $\to 0$ (only C_- or C_2), with ER equal to 1 for the racemate. Furthermore, ERs are not based on a numerical but on a logarithmic scale, which may cause some problems; such as the fact that while the ER of a tenfold amount of the second eluted enantiomer is 0.1, the ER of a tenfold amount of the first enantiomer is 10. Hence the mean value will be $10^{(-1-1)/2} = 10^0 = 1.0$ (rather than the arithmetic mean 5.05). The reciprocal values of ER = 0.5 and 0.4 (ΔER = 1.0) are 2.0 and 2.5 (ΔER = 0.5), and an ER of 2.0 and 2.1 (ΔER = 1.0) corresponds to reciprocal values of 0.5 and 0.48 (ΔER = 0.02) [45]. Therefore, calculation of mean values of ERs in repetitive injections of standards or from different samples should be carried out after transfer to the log scale. Basically, the values of ERs should be greater than one, with an accuracy of ± 0.1. For ERs less than one, it is often necessary to present two values after the decimal point.

Sometimes, the enantiomeric purity is expressed by the enantioexcess (ee), which indicates the excess of one enantiomer that has a higher concentration (C_H) over another that has a lower concentration (C_L), as follows:

$$ee = C_H - C_L/C_H + C_L \qquad (1.2)$$

The value of ee varies from zero to one for the racemic and optically active pure enantiomers. In general, ee is expressed in terms of a percentage, as follows:

$$\% \, ee = 100(C_H - C_L/C_H + C_L) \qquad (1.3)$$

The % ee is compatible with the law of mixing (mixing of samples of different ee) and % ee is equivalent to the optical purity:

$$\text{optical purity} = 100(\alpha_{samp}/\alpha_{enan}) \qquad (1.4)$$

where α_{enan} is the specific rotation of the pure enantiomer and α_{samp} is the specific rotation of the sample [58]. The chromatographic purity or enantiopurity ee enantiofraction is described by the following equation:

$$\text{enantiopurity} = C_1(C_1 + C_2) \qquad (1.5)$$

which has been used to express enantiomer purity in the pharmaceutical, agrochemical, environmental and other analytical sciences [59–61].

The enantiomeric ratio data are transferred into enantiomer fractions (EFs) as a standard descriptor [62] and the EF can be calculated from the ER using the following equation:

$$EF = ER/(ER + 1) \qquad (1.6)$$

This descriptor provides a more meaningful representation of the graphical data than the ER, and is more easily employed in mathematical expressions [62, 63].

1.11 Methods for the Separation of Chiral Pollutants

The splitting of a chiral pollutant into its enantiomers is called separation. Various methodologies have been used for the separation of the enantiomers of drugs and pharmaceuticals. The basic principles of the enantiomeric separation of chiral pharmaceuticals and environmental pollutants are similar and, therefore, the approach used for the enantiomeric separation of pharmaceuticals may be used for the chiral separation of environmental pollutants. The different approaches applied for chiral separation of pharmaceuticals include: the classical approach, using enzymatic degradation of one of the enantiomers; preferential crystallization; and modern technologies, including spectroscopic, electrophoretic and chromatographic methods.

In the enzymatic method, the destruction of one form affects separation by biochemical processes. When certain micro-organisms, such as yeast, moulds and bacteria, are allowed to grow in the solution of racemic mixtures, they assimilated one form selectively, leaving the other one behind in solution. For example, if ordinary mould, *Penicillium glaucum*, is added to a racemic solution of ammonium tartarate, the solution becomes levorotatory due to the destruction of the *dextro-* form [29]. The principle of crystallization is based on the formation of diastereomeric salts by the two enantiomers with the optically pure compound, and these diastereoisomeric salts can easily be separated [21, 64]. In this process, the optically active resolving agent must be of high optical purity. In most cases, after the separation of the desired enantiomers from the diastereoisomeric salts, the resolving agent is recovered and made available for reuse [64–66]. Additionally, mechanical methods of separation (by needle etc.) have also been utilized for separation of the crystals of some racemic compounds, such as sodium ammonium tartarate and quartz, as the crystals of these compounds are mirror images of each other. These classical methods have not been able to achieve the status of routine laboratory practice due to certain drawbacks. The most important drawbacks associated with these methods are the degradation of one enantiomer in the enzymatic method, while the applications of the crystallization method are very limited. Nowadays, chromatographic, electrophoretic, spectroscopic, biosensing and membrane methods are the most common techniques applied in this area [37–40, 67, 68], the chromatographic and electrophoretic methods being very sensitive, reproducible and reliable. Moreover, these methods can easily be used to determine the enantiomeric ratio of chiral pollutants in different matrices. Briefly, the chromatographic and electrophoretic methods for chiral separation are ideal and practical and, therefore, these methods will be discussed herein.

1.11.1 Chromatographic Methods

Nowadays, chromatographic and electrophoretic methods are the most popular techniques applied in this field of work. In chromatographic methods, two approaches are used, the indirect and the direct. The indirect chromatographic separation of racemic mixtures can be achieved by derivatization of the racemic pollutant with a chiral derivatizing agent (CDA), resulting in the formation of a diastereoisomeric complex/salt. Diastereoisomers that have differing physical and chemical properties can be separated from each other by an achiral chromatographic method. A precondition for a successful derivatization is the presence of suitable functional groups in the pollutant. Additionally, to increase the physicochemical differentiation, the derivatization should occur close to the chiral atom. Although the indirect chromatographic approach has the advantage of predetermining the elution order, which can be important for the determination of optical purities, there are some limitations to this technique. The derivatization procedure is tedious and time-consuming, due to the different reaction rates of the individual enantiomers, and a suitable chiral derivatizing agent in a pure form is sometimes poorly available. Moreover, this approach cannot be used easily with environmental samples.

On the other hand, the direct chromatographic approach involves the use of a chiral selector either in the mobile phase, where it is called a chiral mobile phase additive (CMPA), or in the stationary phase, called a chiral stationary phase (CSP). In the later case, the chiral selector is chemically bonded or coated, or allowed to be absorbed on some suitable solid support. Of course, the use of chiral selectors in the form of CMPAs still exists, but there are few publications in the literature on this approach. This is due to the high running cost of the experiment, as a greater amount of chiral selector is required for the preparation of the mobile phase. Besides, there is very little chance of recovery of the chiral selectors and hence a large amount of costly chiral selector is wasted. Contrary to this, CSPs have achieved a great reputation in the chiral separation of enantiomers by chromatography and, today, they are the tools of choice in almost all analytical, biochemical, pharmaceutical and pharmacological institutions and industries. The most important and useful CSPs are available in the form of open and tubular columns. However, some chiral capillaries and thin layer plates are also available, for use in capillary electrophoresis and thin layer chromatography. Chiral columns and capillaries are packed with several chiral selectors.

The chromatographic methods involve the use of gas or liquid separately, as the mobile phases. Therefore, the former kind of chromatography is called gas chromatography (GC), while the later is termed liquid chromatography (LC). Due to some problems associated with gas chromatography, it cannot be accepted as the method of choice for chiral separation of racemic compounds. The major disadvantage of GC is its requirement of the conversion of the racemic compound into volatile species, which is carried out by a derivatization process. Therefore, LC is the best remaining technology for the chiral separation of a wide variety of racemates. The main advantage of LC is its ability to determine the enantiomers in environmental samples directly. Over the course of time, various types of liquid chromatographic approaches have been developed and used in this field of work, the most important methods being high performance liquid chromatography (HPLC), sub- and supercritical fluid chromatography (SFC), capillary electrochromatography (CEC) and thin layer chromatography (TLC). The different modes of chromatography used for the chiral separation of a variety of racemic drugs, pharmaceuticals and pollutants are shown in Figure 1.4.

Among the various liquid chromatographic techniques mentioned above, HPLC remains as the best modality, due to its several advantages in comparison to the other options. High speed, sensitivity and reproducible results make HPLC the method of choice in almost all laboratories. About 90 % of the chiral separation of pharmaceuticals has been carried out using the HPLC mode of chromatography. Due to the wide range of applications of HPLC in chiral separation, several chiral selectors are available in the form of HPLC columns. A variety of mobile phases, including normal, reversed, polar organic and polar ionic modes, are used in HPLC. The composition of the mobile phases may be modified by the addition of various aqueous and nonaqueous solvents. The optimization of the chiral separation is carried out using a number of parameters.

The use of a supercritical fluid (SFC) as the mobile phase for chromatographic separation was first reported more than 30 years ago, but most of the growth in SFCs has occurred recently. A supercritical fluid exists when both the temperature and pressure of the system exceed the critical values; that is, a critical temperature T_c and a critical pressure P_c. Critical fluids have physical properties that lie between those of a liquid and a gas.

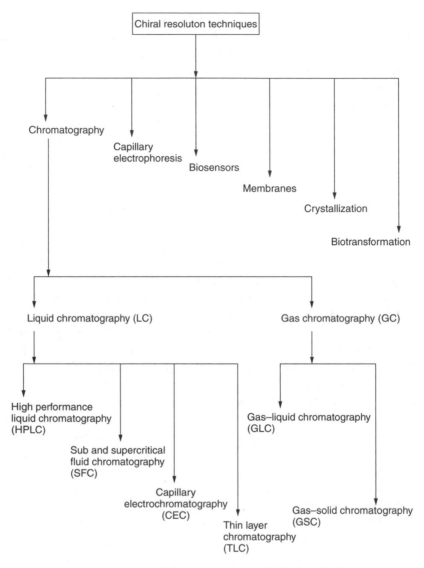

Figure 1.4   The different techniques of chiral resolution.

Like a gas, a supercritical fluid is highly compressible, and the properties of the fluid – including the density and the viscosity – can be maintained by varying the pressure and temperature conditions. In chromatographic systems, the solute diffusion coefficients are often of a higher order of magnitude in supercritical fluids than in traditional liquids. On the other hand, the viscosities are lower than those of liquids [69]. At temperatures below T_c and pressures above P_c, the fluid becomes a liquid. On the other hand, at temperatures above T_c and pressures below P_c, the fluid behaves as a gas. Therefore, a supercritical fluid can be used as part of a liquid–gas mixture [70]. The commonly used supercritical fluids (SFs) are carbon dioxide, nitrous oxide and trifluoromethane [69–71]. Its compatibility with most detectors, low critical temperature and pressure, low toxicity and environmental burden, and low cost make carbon dioxide the supercritical fluid of choice. The main drawback of supercritical carbon dioxide as a mobile phase is its inability to elute more polar compounds. This can be improved by the addition of organic modifiers to the relatively apolar carbon dioxide. Chiral sub-FC and SFC have been carried out in packed and open tubular columns and capillaries [72]. The first report on chiral separation by SFC was published in 1985, by Mourier *et al.* [73]. Since then, several papers and reviews have appeared on the subject [74–79].

Basically, capillary electrochromatography (CEC) is a hybrid technique that works on the basic principles of capillary electrophoresis and chromatography [80]. This mode of chromatography is used either on packed or tubular capillaries/columns. Packed column ECE was first introduced by Pretorius *et al.* [81] in 1974, while open tubular CEC was presented by Tsuda *et al.* [82] in 1983. In 1984, Terabe *et al.* [83] introduced another modification in liquid chromatography – micellar electrokinetic capillary chromatography (MECC). Of course, this mode too depends on the working principles of capillary electrophoresis and chromatography, but it also involves the formation of micelles. CEC and MECC have been used recently in the chiral separation of racemic compounds and hence some publications have appeared on this issue [84–89]. Their high speed, sensitivity, lower limit of detection and reproducible results make CEC and MECC the methods of choice in chiral separation. However, these methods are not yet in common use, as the techniques are not fully developed and research is still under way.

The development of thin layer chromatography (TLC) has a very long history, but its use in chiral separation goes back about 25 years. Most TLC enantioseparations of pharmaceuticals and other compounds have been carried out in the indirect mode; that is, by preparing diastereoisomers and resolving them using TLC. The derivatization of racemic mixtures and their subsequent separation on silica gel or RP TLC plates represents a method of chiral separation. Only a few reports have appeared on direct enantiomeric separation on chiral TLC plates; that is, using CMPAs or CSPs. Among the direct approaches, the use of CSPs is also very limited. Only ligand exchange based chiral thin layer chromatographic plates are commercially available for the chiral separation of racemates. It is worth mentioning here that TLC has not been used for chiral separation of environmental pollutants. However, it can easily be used for this purpose.

It is obvious that various modalities of chromatography have been used for the chiral separation of pharmaceuticals and drugs. Therefore, these approaches can also be used for the chiral separation of pollutants. In view of the importance of chromatography in the chiral separation of pollutants, and to familiarize the reader with chromatographic techniques, it is necessary to set out the chromatographic terms and symbols by which chromatographic separations can be explained. Some of the important terms and equations of chromatographic separations are discussed below. Chromatographic separations are characterized by retention (k), separation (α) and resolution (R_s) factors. The values of these parameters can be calculated using the following standard equations [90]:

$$k = (t_r - t_0)/t_0 \qquad (1.7)$$

$$\alpha = k_2/k_1 \qquad (1.8)$$

$$R_s = 2\Delta t/(w_1 + w_2) \qquad (1.9)$$

where t_r and t_0 are the retention time of the chromatogram of the enantiomers and the dead time (solvent front) of the column, respectively, both in minutes. Δt, w_1 and w_2 are the difference between the retention times of the two peaks of the separated pollutant, and the base widths of peaks 1 and 2, respectively. If the individual values of α and R_s are one or greater, the separation is supposed to be complete. If the individual values of these parameters are lower than one, the separation is considered to be partial or incomplete.

The number of theoretical plates (N) characterizes the quality of a column: the larger the value of N, the more complicated is the sample mixture that can be separated using the column. The value of N can be calculated from the following equations:

$$N = 16(t_r/w)^2 \qquad (1.10)$$

or

$$N = 5.54[(t_r)/w_{1/2}]^2 \qquad (1.11)$$

where t_r, w and $w_{1/2}$ are the retention time (min) of the peak, and the peak widths at base and at half of the height of the peak, respectively. The height equivalent to a theoretical plate (HETP) h is a section of a column in which a solute is in equilibrium with mobile and the stationary phases.

Since a large number of theoretical plates are desired, h should be as small as possible. Naturally, there are no real plates in a column. The concept of a theoretical plate is a variable, the value of which depends on the particle size, the flow velocity, the mobile phase (viscosity) and, especially, on the quality of the packing. h can be calculated using the following equation:

$$h = L/N \qquad (1.12)$$

where L is the length of the column used.

1.11.2 The Capillary Electrophoretic Method

At present, capillary electrophoresis (CE), a versatile technique that offers high speed, a high sensitivity and a lower limit of detection, is a major trend in analytical science, and in the field of chiral separation the number of publications has increased exponentially in recent years [91]. Among the electrophoretic methods of chiral separation, various forms of capillary electrophoresis, such as capillary zone electrophoresis (CZE), capillary iso-tachphoresis (CIF), capillary gel electrophoresis (CGE), capillary isoelectric focusing (CIEF), affinity capillary electrophoresis (ACE) and separation on microchips, have been used. In contrast to the others, the CZE model has frequently been used for this purpose [91]. However, it is necessary to mention here that capillary electrophoresis cannot achieve the status of a routine analytical technique in chiral separation, because of some associated drawbacks. The limited application of these methods is due to the lack of development of modern chiral phases.

Again, it is worth explaining some fundamental aspects of capillary electrophoresis, so that the reader can use this technique in the proper way. The separation mechanism in CE is based on the difference in the electrophoretic mobilities of the pollutants. Under the CE conditions, the migration of the pollutants is controlled by the sum of the intrinsic electrophoretic mobility (μ_{ep}) and the electro-osmotic mobility (μ_{eo}), due to the action of electro-osmotic flow (EOF). The observed mobility (μ_{obs}) of the pollutant is related to μ_{eo} and μ_{ep} by the following equation:

$$\mu_{obs} = E(\mu_{eo} + \mu_{ep}) \qquad (1.13)$$

where E is the applied voltage (kV).

The simplest way to characterize the separation of two components, the resolution factor (R_s), is to divide the difference in the retention times by the average peak width, as follows:

$$R_s = 2(t_2 - t_1)/(w_1 + w_2) \qquad (1.14)$$

where t_1, t_2, w_1 and w_2 are the retention times of peaks 1 and 2 and the widths of peaks 1 and 2, respectively.

The value of the separation factor may be correlated with μ_{app} and μ_{ave} by the following equation:

$$R_s = \tfrac{1}{4}(\Delta\mu_{app}/\mu_{ave})N^{1/2} \qquad (1.15)$$

where μ_{app} is the apparent mobility of the two enantiomers and μ_{ave} is their average mobility. The utility of Equation (1.9) is that it permits independent assessment of the two factors that affect separation, selectivity and efficiency. The selectivity is reflected in the mobility of the analytes, while the efficiency of the separation process is indicated by N. Another expression for N is derived from the following equation:

$$N = 5.54(L/w_{1/2})^2 \qquad (1.16)$$

where L and $w_{1/2}$ are the capillary length and the peak width at half height, respectively.

It is important to point out that it is misleading to discuss theoretical plates in CE: it is simply a carryover from chromatographic theory. In electrophoresis, the separation is governed by the relative mobilities of the analytes in the applied electric field, which are a function of their charge, mass and shape. The theoretical plate in CE is merely a convenient concept to describe the shape of the analyte peaks and to assess the factors that affect separation. The efficiency of the separations on a column is expressed by N, but it is difficult to use this variable to assess the factors that affect efficiency. This is because it refers to the behaviour of a single component during the separation process, and it is not suitable for describing the separation in capillary electrophoresis. However, a more useful parameter is the height equivalent of a theoretical plate (HETP), given as follows:

$$\text{HETP} = L/N = \sigma_{tot}^2/L \qquad (1.17)$$

The HETP may be considered as the function of the capillary occupied by the analyte, and it is more practical to measure separation efficiency compared to N. σ_{tot}^2 is affected not only by diffusion but also by differences in the mobilities, the Joule heating of the capillary and the interaction of the analytes with the capillary wall, and hence σ_{tot}^2 can be represented as follows:

$$\sigma_{tot}^2 = \sigma_{diff}^2 + \sigma_T^2 + \sigma_{int}^2 + \sigma_{wall}^2 \qquad (1.18)$$

1.12 Chiral Selectors in Chromatography and Capillary Electrophoresis

The presence of a chiral phase, called a chiral selector, is essential for the enantiomeric analysis of chiral pollutants by chromatographic and capillary electrophoretic methods. Therefore, several optically active compounds have been used for this purpose. The most important classes of such types of substances are polysaccharides, cyclodextrins, antibiotics, proteins, Pirkle-type CSPs (see below), ligand exchangers, crown ethers and several other types. The basic requirements for a suitable chiral selector are that it should be easily available and inexpensive, that it should have sufficient groups, atoms, grooves, cavities and so on for complexing with chiral pollutants, and that it should be capable of forming diastereomers that are non-UV-absorbing in nature as, generally, the detection in chromatography and capillary electrophoresis is carried out by a UV/visible detector.

Most of the naturally occurring polymers, including the polysaccharides, are chiral and optically active because of their asymmetric structures. These polymers often possess a specific conformation or higher order structure arising from chirality that is essential for the chiral analysis of racemic pollutants [92]. Therefore, the polysaccharides have a potential application in the chiral separation of chiral pollutants by chromatography and capillary electrophoresis [93, 94]. The polysaccharide polymers, such as cellulose, amylose, chitosan, xylan, curdlan, dextran and inulin, have been used for chiral separation in chromatography [95]. However, these derivatives cannot be used as commercial chiral stationary phases (CSPs), because of their poor separation capacity and handling problems [92]. Therefore, derivatives of these polymers have been synthesized in the past two decades [92]. Among the various polymers of polysaccharides, cellulose and amylose are the most readily available naturally occurring forms, and they have been found to be suitable for chiral separations. Therefore, most chiral applications involving chromatography and capillary electrophoresis have been reported as using these two polysaccharides [92, 95].

Cyclodextrins (CDs) are cyclic and nonreducing oligosaccharides, and are obtained from starch. Schardinger [96] identified three different forms of naturally occurring CDs – α-, β- and γ-CDs – and referred to them as Schardinger's sugars. They are also called cyclohexamylose (α-CD), cycloheptamylose (β-CD), cyclooctamaylose (γ-CD), cycloglucans, glucopyranose and Schardinger dextrins. The ability of CDs to form complexes with a wide variety of molecules has been documented [97–102]. The complex formation of CDs and their binding constants have been determined and are controlled by several different factors – hydrophobic interactions, hydrogen bondings and van der Waals interactions. Therefore, CDs and their derivatives have been widely used in separation science since the early 1980s [103, 104]. The evolution of CDs as chiral selectors in chromatographic and capillary electrophoretic separations of enantiomers has become a subject of interest in the past two decades. The presence of a chiral hollow basket/cavity in these molecules makes them suitable for chiral separation of a wide range of chiral pollutants. At present, the use of CDs as chiral selectors for enantiomeric separation by chromatography and capillary electrophoresis is very common. As chiral selectors, CDs have been used in the form of chiral stationary phases (CSPs) and chiral mobile phase additives (CMPs).

Macrocyclic antibiotics are one of the newest and perhaps the most varied classes of chiral selectors [105]. The concept of utilizing macrocyclic glycopeptide as a chiral stationary phase for HPLC was introduced by Dr D. W. Armstrong in 1994 [106]. Since then, their use for chiral analysis in chromatography and capillary electrophoresis has increased exponentially [67, 107]. The antibiotics have been found to have a very good potential for the chiral separation of a wide range of racemates. This may be due to their specific structures and the possibility of using a wide range of mobile phases. Additionally, due to their relatively small size and the fact that their structures are known, basic studies on chiral recognition can be carried out easily and precisely. They are often complementary in the types of compounds they can separate. For example, rifamycin B, an ansamycin, is enantioselective for many positively charged analytes, whereas vancomycin, a glycopeptide, can resolve a variety of chiral compounds containing free carboxylic acid functional groups. In addition, the separation of enantiomers by antibiotics is not very sensitive and hence is highly robust. The antibiotics most commonly used for chiral separation are vancomycin, teicoplanin, teicoplanin aglycon and Ristocetin A, although vancomycin aglycon, thiostrepton, rifamycin, fradiomycin, streptomycin, kanamycin and avoparcin are also used.

Proteins are natural polymers and are made of amino acids – which are chiral molecules, with the exception of glycine – through amide bonds. However, some glycoproteins also contain sugar moieties. The protein polymer remains in the twisted form because of the different intramolecular bondings. These bondings are also responsible for different types of loops/grooves that are present in the protein molecule. This sort of twisted three-dimensional structure of the protein makes it enantioselective in nature. Enantioselective interactions between small molecules and proteins in biological systems are well known [108]. Although all of the protein molecules are complex in structure and enantiospecific, they have not been used as successful chiral selectors yet, because the enantiomeric separation varies from one protein to another. The albumin proteins used as chiral selectors in chromatography and capillary electrophoresis are bovine serum albumin (BSA), human serum albumin (HSA), rat serum albumin (RSA) and guinea pig serum albumin (GPSA), but BSA and HSA have been found to be the successful chiral selectors. However, other protein molecules have been explored for their chiral separation capacities; that is, glycoproteins such as α_1-acid glycoprotein (AGP), ovomucoid (OVM), ovotransferin, avidin and trypsin (CT), and certain enzymes such as chymotrypsin, riboflavin, lysozyme, pepsin, amyloglucosidase and lactoglobulin. Additionally, cellobiohydrase-I (CBH-I), a protein obtained from fungi, has also been used as a chiral selector in HPLC [109].

In 1976, Mikeš *et al.* [110] introduced a new concept by attaching a small chiral molecule to silica gel. In this CSP, the organic groups of the chiral molecule remain directed away from the silica gel, appearing in the form of a brush, and hence this is called a brush type phase. Later on, Pirkle and coworkers developed these types of CSP extensively, and nowadays they are known as Pirkle-type CSPs [111–119]. Normally, the chiral molecule attached to the silica gel contains π electron donors or π electron receptors, or both types of group. Therefore, these CSPs are classified into three groups; π-acidic (with π electron acceptor groups), π-basic (with π electron donor groups), and π acidic–basic (with π electron acceptor and donor groups), respectively. The reciprocality concept put forth by Pirkle has allowed the development of several generations of these types of CSP [113, 116]. The main advantage of these types of phases is that one can choose the type of chiral molecule to be attached to the silica gel. A specific and required chiral molecule (to be attached to the silica gel) can be selected by the reciprocality concept and bonded to the silica gel, and hence the chiral separation of a wide variety of racemic compounds can be performed easily and successfully. Recently, some chiral molecules that have specific groups other than π donors or π acceptors, such as polar and polarizable groups, have been grafted on to the silica gel surface. These types of CSPs have been found to have great potential for the chiral separation of different racemic compounds.

Ligand exchange chiral selectors involve the breaking and formation of coordinate bonds among the metal ions of the complex, the ligands and the chiral pollutants. Therefore, ligand exchange chromatography is useful for the chiral separation of pollutants that contain electron-donating atoms, such as oxygen, nitrogen and sulfur. These types of pollutant contain amino, hydroxy and acid groups. Sometimes, the fast kinetics of the ligand exchange reactions in the metal ion coordination sphere make this technique suitable for the chiral separation of kinetically labile pollutants. Chiral ligand exchange chromatography was developed by Davankov [120, 121] in 1969. Copper(II) has been used as the ligand metal ion in most of the applications of ligand exchange chromatography. However, some other metal ions, such as nickel and zinc, have also been tested [122, 123]. Nowadays, these selectors are also useful for chiral analysis in capillary electrophoresis.

The crown ethers are synthetic macrocyclic polyethers: their name derives from both the crown-like appearance of their molecular structures and their ability to crown selectively with cations. The ether oxygens that are electron donors remain in the inner wall of the crown cavity, and are surrounded by methylene groups in a collar fashion. The IUPAC nomenclature for these ethers is complex, and hence trivial names are commonly used [124]. For example, 2,311,12-dibenzo-1,4,7,10,13,16-hexaoxa-cyclo-octadeca-2,11-diene is called dibenzo-18-crown-6 ether, where dibenzo, 18 and 6 indicate the substituent groups, the total number of atoms in the ring and the number of oxygen atoms, respectively. If the oxygen atoms of the ether are replaced by nitrogen or sulfur atoms, the crown ether is called an aza or a thia crown ether, respectively. Chirality in crown ethers is developed by introducing chiral moieties and, hence, the developed crown ether is called a chiral crown ether (CCE). The most important chiral groups used for this purpose are binaphthyl [125–128], biphenanthryl [129–131], hericene derivatives [132], tartaric acid derivatives [133], carbohydrate moiety [134], a chiral carbon atom with a bulky group directly incorporated in the crown ring [135, 136], aromatic bicyclo derivatives [3.3.1], nonane derivatives [137, 138] hexahydrochrysese or tetrahydroindenoinden [139], and 9,9'-spiro-bifluorene groups [140]. The capability of these CCEs of crowning selectively, and their stereospecific configurations, make them suitable chiral selectors in chromatography and capillary electrophoresis. The application of these chiral selectors is limited, as they can only be used for the analysis of chiral pollutants containing amine and amide groups.

The new strategy for chiral separation in chromatography and capillary electrophoresis is the development of molecularly imprinted polymers. First of all, Wulf *et al.* [141] presented the idea of a molecularly imprinted polymers technique. This involves the incorporation of a target molecule (an imprint molecule) into a polymer and the removal of the print molecule, to leave a substrate selective site or cavities. This may be achieved either by a print molecule (template) bonded to functionalized monomers in solution, and where after copolymerization with an excess of a crosslinker the print molecule is chemically cleaved off the polymer, or by the imprint molecule being mixed with functional monomers and undergoing electrostatic (ionic and hydrophobic) interactions with the monomer prior to polymerization in solution. After polymerization, hydrogen bonds are formed between the template and the carboxylic groups of the recognition sites (polymeric). Then the imprint molecule is removed by washing the polymer with an acidic organic solvent. The most commonly used monomers are methacrylic acid, 2-vinylpyridine and 4-vinylpyridine. This technique has not been popular up to the present day, but it is promising for the near future, as the required imprinted polymer can be synthesized and chiral analysis can be achieved easily and successfully. However, the synthesis of these chiral selectors suffers from certain drawbacks. To make a successful polymer of a chiral compound, the proper amount of the enantiopure imprint molecule is required. In some cases, a molecularly imprinted polymer can be so selective for a certain molecule that it would become necessary to make an imprinted polymer for each analyte to be tested [142]. Additionally, a small amount of the print molecule can be irreversibly incorporated into the polymer matrix, which would exclude the use of molecularly imprinted polymers for the determination of enantiomeric purity [143].

1.13 Detection in Chromatography and Capillary Electrophoresis

UV detection is used in most chiral analysis by HPLC and other liquid chromatographic modalities. However, some other detectors, such as conductivity, fluorescent and refractive index types, are also used. The choice of detector depends on the properties of the racemic compound to be resolved [41, 144]. Chiroptical detectors, which are based on the principle of polarimetry [145] or circular dichroism [146, 147], are also available. The enantiomer (+)- or (−)-notation is determined by these detectors. Some organochlorine pesticides are not UV-sensitive, and hence they are difficult to detect in liquid chromatography. The detection of these types of pollutant can be achieved by using a mass spectrometry (MS) detector, and therefore LC–MS instruments are now being put on the market for routine use [148, 149].

Due to the presence of halogen atoms in some pesticides, an electron capture detector is considered to be one of the most efficient detection methods in gas chromatography. However, other detectors, such as conductivity, flame ionization, nitrogen and phosphorus types, may also be used in GC. Most chiral separation of these types of pollutants is carried out using gas chromatography. Also, GC–MS coupling is used to identify the separated enantiomers in real samples.

In TLC, detection is carried out using a number of approaches. The UV-sensitive enantiomers are observed in a UV cabinet, while some other non-UV-absorbing chiral molecules may be detected by developing a colour using a suitable reagent; for example, compounds containing amino acids and amines are located by developing a colour on the TLC plate using ninhydrin [150]. Moreover, iodine vapour is used as a universal detection method on TLC plates. The separated enantiomers adsorb the iodine vapour and become yellow in colour. The detection limits for chiral pollutants depend on a number of factors, such as the properties of the molecules, the mobile phase, the chiral stationary phase, the type of chiral selector and the sensitivity of the detector. The lowest detection limits have been reported at the milligram to nanogram levels.

1.14 Other Methods of Separation of Chiral Pollutants

Besides chromatographic and capillary electrophoretic methods, some other alternatives are also available for the chiral separation of environmental pollutants. These include spectroscopic, sensor, simulating moving bed adsorption and membrane methods [37–40, 67, 68, 151–153]. Optical rotation measurements, nuclear magnetic resonance (NMR) and infrared (IR) spectroscopic methods have been reported to distinguish enantiomers. In IR, differential scanning colorimetery (DSC) only distinguishes racemic mixtures ($\pm$) and individual enantiomers. In NMR, chiral solvating agents (CSA) can be utilized to promote a change in the chemical shift of the chiral (carbon) atoms. However, while the spectroscopic and DSC methods are sensitive to interference by impurities of a chiral or nonchiral kind in the sample of interest, they can lack sensitivity and accuracy, since the differences between the isomers and/or the isomer mixtures may be very small.

Biosensors have been used for the discrimination of industrial, medicinal and environmental chiral molecules [68]. The molecularly imprinted polymer based biomimetic sensors are useful for this type of chiral analysis [151]. The advantages of molecularly imprinted polymers over other biomolecules are their high stability and their resistance to the strong medium. Gas sensors have also been used for chiral analysis, this technique involving the use of chiral amide (octyl-Chirasil-Val®) attached to a polysiloxane chain and coated on a solid surface such as quartz, glass and so on [154]. In this method, one enantiomer undergoes a greater interaction with the incorporated chiral selector in the polymer coating as compared to the other antipode, and this difference leads to the enantiodifferentiation. Chiral discrimination of the vaporized racemates results in a small change in the thickness of the polymer film after absorption of the pollutants into the coating, which causes a shift in the reflection of monochromatic light through the transparent film on the glass substrate surface.

Chiral discrimination has been achieved by the use of membranes, this technique involving the principle of enrichment of enantiomers in organic solutions on both sides of a membrane, which prevents the two organic phases from mixing together [152]. These membranes are permeable for the enantiomers, with a preferred passage of only one enantiomer in comparison to its antipode, which results in the enrichment of only one enantiomer. Various types of membranes are available, including polymers, solid supported liquid membranes [155], emulsion liquid membranes [156] and bulk liquid membranes [157]. In spite of the good capacity of these membranes, the method is not yet popular, as it is still under development.

The simulated moving bed adsorption technique is based on the movement of the stationary phase. The front and back ends of a series of columns are connected to form a circle, and during rotation of the columns a countercurrent movement of the phase relative to the liquid stream in the system is developed [158]. Injections of the fresh chiral analyte and the solvent are made at various connecting points, and the separated enantiomers are withdrawn simultaneously at certain time intervals. This is a continuous process that provides certain advantages for enantiodiscrimination. The chiral selectors used in this technique are the same as those utilized in liquid chromatography and capillary electrophoresis.

In view of the importance of the enantiomeric separation of chiral pollutants, attempts are made in this book to explain the art of enantiomeric separation of pollutants by chromatography and electrophoresis. Following this introduction, the remaining nine chapters discuss enantioselective toxicities, metabolism and biotransformation, extraction, purification and pre-concentration, and the determination of chiral pollutants by chromatography and capillary electrophoresis. In this book, the art of the chromatographic and electrophoretic analysis of chiral pollutants is described in a well defined, systematic and scientific way, including discussions on optimization of the methods and the separation mechanisms. We hope that this book will be useful to academicians, scientists, workers in other regulatory authorities and students dealing with the analysis of chiral pollutants.

References

1. L. B. Franklin, *Wastewater Engineering, Treatment, Disposal and Reuse*, McGraw-Hill, New York, 1991.
2. R. L. Droste, *Theory and Practice of Water and Wastewater Treatment*, Wiley, New York, 1997.
3. J. De Zuane, *Handbook of Drinking Water Quality, Standards and Controls*, Van Nostrand Reinhold, New York, 1990.
4. *Toxic Substance Control Act*, US EPA, III 344 (1984).
5. J. W. Moore and S. Ramamoorthy, *Heavy Metals in Natural Waters, Applied Monitoring and Impact Assessment*, Springer, New York, 1984.
6. *IARC Monographs, Suppl. 7, IARC, Lyon* **54**, 40 (1987).
7. J. Dich, S. H. Zahm, A. Hanberg and H. O. Adami, *Cancer Causes & Control* **8**, 420 (1997).
8. D. Brusick, *Toxicol. Indust. Health* **9**, 223 (1993).
9. D. Kaniansky, E. Kremova, V. Madajova and M. Maser, *Electrophoresis* **18**, 260 (1997).
10. D. H. Hutson and T. R. Roberts, *Environmental Fate of Pesticides*, Wiley, New York, vol. 7, 1990.
11. I. Ali and C. K. Jain, *Curr. Sci.* **75**, 1011 (1998).
12. C. K. Jain and I. Ali, *Int. J. Environ Anal. Chem.* **68**, 83 (1997).
13. R. Kallenborn and H. Hühnerfuss, *Chiral Environmental Pollutants, Trace Analysis and Ecotoxicology*, Springer-Verlag, Berlin, 2001, p. 3.
14. G. A. Cutter, *Marine Chem.* **40**, 65 (1992).
15. I. Ali, V. K. Gupta and H. Y. Aboul-Enein, *Curr. Sci.* **84**, 152 (2003).
16. K. Mislow, *Introduction to Stereochemistry*, Benjamin, New York, 1965.
17. E. L. Eliel, S. H. Wilen and L. N. Mander, *Stereochemistry of Organic Compounds*, Wiley, New York, 1994.
18. P. Curie, *J. Phys.* **3**, 393 (1894).
19. P. Drude, *The Theory of Optics*, Dover, New York, 1900.
20. T. M. Lowry, *Optical Rotatory Power*, Longmans Green, London, 1935.
21. R. Hegstrom and D. K. Kondepudi, *Sci. Am.* **262**, 108 (1990).
22. D. Kondepudi and D. J. Durand, *Chirality* **13**, 351 (2001).
23. J. Bailey, A. Chrysostomou, J. H. Hough, T. M. Gledhill, A. McCall, S. Clark, F. Menard and M. Tamura, *Science* **281**, 672 (1998).
24. R. J. Haüy, *Tableaux comparatif des resultats de la crystallographie et de l'analyse chimique relativement a la classification des mineraux*, Paris, 1809, p. VII.
25. E. Mitscherlich, *Abh. Dt. Akad. Wiss., Berlin*, 427 (1819).
26. J. W. F. Herschel, *Trans. Camb. Phil. Soc.* **1**, 43 (1822).
27. A. Fresnel, *Bull. Soc. Philomath.* 147 (1824).
28. M. Mitscherlich, *C. R. Acad. Sci.* **19**, 719 (1844).
29. L. Pasteur, *C. R. Acad. Sci.* **26**, 535 (1848).

30. J. A. Le Bel, *Bull. Soc. Chim. Fr.* **22**, 337 (1874).
31. J. H. Van't Hoff, *Arch. Neerland Sci. Exactes et Naturelles* **9**, 445 (1874).
32. R. A. Aitken, D. Parker, R. J. Taylor, J. Gopal and R. N. Kilenyi, *Asymmetric Synthesis*, Blackie, New York, 1992.
33. G. Blaschke, H. P. Kraft, K. Fickentscher and F. Köhler, *Drug Res.* **29**, 1640 (1979).
34. B. Knoche and G. Blaschke, *J. Chromatogr. A* **666**, 235 (1994).
35. B. Testa and P. A. Carrupt, *Chirality* **5**, 105 (1993).
36. E. J. Ariëns and E. W. Wuis, *J. R. Coll. Physns (Lond.)* **28**, 395 (1994).
37. H. Y. Aboul-Enein and I. Ali, *Chiral Separations by Liquid Chromatography and Related Technologies*, Dekker, New York, 2003.
38. S. Allenmark, *Chromatographic Enantioseparation, Methods and Applications*, 2nd edn., Ellis Horwood, New York, 1991.
39. G. Subramanian, ed., *A Practical Approach to Chiral Separations by Liquid Chromatography*, VCH, Weinheim, 1994.
40. H. Y. Aboul-Enein and I. W. Wainer, eds, *The Impact of Stereochemistry on Drug Development and Use*, *Chemical Analysis*, vol. 142, Wiley, New York, 1997.
41. T. E. Beesley and R. P. W. Scott, *Chiral Chromatography*, Wiley, New York, 1998.
42. S. C. Stinson, *Chem. Eng. News* **72**, 38 (1994).
43. E. Y. Spencer, *Guide to Chemicals Used in Crop Protection*, 7th edn, Ottawa, Canadian Govt Publ. Centre, 1981.
44. D. L. Lewis, A. W. Garrison, K. E. Wommack, A. Whittemore, P. Steudler and J. Melillo, *Nature* **401**, 898 (1999).
45. W. Vetter, *Food Rev. Intl* **17**, 113 (2001).
46. R. S. Cahn, C. K. Ingold and V. Prelog, *Experientia* **12**, 81 (1956).
47. S. Hauptmann, in S. Hauptmann, ed., *Organische Chemie*, VEB Deutscher Verlag für Grundstoffindustrie, 1988.
48. J. Caldwell, *J. Chromatogr. A* **694**, 48 (1995).
49. E. J. Ariens, J. J. S. van Rensen and W. Welling, eds, *Stereoselectivity of Pesticides, Biological and Chemical Problems*, Chemicals in Agriculture, Vol. 1, Elsevier, Amsterdam, 1988.
50. H. R. Buser, M. D. Muller and C. Rappe, *Environ. Sci. Technol.* **26**, 1533 (1992).
51. V. Schurig, A. Glausch and M. Fluck, *Tetrahedron Asymm.* **6**, 2161 (1995).
52. G. Weseloh, C. Wolf and W. A. König, *Chirality* **8**, 441 (1996).
53. C. Wolf, D. H. Hochmuth, W. A. König and C. Russel, *Liebigs. Ann.* 357 (1996).
54. M. T. Harju and P. Haglund, *J. Anal. Chem.* **364**, 219 (1999).
55. R. Kallenborn, H. Hühnerfuss and W. A. Köning, *Angew. Chem.* **103**, 328 (1991).

56. J. Faller, H. Hühnerfuss, W. A. Köning, R. Krebber and P. Ludwig, *Environ. Sci. Technol.* **25**, 676 (1991).
57. W. Vetter and V. Schurig, *J. Chromatogr. A* **774**, 143 (1997).
58. V. Schurig and W. Houben, *Methods of Organic Chemistry, Stereochemistry and Stereoselective Synthesis*, Thiem, Stuttgart and New York, 1995.
59. W. Vetter, K. Hummert, B. Luckas and K. Skirnisson, *Sci. Total Environ.* **170**, 159 (1995).
60. H. J. De Gues, H. Besseling, A. Brouwer, J. Klungsöyr, B. McHugh, E. Nixon, G. C. Rimkus, P. G. Wester and J. de Boer, *Environ. Health Perspect.* **107**, 115 (1999).
61. A. T. Fisk, R. J. Norstrom, K. A. Hobson, J. Moisey and N. J. Karnovsky, *Organohal. Compds* **40**, 413 (1999).
62. T. Harner, K. Wiberg and R. Norstrom, *Environ. Sci. Technol.* **34**, 218 (2000).
63. W. J. M. Hegeman and W. P. M. Laane, *Rev. Environ. Toxicol.* **173**, 85 (2002).
64. FDA's policy statement for the development of new stereoisomeric drugs, *Chirality* **4**, 338 (1992).
65. S. C. Stinson, *Chem. Eng. News* **75**, 28 (1997).
66. S. C. Stinson, *Chem. Eng. News* **73**, 44 (1995).
67. N. M. Maier, P. Franco and W. Lindner, *J. Chromatogr. A* **906**, 3 (2001).
68. R. I. Stefan, J. F. van Staden and H. Y. Aboul-Enein, *Electrochemical Sensors in Bioanalysis*, Dekker, New York, 2001.
69. R. C. Weast, ed., *Handbook of Chemistry and Physics*, 54th edn, CRC Press, Cleveland, 1973.
70. T. A. Berger, in R. M. Smith, ed., *Packed Column SFC*, The Royal Society of Chemistry, Chromatography Monographs, Cambridge, 1995.
71. P. J. Schoenmakers, in R. M. Smith, ed., *Packed Column SFC*, Royal Society of Chemistry, Chromatography Monographs, Cambridge, 1988.
72. G. Terfloth, *J. Chromatogr. A* **906**, 301 (2001).
73. P. A. Mourier, E. Eliot, R. H. Caude, R. H. Rosset and A. G. Tambute, *Anal. Chem.* **57**, 2819 (1985).
74. P. Macaudiere, M. Caude, R. Rosset and A. Tambute, *J. Chromatogr.* **405**, 135 (1987).
75. C. R. Lee, J. P. Porziemsky, M. C. Aubert and A. M. Krstulovic, *J. Chromatogr.* **539**, (1991).
76. P. Biermanns, C. Miller, V. Lyon and W. Wilson, *LC–GC* **11**, 744 (1993).
77. P. Pettersson and K. E. Markides, *J. Chromatogr. A* **666**, 381 (1994).
78. N. Bargmann-Leyder, A. Tambute and M. Caude, *Chirality* **7**, 311 (1995).
79. K. L. Williams, L. C. Sander and S. A. Wise, *J. Chromatogr. A* **746**, 91 (1996).
80. J. R. Cronin and S. Pizarello, *Science* **275**, 95 (1997).
81. V. Pretorius, B. J. Hopkins and J. D. Schieke, *J. Chromatogr.* **99**, 23 (1974).
82. T. Tsuda, K. Nomura and G. Nagakawa, *J. Chromatogr.* **248**, 241 (1982).

83. S. Terabe, K. Otsuka, K. Ichikawa, Λ. Tsuchiya and T. Ando, *Anal. Chem.* **56**, 111 (1984).
84. V. Schurig and D. Wistuba, *Electrophoresis* **20**, 2313 (1999).
85. S. Fanali, P. Catarcini, G. Blaschke and B. Chankvetadze, *Electrophoresis* **22**, 3131 (2001).
86. A. S. Cohen, A. Paulus and B. L. Karger, *Chromatographia* **24**, 15 (1987).
87. M. L. Marina, I. Benito, J. C. Diez-Masa and M. J. Gonzalez, *J. Chromatogr. A* **752**, 265 (1996).
88. A. L. Crego, M. A. Garcia and M. L. Marina, *J. Microcol. Sepn* **12**, 33 (2000).
89. W. C. Lin, C. C. Chang and C. H. Kuei, *J. Microcol. Sepn* **11**, 231 (1999).
90. V. R. Meyer (ed.), *Practical High Performance Liquid Chromatography*, Wiley, New York, 1993.
91. B. Chankvetadze, *Capillary Electrophoresis in Chiral Analysis*, Wiley, New York, 1997, p. 353.
92. Y. Okamoto and E. Yashima, in K. Hatada, T. Kitayama and O. Vogl, eds, *Molecular Design of Polymeric Materials*, Dekker, New York, 1997, p. 731.
93. Y. Okamoto, *Chemtech.* **17**, 176 (1987).
94. Y. Okamoto and R. Aburatani, *Polym. News* **14**, 295 (1989).
95. E. Yashima and Y. Okamoto, in H. Y. Aboul-Enein and I. W. Wainer, eds, *The Impact of Stereochemistry on Drugs Development and Use*, Wiley, New York, 1997, p. 345.
96. F. Schardinge, *Zentr. Bacteriol. Parasitenk Abt. II* **29**, 188 (1911).
97. J. Michon and A. Rassat, *J. Am. Chem. Soc.* **101**, 4337 (1979).
98. S. M. Han, W. M. Atkinson and N. Purdie, *Anal. Chem.* **56**, 2827 (1984).
99. D. A. Lightner, J. K. Gawronski and J. Gawronska, *J. Am. Chem. Soc.* **107**, 2456 (1985).
100. Y. Ihara, E. Nakashini, K. Mamuro and J. Koga, *Bull. Chem. Soc. Jpn* **59**, 1901 (1986).
101. D. W. Armstrong, T. J. Ward, R. D. Armstrong and T. E. Beesley, *Science* **222**, 1132 (1986).
102. J. A. Hamilton and L. Chen, *J. Am. Chem. Soc.* **110**, 5833 (1988).
103. S. M. Han and D. W. Armstrong, in A. M. Krstulovic, ed., *Chiral Separations by HPLC*, Ellis Horwood, Chichester, 1989, p. 208.
104. A. M. Stalcup, in G. Subramanian, ed., *A Practical Approach to Chiral Separations by Liquid Chromatography*, VCH, Weinheim, 1994, p. 95.
105. D. W. Armstrong, Y. Tang, S. Chen, Y. Zhou, C. Bagwill and R. Chen, *Anal. Chem.* **66**, 1473 (1994).
106. *Chirobiotic Handbook, Guide to Using Macrocyclic Glycopeptide Bonded Phases for Chiral LC Separations*, 2nd edn, Advanced Separation Technologies, Inc., Whippany, NJ, 1999.
107. H. Y. Aboul-Enein and I. Ali, *Chromatographia* **52**, 679 (2000).
108. C. R. Lowe and P. G. D. Dean, *Affinity Chromatography*, Wiley, London, 1974.

109. J. Haginaka, *J. Chromatogr. A* **906**, 253 (2001).
110. F. Mikeš, G. Boshart and E. Gil-Av, *J. Chromatogr.* **122**, 205 (1976).
111. W. H. Pirkle and D. L. Sikkenga, *J Chromatogr.* **123**, 400 (1976).
112. W. H. Pirkle and D. W. House, *J. Org. Chem.* **44**, 1957 (1979).
113. W. H. Pirkle, D. W. House and J. M. Finn, *J. Chromatogr.* **192**, 143 (1980).
114. W. H. Pirkle and J. M. Finn, *J. Org. Chem.* **46**, 2935 (1981).
115. W. H. Pirkle and C. J. Welch, *J. Org. Chem.* **49**, 138 (1984).
116. W. H. Pirkle and R. Däppen, *J. Chromatogr.* **404**, 107 (1987).
117. J. M. Finn, in M. Zief and L. J. Crane, eds, *Chromatographic Chiral Separations*, Chromatographic Science Series, vol. 40, Dekker, New York, 1988.
118. P. Macaudiere, M. Lienne, A. Tambute and M. Caude, in A. M. Krstulovic, ed., *Chiral Separations by HPLC*, Ellis Horwood, New York, 1989.
119. B. A. Persson and S. Andersson, *J. Chromatogr. A* **906**, 195 (2001).
120. German Pat. 1932190 (1969), S. V. Rogozhin and V. A. Davankov.
121. V. A. Davankov, J. D. Navratil and H. F. Walton, *Ligand exchange chromatography*, CRC Press, Boca Raton, FL 1988.
122. B. Feibush, M. J. Cohen and B. L. Karger, *J. Chromatogr.* **282**, 3 (1983).
123. Y. A. Zolotarev, N. F. Myasoedov, V. I. Penkina, O. R. Petrenik and V. A. Davankov, *J. Chromatogr.* **207**, 63 (1981).
124. C. J. Pedersen, *J. Am. Chem. Soc.* **89**, 7017 (1967).
125. E. P. Kyba, J. M. Timko, J. L. Kaplan, F. de Jong, G. W. Gokel and D. J. Cram, *J. Am. Chem. Soc.* **100**, 4555 (1978).
126. S. C. Peacock, L. A. Domeier, F. C. A. Gaeta, R. C. Helgeson, J. M. Timko and D. J. Cram, *J. Am. Chem. Soc.* **100**, 8190 (1978).
127. S. C. Peacock, D. M. Walba, F. C. A. Gaeta, R. C. Helgeson, J. M. Timko and D. J. Cram, *J. Am. Chem. Soc.* **102**, 2043 (1980).
128. D. S. Linggenfelter, R. C. Helgeson and D. J. Cram, *J. Org. Chem.* **46**, 393 (1981).
129. K. Yamamoto, H. Yumioka, Y. Okamoto and H. Chikamatsu, *J. Chem. Soc. Chem. Commun.*, 168 (1987).
130. K. Yamamoto, T. Kitsuki and Y. Okamoto, *Bull. Chem. Soc. Jpn* **59**, 1269 (1986).
131. K. Yamamoto, K. Noda and Y. Okamoto, *J. Chem. Soc. Chem. Commun.*, 1065 (1985).
132. M. Nakazaki, K. Yamamoto, T. Ikeda, T. Kitsuki and Y. Okamoto, *J. Chem. Soc. Chem. Commun.*, 787 (1983).
133. J. P. Behr, J. M. Girodeau, R. C. Hayward, I. M. Lehn and J. P. Sauvage, *Helv. Chim. Acta* **63**, 2096 (1980).
134. D. Gehin, P. D. Cesare and B. Gross, *J. Org. Chem.* **51**, 1906 (1986).
135. R. B. Davidson, J. S. Bradshaw, B. A. Jones, N. K. Dalley, J. J. Christensen, R. M. Izatt, F. G. Morin and D. M. Grant, *J. Org. Chem.* **49**, 353 (1984).

136. J. D. Chadwick, I. A. Cliffe, I. O. Sutherland and R. F. Newton, *J. Chem. Soc. Perkin Trans.* **I**, 1707 (1984).
137. K. Naemura and R. Fukunaga, *Chem. Lett.*, 1651 (1985).
138. K. Naemura, R. Fukunaga and M. Yamanaka, *J. Chem. Soc. Chem. Commun.*, 1560 (1985).
139. K. Naemura, M. Komatsu, K. Adachi and H. Chikamatsu, *J. Chem. Soc. Chem. Commun.*, 1675 (1986).
140. A. P. Thoma, A. Viviani-Nauer, K. H. Schellenberg, D. Bedekovic, E. Pretsch, V. Prelog and W. Simon, *Helv. Chim. Acta*, **62**, 2303 (1979).
141. G. Wulf, A. Sarhan and K. Zabrocki, *Tetra. Lett.* **44**, 4329 (1973).
142. M. Kempe and K. Mosbach, *J. Chromatogr. A* **694**, 3 (1995).
143. L. I. Andersson, D. J. O'Shannessy and K. Mosbach, *J. Chromatogr. A* **513**, 167 (1990).
144. R. P. Scott, *Chromatography Detectors*, Chromatography Science Series, vol. 73, Dekker, New York, 1996.
145. B. Baraj, L. F. Niencheski, J. A. Soares and M. Martinez, *Fresenius J. Anal. Chem.* **367**, 12 (2000).
146. P. Salvadori, C. Bertucci and C. Rosini, *Chirality* **3**, 376 (1991).
147. G. Brandl, F. Kastner, A. Mannschreck, B. Nölting, K. Andert and R. Wetzel, *J. Chromatogr.* **586**, 249 (1991).
148. C. J. L. Bugge, I. Crun, A. Ljungqvist, M. Vatankhan, D. B. Garci, H. B. Warren and S. Gupta, Abstracts, American Association of Pharmaceutical Scientists Annual Meeting, Seattle, Washington, 1996.
149. L. Ramos, R. Bakhtiar, T. Majumdar, M. Hayes and F. Tse, *Rapid Commun. Mass Spectrom.* **13**, 2054 (1999).
150. E. Stahl, *Thin Layer Chromatography*, 2nd edn, Springer-Verlag, Berlin, 1969.
151. D. Kriz, O. Ramström and K. Mosbach, *Anal. Chem. News Features*, 345A, (1997).
152. J. T. F. Keurentjes and F. J. M. Voermans, in A. N. Collins, G. N. Sheldrake and J. Crosby, eds, *Chirality in Industry II. Developments in the Commercial Manufacture and Applications of Optically Active Compounds*, Wiley, New York, 1997, Ch. 8.
153. S. C. Stinson, *Chem. Eng. News* **72**, 38 (1994).
154. K. Bodenhöfer, A. Hierlemann, J. Seemann, G. Gauglitz, B. Christian, B. Koppenhoefer and W. Göpel, *Anal. Chem.* **69**, 3058 (1997).
155. J. T. F. Keurentjes, *Cemische Magazine*, 352 (1994).
156. P. J. Pickering and J. B. Chaudhuri, *Chirality* **9**, 261 (1997).
157. W. H. Pirkle and W. A. Bowen, *Tetrahedron Asymm.* **5**, 773 (1994).
158. M. Negawa and F. Shoji, *J. Chromatogr. A* **590**, 113 (1992).

Chapter 2

Chiral Pollutants: Sources and Distribution

2.1 Introduction

At present, about 60 000 organic substances are used by human beings, and presumably some of these compounds are toxic and are contaminating our environment. These toxic organic compounds enter into the human body through the various stages of the food chain. It is well known that mostly chirality exists in organic compounds and, therefore, most chiral pollutants are organic compounds. Some pesticides, phenols, plasticizers and polynuclear aromatic hydrocarbons are chiral toxic pollutants. There are also some chiral drugs that have a toxic enantiomer, and such drug enantiomers may be considered as chiral pollutants. The best example is thalidomide which, was put on the market in the late 1950s as a sedative, in the racemic form. Even when applied in the therapeutic and harmless (+)-form, the *in vivo* interconversion of the (+) enantiomer into the harmful (−)-isomer was found to be responsible for the disastrous malformations of embryos when thalidomide was administered to women during pregnancy [1–3]. It has been reported that about 25 % of pesticides are chiral in nature [4]. In addition, some nonchiral pollutants degrade into the environment and their degradation products are toxic and chiral in nature. γ-Hexachlorocyclohexane (γ-HCH) is an achiral pesticide, but it degrades into chiral and toxic γ-pentachlorocyclohexene (γ-PCCH)

Chiral Pollutants. I. Ali and H. Y. Aboul-Enein
© 2004 John Wiley & Sons, Ltd ISBN: 0-470-86780-9

enantiomers. The present chapter describes the sources of pollution by chiral xenobiotics. Attempts have also been made to describe the distribution of these xenobiotics in water resources, soils, sediments, air, animals and plants. Some of the chiral pollutants detected in the environment are listed in Table 2.1. The chemical structures of some commonly reported chiral pollutants are given in Figure 2.1.

Table 2.1   A list of some common chiral pollutants

---

o,p-DDT [1,1,1-trichloro-2-(*o*-chlorophenyl)-2-(*p*-chlorophenyl)] ethane
o,p-DDD [1,1-dichloro-2-(*o*-chlorophenyl)-2-(*p*-chlorophenyl)] ethane
α-, β-, γ- and δ-Hexachlorocyclohexanes (HCHs)
HHCB (1,3,4,6,7,8-hexahydro-4,6,6,7,8,8-hexamethylcyclopenta[g]-2-benzopyran);
    galaxolide and metabolites
AHTN [1-(5,6,7,8-tetrahydro3,5,5,6,8,8-hexamethyl-2-naphthalenyl) ethanone; tonalide]
ATTI (1-[2,3-dihydro-1,1,2,6-tetramethyl-3-(1-methyl ethyl)-1H-inden-5-yl]ethanone;
    traseolide
AHDI [1-(2,3-dihydro-1,1,2,3,3,6-hexamethyl-1H-inden-5-yl) ethanone phantolide]
4-Tolylethyl sulfoxide
Chlorinated bis(propyl)ether
Chlordane (*cis-, trans-* and other congeners)
Oxychlordane
Chlordene and metabolites
Photochlordane
Heptachlor
Heptachlor *exo*-epoxide
Photoheptachlorepoxide
Deltamethrin
Tetrodotoxin
Saxitoxin
Anatoxin-a
Homoanatoxin-a
Bromocyclen
Toxaphene
Polychlorinated biphenyls (PCBs)
Polychlorinated biphenyls (PCBs) – methyl sulfones
Bromocyclane (Bromodan)
MCCP [2-(4-chloro-2-methyl-phenoxy) propionic acid]
DCPP [dichlorprop or 2-(2,4-dichlorophenoxy) propionic acid]
MDCPP [methyl dichlorprop or methyl 2-(2,4-dichlorophenoxy) propionic acid]
Clofibric acid
Thalidomide
Naloxone
Ibuprofen and metabolites

Table 2.1   *(continued)*

Cruformate
Biollethrin
Bromocyclen
Endosulfan
Methamidophos
Acephate
Tricholfon
Malaxon
Bromoacil
Crotoxyphos
Dialifor
Fonofos
Fenamiphos
Fensulfothion
Isofenphos
Malathion
Isomalathion
Methamidophos
Profenofos
Crufomate
Prothiophos
Trichloronate
Ruelene (*tert*-butyl-2-chlorophenylmethyl-*N*-methylphosphoramidate)

2.2 Sources of Contamination

To control and minimize pollution, it is essential to understand the sources of contamination. Knowledge about pollution sources is also helpful in explaining the transportation behaviour of the pollutants. Several books and articles [5, 7–10] have appeared on this issue, and interested readers should consult them. However, the main sources of chiral pollutant contamination are discussed herein in brief. The sources of contamination can be categorized into point and nonpoint classes. The point sources include aspects of some industrial and domestic activities (lawns, ornaments, pest control etc.), while the nonpoint sources are agricultural, forestry and some other related activities. Nonpoint-source pollution comes from many diffuse sources, and is caused by rainfall or snowmelt moving over and through the ground. Nonpoint-source pollutants include excess agricultural and residential fertilizers, herbicides, pesticides, toxic chemicals, sediments and nutrients from animal waste. These pollutants take the form of liquid and

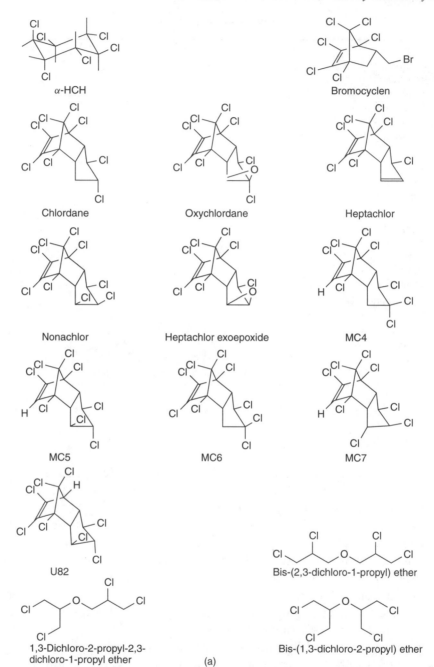

α-HCH

Bromocyclen

Chlordane

Oxychlordane

Heptachlor

Nonachlor

Heptachlor exoepoxide

MC4

MC5

MC6

MC7

U82

Bis-(2,3-dichloro-1-propyl) ether

1,3-Dichloro-2-propyl-2,3-
dichloro-1-propyl ether

Bis-(1,3-dichloro-2-propyl) ether

(a)

Figure 2.1   The chemical structures of some chiral pollutants. (a) Aliphatic organochlorine pesticides; (b) aromatic organochlorine pesticides; (c) phosphorous pesticides.

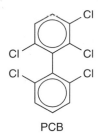

PCB

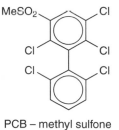

PCB – methyl sulfone

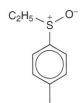

(S)-4-tolyl ethyl sulfoxide

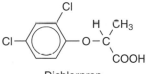

Dichlorprop

Methyl dichlorprop

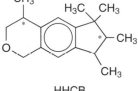

HHCB
(Galaxolide®)

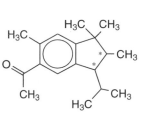

ATII
(Traseolide®)

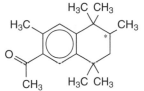

AHTN
(Tonalide®)

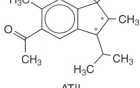

AHDI
(Phantolide®)

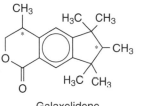

DPMI
(Cashmeran®)

Galaxolidone

(b)

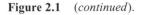

Figure 2.1 (*continued*).

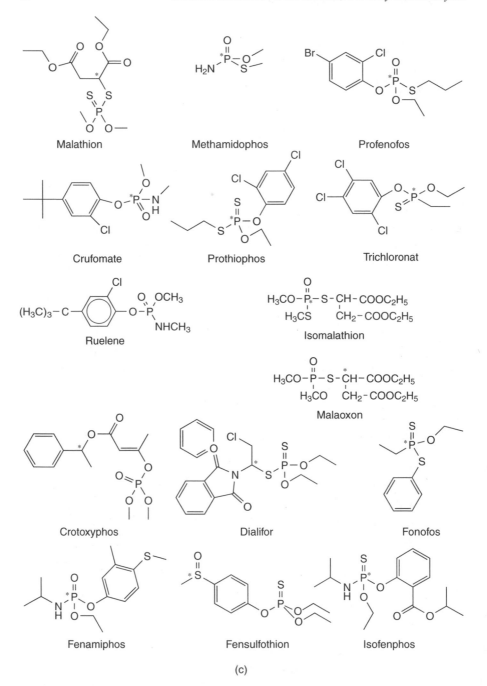

Malathion

Methamidophos

Profenofos

Crufomate

Prothiophos

Trichloronat

Ruelene

Isomalathion

Malaoxon

Crotoxyphos

Dialifor

Fonofos

Fenamiphos

Fensulfothion

Isofenphos

(c)

Figure 2.1 *(continued)*.

solid wastes from all types of farming, including run-off from pesticides, fertilizers and feedlots; erosion and dust from ploughing; animal manure and carcasses; and crop residues and debris. Agriculture is generally recognized as the leading nonpoint source of water pollutants, such as sediments, nutrients and pesticides.

There are many insecticidal, chemical and pharmaceutical industries that are discharging chiral pollutants into the environment. Many chiral pesticides are used to control pests in houses and other buildings, and in turn contaminate the environment. Chiral pesticides from agricultural and forestry soils contaminate the surface and ground water in the rainy season and during irrigation of the land. The basic route for environmental contamination by these pollutants follows the waste water from industrial and housing complexes, which contaminates rivers, ocean and ground water, sediments and soils. Pesticide sprays are also a source of atmospheric contamination. Therefore, the environment is continuously being contaminated by chiral pollutants from point and nonpoint sources.

Various pesticides – including organochlorines, which are very persistent in the environment – are used in agriculture and forestry to control pests. For example, chlordane and heptachlor have been banned since 1988 in many parts of the world, but they are still present in the environment. Some authors have reported the volatilization of these organochlorines and their metabolites, which has resulted in the contamination of the atmosphere [11–13]. These authors have made claims about the volatilization of chlordane, heptachlor and heptachlor *exo*-epoxide pesticides, which have accumulated in the atmosphere. They have also reported that the spray, used to control termites, is the main cause of the atmospheric contamination in the southern United States.

Bidleman *et al.* [12–15] carried out extensive studies of the air/water exchange of α-HCH pesticides by determining the levels of these pesticides in water, rain and air samples near Lake Ontario. The authors reported the contamination of air due to the volatilization of α-HCH from the lake water. Furthermore, Jantunen and Bidleman [16] discussed the α-HCH concentrations in water and air using the fugacity ratio: the concentrations of α-HCH in water and air are the same if the fugacity ratio equals one. In 1996, Harner *et al.* [17] carried out air–water exchange studies of α-HCH in the Barents Sea and the eastern Arctic Ocean. The authors reported 0.74–0.94 as the enantiomeric ratio of α-HCH in surface water, indicating selective transformation of this pesticide. The range of the enantiomeric ratio in air was 0.87–1.00, which indicates a high level of volatilization of α-HCH from the surface of the water.

2.3 The Distribution of Chiral Pollutants

The basic sources of chiral pollutants are point and nonpoint in origin. The point and nonpoint sources have already been discussed above. Mostly, chiral pollutants contaminate water resources, sediments and soils. However, these pollutants have also been reported in the atmosphere, and in animal and plant bodies [5, 6]. The percentage contamination of air is very low in comparison to that of water, sediments and soils [5]. Recently, Hegeman and Laane [7] reviewed the distribution of chiral pollutants in the environment. Briefly, chiral pollutants enter into the food chain and are transported into vegetables, fruit and other edible foodstuffs, and finally into the human body. This results in different types of disease, which will be discussed in Chapter 4.

2.3.1 Distribution in Water

Water is the most important and essential component of the universe and it plays a vital role in the proper functioning of the Earth's ecosystems. The quality of the water resources is deteriorating exponentially due to contamination by xenobiotics, especially chiral xenobiotics. Both point and nonpoint sources are responsible for the pollution of water resources. The ground and surface waters in many places have been reported to be contaminated by chiral pollutants. Briefly, water pollution due to these notorious xenobiotics is a serious issue, as it affects all living creatures, and it can negatively affect the use of water for drinking, household purposes, recreation, fishing, transportation and commerce. Under such circumstances, we will be in great difficulty after a few decades.

Since the 1970s, water quality regulations have focused on a select subset of chemicals that are deemed representative of the human impact on the environment. These regulated – so-called 'priority' – pollutants come from agrochemical and industrial manufacturing sources. However, for the past decade, European data have suggested that a wide range of chemicals from consumer activities could be trickling through our waterways. The US Geological Survey have released a report that shows that US waters are awash with trace amounts of pharmaceuticals, including contraceptives and antibiotics, as well as cholesterol, insecticides and caffeine, that have made the journey from the human body through sewage plants to our streams. These substances are largely unregulated and are present at such low levels that they elude the typical biodegradation processes at municipal wastewater treatment plants. According to Christian Daughton at the Environmental Chemistry of the US Environmental Protection Agency in Las Vegas, this is actually the first formal scan across the country of pollutants that result from people's actions and activities. The results related to the presence of pharmaceuticals in water were presented at the 3rd International Conference on Pharmaceuticals and Endocrine Disrupting Chemicals in Water, held by the National Ground Water Association on 19–21 March 2003, in Minneapolis, Minnesota. The Food and Drug Administration (FDA) has welcomed these results: according to Stephen Sundlof at the FDA's Center of Veterinary Medicine, the FDA will study the results of the US Geological Survey's national reconnaissance survey and, if necessary, take the appropriate actions to protect public health.

More than 700 pollutants have been reported in water, and these include substances that are both inorganic and organic in origin. Microbial populations are also contributing to the pollution of water resources to some extent. Water is a universal solvent for most polar organic compounds and so the presence of chiral pollutants in water is a common phenomenon. Therefore, almost every type of chiral polar pollutant is found in the various water bodies, as reported in the literature [5–7, 18]. The main water bodies contaminated by chiral pollutants are the oceans, rivers, lakes and ground water. In view of these points, the contamination of various kinds of water bodies is discussed in what follows.

Sea Water

Basically, the sea is the ultimate destination of any river, culvert or tributary and, therefore, it is very common to have chiral xenobiotics in marine water. Many workers have carried out qualitative and quantitative analysis of chiral pollutants in sea water. To make the concepts clear to readers, some examples are summarized here.

Jantunen and Bidleman [19] described the presence of organochlorine pesticides in Arctic Ocean water. The authors reported heptachlor, *exo*-epoxide (a metabolite of heptachlor), α-HCH and toxaphene in the sea water. They collected seawater samples from the surface layer (40–60 m) in the summers of 1993 and 1994. The concentrations of organochlorine pesticides that were determined indicated a spatial distribution of these pesticides in the Arctic Ocean from the Bering and Chukchi Seas to the North Pole, to a station north of Spitsbergen, and then south into the Greenland Sea. *Exo*-epoxide and α-HCH increased from the Chukchi Sea to the pole, and then decreased towards Spitsbergen and the Greenland Sea. Chlorinated bornanes (toxaphene) followed a similar trend, but levels were also high near Spitsbergen and in the Greenland Sea. A reverse trend was found for endosulfan, with lower concentrations in the ice-covered regions. Little variation was seen in the chlordane concentrations, although the *cis/trans*-chlordane ratio decreased at high latitudes. Enrichment of (+)-heptachlor *exo*-epoxide was found in all regions. Furthermore, the same group [14, 15, 20–22] have carried out investigations related to the analysis of α-HCH in coastal bays and marine waters of various regions all over the world.

Bethan *et al.* [23] reported the presence of α-HCH in the sea water of the North Sea. Enantioselective analysis of the samples indicated an unequal distribution of the enantiomers of this pesticide in the sea water. Due to the warmer water surface during late summer to early autumn, the equilibrium of α-HCH between air and water is dominated by volatilization rather than deposition. The volatilization of nonracemic α-HCH, as is known to occur from sea water, changes the enantiomeric ratio in the sea water. Moisey *et al.* [24] determined the concentrations of α-, β- and γ-HCH isomers and enantiomer fractions (EFs) of α-HCH in the water of the North Sea. The authors reported different distributions of these pesticides in the water. Hühnerfuss *et al.* [25] detected α-HCH, β-PCCH, γ-PCCH, chlordane, octachlordane, oxychlordane and heptachlor pesticides in marine water. Faller *et al.* [26] reported the enantiomeric ratios of α-HCH in the eastern North Sea, including German Bight and Skagerrak water. In 1992, the same group [27] detected (+)-α-HCH in the eastern part of the North Sea. Ludwig [28] has detected phenoxyalkanoic acid herbicides in the German Bight. The pesticides detected were 2,4-dichlorophenoxyacetic acid, (2,4-*D*)-2-(2,4-dichlorophenoxy) propionic acid (dichlorprop or DCPP) and 2-(4-chloro-2-methylphenoxy) propionic acid (MCPP). Gaul and Ziebarth [29] collected water samples from the North Sea and reported an enantiomeric ratio of α-HCH. Similarly, Hühnerfuss *et al.* [30] have reported α-HCH in the eastern part of the North Sea by collecting water samples from the German Bight and the Baltic Sea. Synthetic musks (nitro and polycyclic musks) are a group of chemicals that offer a wide range of important properties for environmental monitoring programs. They are produced as odorous chemicals and added to a wide variety of perfumes, toiletries and other household products. As such, they are directly applied to the human body in considerable concentrations in cosmetic products or in washed textiles, and they accumulate owing to dermal resorption. In addition, synthetic musks also enter the environment via wastewater treatment. Several polycyclic musks are chiral and have also been isolated from plants, from angelica root, Ambrette seed oil and so on. Musks have frequently been used in the perfume industry for a long time, and consequently they contaminate our environment. These perfume ingredients have been detected in marine water [31, 32].

River Water

Pesticides and agrochemicals are mostly used in agriculture and in forestry activities. These chiral and nonchiral xenobiotics contaminate rivers during the course of rainfall and irrigation processes. However, some chiral pesticides have been reported at the poles and, hence, found in river water which comes from melting of snow at the poles. Another possibility for the presence of chiral pollutants in river water may be the dynamic exchange of pollutants between river water and the atmosphere (near agricultural fields). In view of this, some workers have tried to detect chiral pesticides in river water.

Wong *et al.* [33] studied the enantiomeric composition of polychlorinated biphenyl (PCB) enantiomers in rivers from selected sites throughout the United States. Nonracemic enantiomeric fractions (EFs) were observed for PCBs 91, 95, 136 and 149 in aquatic samples. Bromocyclen, a new insecticide with a low toxicity to mammals, is currently in use in Europe for the treatment of domestic animals against ectoparasites. Therefore, bromocyclen has been reported in the waters of the Stör river, a tributary of the River Elbe in northern Germany [34–36]. Franke *et al.* [37] reported chlorinated bis(propyl)ethers in the waters of the River Elbe. Three isomers of bis(propyl)ethers have been reported in different parts of the river, and the authors identified an epichlorohydrin production site, close to the Czech border, as the source of this contamination. Many toxic pesticides, such as BHC, aldrin, dieldrin, DDT and so on, have been detected in the Yamuna River, which is a very important river in India [38]. Similarly, DDT, BHC, aldrin, dieldrin, malathion and so on have also been detected in another very famous river, the Ganges [39]. Recently, Ali and Jain [40] have reported the presence of lindane, malathion, BHC, p,p'-DDD, o,p'-DDT and methoxychlor in the Hindon River, also in India.

Eschke *et al.* [41, 42] reported polycyclic musks in the waters of the River Ruhr. Franke *et al.* [43] also detected polyaromatic musks in the waters of the Orda and the Elbe. Winkler and coworkers [44] reported clofibric acid and ibuprofen in the Elbe and the Sale. The authors also detected hydroxy-ibuprofen, a metabolite of ibuprofen, in Elbe river water. Ternes [45] also detected ibuprofen, with several other pharmaceuticals – antiphlogistics, lipid regulators, psychiatrics, anti-epileptics, beta blockers and β_2-sympathomimetics – in river water.

Lake Water

Lakes collect their water from agricultural and domestic run-off during the rainy season, and hence the presence of chiral pollutants in lake water is a common phenomenon. Chiral pollutants have been reported in lakes in several parts of the world. Bidleman *et al.* [14, 15, 20, 21] reported the presence of α-HCH in lake water. The enantiomeric ratios of α-HCH in the Amituk and Ontario lakes of Canada ranged from 0.65 to 0.99. Wiberg *et al.* [46] found nonracemic forms of heptachlor *exo*-epoxide in lake water. Many other chiral pesticides and agrochemicals, such as DDT, lindane and other phosphorous pesticides, have also been detected in lake water [18].

Ground Water

Ground water is thought to be the least contaminated of the various water bodies. However, due to pollution overload, it is being contaminated on a daily basis. Ground water is not untouched by the notorious chiral xenobiotics: it is polluted due to the infiltration of agricultural water during the rainy season and irrigation processes on agricultural and forestry soils. However, the ground water near rivers, canals and culverts has been found to be heavily contaminated, due to the seepage and infiltration of river water. Therefore, in brief, ground water pollution by chiral pesticides is a common process, as these pesticides leach into the ground water from soil, sediment or surface water. Many reports have been published on ground water contamination by pesticides and other toxic organic pollutants [47–51]. The chiral pesticides detected in ground water are DDT, chlordane, heptachlor, HCH, alachlor, metaclor, acetochlor and endosulfan. Buser *et al.* [52] reported the presence of ibuprofen in drinking water. Recently, Kümmerer [53] has reported the presence of several drugs in surface, ground and drinking water.

Waste Water

Its well known that waste water from housing, industry and other activities contains many toxic substances. Chiral xenobiotics have also been detected in waste water sampled in various parts of the world. Inspite of this, little work has been carried out on the detection of the notorious chiral xenobiotics in waste water. The analysis of sewage effluent from Saskatoon city, discharged to the South Saskatchewan River, has indicated the presence of clofibric acid and two metabolites, hydroxy- and carboxy-ibuprofen. Kümmerer *et al.* [54] reported some pharmaceuticals, such as antineoplastics, carcinogenics, mutagenics, teratogenics and fetotoxics, in wastewater samples. Gatermann *et al.* [31, 55–57] reported the presence of musks in waste water. Yamagishi *et al.* [58, 59] also identified nitro musks in waste water. Eschke *et al.* [41, 42] reported the presence of polycyclic musks at some municipal sewage treatment plants. Weigel [60] has reported the presence of several drugs in waste water. Subsequently, Buser *et al.* [52] identified ibuprofen, a nonsteroidal anti-inflammatory drug, in waste and river waters. Furthermore, the authors also reported a high concentration of the active *S*-enantiomer of ibuprofen, and they detected two metabolites of ibuprofen, namely hydroxy- and carboxy-ibuprofen, in waste water. Ternes [45] also detected ibuprofen, with several other pharmaceuticals, in waste water. Bethan *et al.* [36] reported the presence of bromocyclen in waste waters that are contaminating the River Elbe in northern Germany. Buser and Müller [61, 62] reported four isomers of HCH (i.e. α-, β-, γ- and δ-isomers) in waste water and sewage sludge. The distribution of chiral pollutants in water resources is given in Table 2.2.

2.3.2 Distribution in Sediment

Chiral pollutants in water remain in dynamic equilibrium with the sediment in a particular water body. Therefore, the presence of chiral pollutants in sediment is a natural situation. The concentration of these pollutants in the sediment depends on the various environmental factors and the adsorption capacity of the sediment itself. Vetter *et al.* detected two major compounds of technical toxaphene in Canadian lake sediment [63], and chloroborane congeners in the sediment from a toxaphene-treated Yukon lake [64]. The authors analysed sediment samples collected over almost 60 years, from 1935 to 1992. The results are given in Table 2.3, which

Table 2.2 The distribution of some chiral pollutants in various water resources

Chiral pollutant	Water resource	References
α-HCH	Sea water	26, 27, 153–155
	Sea water (Great Britain)	26
	Rain water	25
β-PCCH	Rain water	25
γ-PCCH	Sea water	27, 155, 156
DCPP	Sea water	61, 157
Clofibric acid	River water	44
Ibuprofen	River water	44
Polyaromatic musks	River water	41–43
HCH	River water	47–51
Chlordane	River water	47–51
Heptachlor	River water	47–51
Acetachlor	River water	47–51
Metachlor	River water	47–51
Endosulfan	River water	47–51

Table 2.3 The enantiomeric ratios of toxaphene congeners B7-1001 and B7-923 in sediment from Hanson Lake, Canada [63, 64]

Sampling date	B7-1001	B6-923
1935	< 1.0	0.97
1946	0.80	1.00
1954	0.81	1.01
1959	0.82	0.97
1964	0.82	1.06
1968	0.81	0.98
1973	0.78	0.98
1979	0.77	0.98
1984–7	–	0.99
1992	0.71	0.96

indicates the different enantiomeric ratios of these pesticides from 1935 to 1992. Another study of chiral pesticides in sediment has been carried out by Rappe *et al.* [65], in Baltic Sea sediment. The sampling was carried out 150 km off the coast in 1985–6. The authors detected the presence of hexachloroborane in the sediment. Benicka *et al.* [66] also identified PCBs 95, 91 and 84 in river sediment. A most comprehensive study of the distribution of chiral pollutants, specially musks, in sediment obtained from wastewater plants has been carried out by Gatemann *et al.* [31, 55–57].

Wong *et al.* [67] measured the enantiomeric ratios for eight polychlorinated biphenyl enantiomers in aquatic sediments from selected sites throughout the United States. Nonracemic enantiomeric ratios (ERs) for PCBs 91, 95, 132, 136, 149, 174 and 176 have been found in sediment cores from Lake Hartwell. Nonracemic ERs for many of the enantiomers have also been found in riverbed sediment samples from the Hudson and Housatonic Rivers. Patterns in enantiomeric ratios among congeners were consistent with known reductive dechlorination patterns in both river sediment basins. The enantioselectivity of PCB 91 was found to be reversed between the Hudson and Housatonic River sites. Moisey *et al.* [24] determined the concentrations of α-, β- and γ-hexachlorocyclohexane isomers and enantiomer fractions of α-HCH in sediment samples obtained from the North Sea. The authors have reported different distributions of these pesticides in the sediment samples.

2.3.3 Distribution in Soil

Agricultural and forestry soils are the direct victims of the notorious chiral pesticides, which are used directly on the soil in agricultural and forestry pest control measures. Therefore, these pesticides have inevitably been detected. Aigner *et al.* [68] reported the enantiomeric ratios of chlordane pesticides in the soil of the Midwestern United States. The soil samples were collected from 38 farmsteads and eight cores in Ohio, Pennsylvania, Illinois and Indiana, from 1995 to 1997. The pesticides detected in these samples were chlordane, heptachlor and heptachlor *exo*-epoxide. Differing enantiomeric ratios of these pesticides were reported in soil samples obtained from different states. Lee *et al.* [69] reported technical chlordane in compost. The authors undertook the first comprehensive examination of technical chlordane residues in a variety of composts – specifically, 13 commercial and 39 municipal compost products – to both characterize and quantify the magnitude of this point source of contamination. The concentrations and the compositional and enantiomeric profiles of the chlordane components were determined. Of the 13 commercial products, nine contained detectable chlordane concentrations, ranging from 4.7 to 292 $\mu g\,kg^{-1}$ (dry weight), while all 39 municipal products contained chlordane residues ranging from 13.9 to 415 $\mu g\,kg^{-1}$ (dry weight). The residue concentrations and profiles suggested possible feedstock sources for the chlordane in the finished compost products. The data also support the conclusion that some composts contribute to anthropogenic cycling of chlordane through the biosphere. Eitzer *et al.* [70] described the presence of chlordane pesticides in soil samples collected from the USA, and Fingerling *et al.* [71] detected chlorinated boranes in soil samples.

Wiberg *et al.* [72] reported the presence of organochlorine pesticides in 32 agricultural and three cemetery soils from Alabama. The enantiomeric signatures were similar to those from other soils in Canada and the USA. The enantiomer fractions (EFs) of o,p'-DDT showed great variability, ranging from 0.41 to 0.57, while the EFs of chlordanes and chlordane metabolites were less variable and, in general, differed significantly from racemic. The enantioselective depletion of (+)-*trans*-chlordane and (+)-*cis*-chlordane and the enrichment of (+)-heptachlor-*exo*-epoxide and (+)-oxychlordane was found in a large majority of the samples. The enantiomeric composition of the α-hexachlorocyclohexane was racemic or close to racemic. Lewis *et al.* [4] reported the presence of dichlorprop pesticide in Brazilian soils. Recently, White *et al.* [73] have reported *cis*- and *trans*-chlordanes in the soil of a greenhouse unit. The authors have reported the enantiomeric ratios of *cis*- and *trans*-chlordanes in different soil samples with higher concentrations of the (−)- and (+)-enantiomers of *trans*- and *cis*-chlordanes, respectively.

Archived background soils (Broadbalk, 1944–86) and sludge amended soils (Luddington, 1968–90) were collected from long term agricultural experiments in the UK by Meijer *et al.* [74]. The authors analysed these samples for organochlorine pesticides, to establish trends over time. The concentrations typically ranged from 0.1 to 10 ng g^{-1} of soil (dry weight), with γ-HCH, dieldrin and p,p'-DDE consistently having the highest concentrations. The trends in the Broadbalk background soils are largely consistent with usage patterns, with peak concentrations occurring in the 1960s for DDTs and between the 1960s and the 1980s for the other organochlorine compounds (OCs). In the Luddington control and sludge amended soils, several of the OCs showed a significant decline in concentrations from the late 1960s to 1990, with half-lives ranging from approximately 7 years (α-HCH) to approximately 25 years (dieldrin). The sludge amended plot received 125 t of sludge per hectare in 1968, which was mixed in to a depth of 15 cm. It appears that the sludge treatment had little effect on the concentrations in the soil, with no significant difference between the control soil and the sludge amended soil for most compounds, except for HCB, p,p'-DDE and dieldrin. The enantiomeric fractions (EFs) of some chiral pesticides (α-HCH, *cis*- and *trans*-chlordane, and o,p'-DDT) were determined in the Luddington soils. The results revealed that enantioselective degradation of OC pesticides was occurring in these soils for *trans*-chlordane (TC) and *cis*-chlordane (CC). However, the depletion over time was not statistically significant, and there was no statistically significant difference between the EFs in the control soil and sludge amended soil. This indicates that enantioselective microbial degradation was not consistent over time, and that the addition of sludge to soil did not significantly alter the enantiomeric preference of the microbial community.

Li *et al.* [75] reported phenthoate and its enantiomeric ratio (ER) in three soil samples. The recoveries of phenthoate from three different types of soil, fortified at levels of 0.1, 1.0, 10.0 μg g^{-1}, ranged from 75 % to 94 %, with relative standard deviations (RSDs) of 1.5–6.5 %. Hutta *et al.* [76] described epoxiconazole in soil samples with two enantiomers of commercially available triazole fungicide epoxiconazoles. Buser *et al.* [77] reported metalaxyl – which is currently being replaced in many countries by metalaxyl-M, the same fungicide enantiomerically enriched with the biologically active *R*-enantiomer – in soil samples. Leone *et al.* [78] reported several chiral organochlorine pesticides in soil samples collected from the Cornbelt region of the USA. The results are given in Table 2.4.

Table 2.4 The enantiomeric ratios of *cis*- and *trans*-chlordane pesticides in soil samples [78]

Soil sample	Enantiomeric ratio of *cis*-chlordane	Enantiomeric ratio of *trans*-chlordane
Pre-bulk (prior to plantation)	1.22	0.861
Post-bulk (after plantation)	1.25	0.872
Near plant root	1.24	0.852

2.3.4 Distribution in the Air

Qualitative and quantitative air pollution due to chiral pesticides is quite low, because only volatile chiral pollutants can contaminate the air, and these are quite low in number. However, some chiral pollutants have been detected in the air. The concentrations of chiral pollutants in the air vary from one place to other and also depend on the local activities. Aigner *et al.* [68] reported the enantiomeric ratios of chlordane pesticides in the air of the Midwestern United States. The air samples were collected from 1995 to 1997, in Ohio, Pennsylvania, Illinois and Indiana, from ambient, above-soil and indoor air. The pesticides detected in these samples were chlordane, heptachlor and heptachlor *exo*-epoxide. Differing enantiomeric ratios of these pesticides were reported in these air samples. The enantiomeric ratios for ambient air samples taken in rural and agricultural areas in the Cornbelt show the same trend as was seen in the soil and in above-soil air samples. Similarly, Bidleman *et al.* [11, 12, 79] collected air samples from the Cornbelt, in South Carolina and Alabama. The samples were again collected from above-soil, ambient and indoor air. The authors reported the presence of *cis*-chlordane, *trans*-chlordane, heptachlor and heptachlor *exo*-epoxide in these samples. Jantunen *et al.* [80] collected air samples from the Arctic and determined the levels of chiral pesticides. The authors reported the presence of α-HCH pesticide, the enantiomeric ratios of which varied from 0.93 to 1.07 and showed a slight variation with latitude. Furthermore, the authors reported a depletion of $(-)$-α-HCH at low latitude and of $(+)$-α-HCH at high latitude, respectively. The enantiomeric ratios of α-HCH at different latitudes are shown in Figure 2.2. It is well known that the pesticides volatilize from water resources and from the soil surface and then contaminate the air. These pesticides have rolled towards the Earth's poles as they have moved from a warmer atmosphere to cooler air. Therefore, the pesticides are also found in the atmosphere and in snow at the poles and, consequently, they move down across the Earth again as the snow melts. The results of these studies are given in Table 2.5.

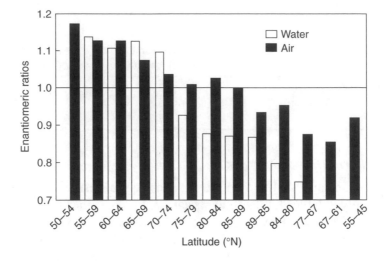

Figure 2.2   The enantiomeric ratios of α-HCH at different latitudes [6].

Table 2.5   The enantiomeric ratios of chlordane, heptachlor and heptachlor *exo*-epoxide in air from the Cornbelt region, South Carolina and Alabama [11, 12]

Location	*trans*-Chlordane	*cis*-Chlordane	Heptachlor	Heptachlor *exo*-epoxide
Cornbelt				
Above	0.74	1.11	–	–
Ambient	0.93	1.04	–	–
Indoor	0.99	0.98	–	–
Alabama				
Ambient	0.98	1.01		
Indoor	0.98	1.00		
Columbia	–	–	0.99	1.51

Dua [81] has detected the α-, β-, γ- and δ-isomers of HCH in rain water in India. It may be concluded that these pesticides were present in the air and hence contaminated the rain water. Ridal *et al.* [21] have detected α-HCH in air samples collected above the surface of Lake Ontario. In one study, Ulrich [82] reported the existence of chlordane in an air sample near the Great Lakes. The sampling (48 samples) was carried out between 1994 and 1995, and the main sampling sites were Lakes Erie, Michigan and Superior. The overall enantiomeric ratio of *trans*-chlordane was 0.88. Other authors who have described the presence of chiral pesticides in air samples include Wiberg *et al.* [46] (chlordane) and Buser and Müller [83] (heptachlor and chlordane).

The exchange of α-HCH between air and water has been studied by Ridal *et al.* [21]. Hühnerfuss and Garrett [84] reported the air–sea exchange of several pesticides. Similarly, Falcorner *et al.* [14] studied the air–water exchange of hexachlorocyclohexanes: the authors collected water and air samples from Amituk Lake, Cornwallis Island (Canada), and off the coast of Cornwallis Island in Resolute Bay. The concentrations of the pesticides varied from one location to another. Jantunen and Bidleman [22] reported the air–water exchange of α-HCH in the Arctic Ocean and near Nova Scotia in the North Atlantic. These authors advocated that this sort of exchange is a two-way process. At equilibrium, the net flux is zero but volatilization and back deposition occur at equal rates for chiral pesticides. Bethan *et al.* [23] reported dynamic equilibrium between the concentrations of α-HCH in the air and water of the North Sea. Enantioselective analyses of the sample extracts yielded evidence of the change in the direction of the net air–water gas transfer of the contaminant, depending on the season. As previously mentioned, due to the warmer water surface during late summer to early autumn, the equilibrium of α-HCH between air and water is dominated by volatilization rather than deposition. The volatilization of nonracemic α-HCH, as is known to occur from sea water, changes the enantiomeric ratio in the air, which is reflected in the observed ratio in rainfall that passes through the air column.

2.3.5 Distribution in Aquatic and Amphibian Biota

In addition to water, sediments, soils and air, chiral pollutants have been detected in some animals and plants. In 1991, König *et al.* [85] reported α-HCH in organisms at different trophic levels. The authors reported the presence of α-HCH in the liver, kidney and muscles of eider duck (*Somateria mollissima*). Subsequently, in 1994, Möller *et al.* [86] also reported α-HCH in the brain of the same duck. Kallenborn *et al.* [87] detected α-HCH in eider duck. In another study, Kallenborn [88] has also identified α-HCH in eider duck, whose diet was molluscs: the pesticide was found in various organs, such as the liver, kidney, muscles and so on. Similarly, several authors have reported α-HCH in different animals, such as seals [25, 86, 89–93], whales [91], polar bears [94] and so on. Müller *et al.* [92] reported the enantiomers of oxychlordane in harbour seals (*Phoca vitulina*) and grey seals (*Halichoerus grypus*). The authors reported higher concentrations of (−)- and (+)-enantiomers in harbour and grey seals, respectively. Wiberg *et al.* [94] also detected α-HCH in the Arctic marine food chain, placing special emphasis on polar bears (*Ursus maritimus*). The food for the polar bears was the blubber of ringed seals (*Phoca hispide*), Arctic cod (*Boreogadus Saida*) and other amphibians. An Arctic cod tissue extract showed α-HCH, which indicated that bioaccumulation took place with or without selective metabolism. This conclusion was in agreement with results obtained by other authors [90, 95]. Furthermore, the same group [96] have studied the concentrations and enantiomer fractions (EFs) of organochlorine compounds (OCs) in the tissues of grey seal and salmon (*Salmo salar*) originating from the Baltic Sea. The selected seal specimens ranged from starved to nonstarved animals, and some of them suffered from a disease complex; while the salmon samples originated from individuals that were known to produce offspring with and without the M74 syndrome. Significant differences in residue levels and EFs were found between the seal groups, but not between the M74 and non-M74 salmon. The relations hips between the chemical and biological variables of the seal samples were investigated using multivariate statistics.

Moisey *et al.* [24] established the concentrations of α-, β- and γ-hexachlorocyclohexane (HCH) isomers and enantiomer fractions (EFs) of α-HCH in pelagic zooplankton (six species), Arctic cod, sea birds (seven species) and ringed seals. Hühnerfuss *et al.* [25] detected α-HCH, β-PCCH, γ-PCCH, chlordane, octachlordane, oxychlordane and heptachlor pesticides in some marine biota. Kallenborn *et al.* [97] and Buser and Müller [98] studied the blubber of a harbour seal for the detection of chloroborane pesticides. Subsequently, Hühnerfuss *et al.* [85, 95, 99] reported the presence of chlordane, oxychlordane and heptachlor *exo*-epoxide in various biota samples. Furthermore, the same group [95, 99] has performed systematic investigations on the enantiomeric distribution of oxychlordane in seagull eggs. Similarly, Müller *et al.* [92] also indicated the presence of oxychlordane in harbour seals and grey seals. The authors reported a high concentration of ($-$)-enantiomer in the harbour seals, while a higher concentration of ($+$)-enantiomer was found in the grey seals. Out of 209 PCBs, 78 possess axial chirality and occur in enantiomeric form. It has been reported that 19 PCBs, which are present in commercial PCB formulations, exist as stable isomers at ambient temperature [100, 101]. Jansen and coworkers [102] reported environmental contamination by PCBs, including their effects on air, water, sediment, fish, wildlife and human beings. Several authors have subsequently reported the presence of PCBs in sea water and at different trophic levels of the marine ecosystem [103–112]. Blanch *et al.* [111] reported atropisomers of three PCBs in shark (*Centroscymnus coelolepis*) liver samples. Vetter *et al.* [113] reported enantio-enrichment of PCB 149 in the blubber of an adult female harbour seal. Reich *et al.* [112] determined enantiomeric ratios of PCBs in dolphins (*Stenella coeruleolba*) from the Mediterranean Sea. Wiberg *et al.* [114] reported the presence of methylsulfonyl PCBs in the adipose tissues of Arctic ringed seals and polar bears. For example, the enantiomeric ratios of 3-methylsulfonyl PCBs (MeSO$_2$-CB149) were 0.32 and < 0.1 in ringed seal blubber and polar bear fat, respectively. Wong *et al.* [33] studied the enantiomeric composition of polychlorinated biphenyl (PCB) enantiomers in riparian biota – fish, bivalves, crayfish, water snakes and barn swallows (*Hioundo rustica*) – from selected sites throughout the United States. Non-racemic enantiomeric fractions (EFs) were observed for PCBs 91, 95, 136 and 149 for aquatic and riparian biota from Lake Hartwell, a reservoir heavily contaminated with PCBs, and for these congeners and PCBs 132, 174, 176 and 183 in river fish. Species-dependent patterns in PCB EFs were observed, which suggested differences in the ability of different species to accumulate PCBs enantioselectively. Furthermore, the same group [115] determined the enantiomeric ratios (ERs) and enantiomeric fractions (EFs) of a number of chiral organochlorine pesticides and PCB atropisomers in five standard (SRM) and certified (CRM) reference materials; that is, SRM 1588a (organics in cod liver oil), SRM 1945 (organics in whale blubber), Marine Mammal Quality Assurance Exercise Control Material IV (NIST IV, organics in whale blubber), CRM trout and CRM EC-5 (sediment). The target analytes were *cis-* and *trans-*chlordane, heptachlor *exo-*epoxide, oxychlordane, U82, MC5, MC6, MC7, *o,p'-*DDT and PCB congeners 91, 95, 136, 149, 174, 176 and 183. These values should contribute to the quality assurance/quality control methodologies for chiral environmental chemistry using standardized reference materials. Wong *et al.* [116] described chiral organochlorine compounds – that is, α-HCH, *trans-*chlordane and chlorobiphenyls (CBs) 95 and 136 – in immature rainbow trout (*Oncorhynchus mykiss*).

Similarly, Hühnerfuss *et al.* [107, 117] detected PCBs in marine and limnic biota tissues. The authors reported the presence of these xenobiotics in blue mussels (*Mytilus edulis*). Toxaphene pesticide was used in the southeastern US states for cotton and soya bean crops, but was subsequently banned by the EPA in 1986, due to its long persistence. Hoekstra *et al.* [118] collected blubber and liver samples from the bowhead whale (*Balaena mysticetus*) in Canada in 1997–8. The authors reported the presence of eight chiral PCB congeners (PCBs 91, 95, 135, 136, 149, 174, 176 and 183) in these samples. De Gues *et al.* [119] reported the presence of toxaphene, produced in large quantities similar to those of polychlorinated biphenyls, in high levels in fish from the Great Lakes and in Arctic marine mammals (up to 10 and 16 μg g^{-1} lipid). Because of the large variability in total toxaphene data, few reliable conclusions were drawn about trends or the geographical distribution of the toxaphene concentrations.

To examine the influence of diet and age on organochlorine (OC) contaminant concentrations in two closely related ringed seal populations, enantiomeric fractions of chiral contaminants and stable isotopes of nitrogen (^{15}N) and carbon (^{13}C) were measured along with OCs in ringed seals collected from the east and west sides of the Northwater Polynya by Fisk *et al.* [120]. Seals from these two locations were allowed to feed at the same trophic level, based on ^{15}N values in muscle, but had slightly different sources of carbon, based on ^{13}C measurements in muscle. After removal of the influence of age, sex and blubber thickness, the OC concentrations did not vary between the ringed seals from the east and west sides of the polynya. PCBs, DDT and chlordane were found to increase with age for both male and female seals. The inclusion of older (> 20 years) female seals, which may have a reduced reproductive effort, may influence the relationships in females. Stable isotopes failed to describe OC concentrations in ringed seals, which suggests that diet was not a major factor in the variation of the OC concentrations within this ringed seal population. It has also been reported that *cis*- and *trans*-chlordane, oxychlordane and heptachlor *exo*-epoxide were all nonracemic in the ringed seal blubber, but did not vary with the age, sex or collection site. EF values in the ringed seals varied considerably from those of other Arctic marine mammals and sea birds. Furthermore, the same authors [121] have determined the concentrations of chlordane in seven species of sea birds – that is, dovekie (*Alle alle*, DOVE), thick-billed murre (*Uria lomvia*, TBMU), black guillemot (*Cepphus grylle*, BLGU), black-legged kittiwake (*Rissa tridactyla*, BLKI), ivory gull (*Pagophila eburnea*, IVGU), glaucous gull (*Larus hyperboreus*, GLGU) and northern fulmar (*Fulmaris glacialis*, NOFU) – found in the Arctic Northwater Polynya between Ellesmere Island and Greenland, which has a high biological productivity compared with other Arctic marine areas. To determine the concentrations of chlordane in liver and fat, and to examine species differences, the enantiomeric fractions (EFs) of chiral chlordane were studied: these varied by over an order of magnitude among the species investigated, from a low of 176 ± 19 ng g^{-1} (lipid corrected) in TMBU liver to a high of 3190 ± 656 ng g^{-1} (lipid corrected) in NOFU liver. The lipid concentrations of chlordane did not vary between the sexes for any species or between fat and liver, except for the DOVE, which had fat concentrations significantly greater than those of the liver. The ^{15}N values described a significant percentage of the variability of concentrations for most chlordane components, although less than has been reported for whole food chains. Slopes of ^{15}N values versus the concentrations of chlordane components and chlordane were similar, with the exception of those that were metabolized (*trans*-chlordane) or formed through biotransformation (oxychlordane). The relative proportions of the chlordane components in the sea birds were related to phylogeny; the procellariid (NOFU) had the greatest percentage of oxychlordane (> 70 %), followed by the larids (BLKI, IVGU and GLGU; 40–50 %) and the acids (DOVE and BLGU; 10–20 %). The exception was TBMU, an acid, where oxychlordane made up > 40 % of its chlordane.

Gao *et al.* [122, 123] studied the uptake of organophosphorus (OP) – that is, malathion, demeton-S-methyl, and crufomate and DDT pesticides – *in vitro*, using the axenically aquatic cultivated plants parrot feather (*Myriophyllum aquaticum*), duckweed (*Spirodela oligorrhiza*) and elodea (*Elodea canadensis*). The results of this study show that the selected aquatic plants have the potential to accumulate chiral pesticides. This study provides new information on plant biochemistry as related to chiral pollutant accumulation. Garrison *et al.* [124] studied the concentration of *o,p*-DDD in fish tissues. The fish species collected were channel catfish (*Ictalurus punctatus*), buffalo (*Ictiobus cyprinellus*) and largemouth bass (*Micropterus salmoides*). The enantiomeric ratio in these fish ranged from 0.40 to 0.79. Gaterman *et al.* [55, 56], Eschke *et al.* [32] and Rimkus *et al.* [125] reported the presence of nitro musks in fish, mussels and shrimps. The authors have also detected these pesticides, along with some polycyclic musks, in human adipose tissue and milk [126–128]. In another study, Eschke *et al.* [41, 42] detected polycyclic musks in fish from the River Ruhr. The compounds HHCB (Galaxolide™) and AHTN (Tonalide™) mainly represent synthetic polycyclic musk fragrances. Because of their volume of use and their bioaccumulation potential, there is concern with respect to their environmental safety. HHCB and AHTN are chiral compounds, the highest lipid concentrations of which were observed in mussels (*Dreissena polymorpha*), tench (*Tinca tinca*) and crucian carp (*Carassius carassius*) [43]. Furthermore, Gatermann *et al.* [129] have reported the chiral polycyclic musks HHCB, AHTN, AHDI and ATII in 18 fish samples (rudd, tench, crucian carp and eel) and in one pooled zebra mussel sample from the pond of a municipal sewage treatment plant. Hühnerfuss *et al.* [34–36] detected the new pesticide bromocyclen in the tissues of some fish. The fish were collected for this purpose, from a fish farm in Denmark and from the River Stör, were rainbow trout (*Oncorhynchus mykiss*), orfe (*Leuciscus idus*), bream (*Abramis brama orientalis*) and pike (*Esox lucius*). The values of the pesticide reported ranged from 0.093 to 1.20 mg kg^{-1} of fat.

2.3.6 Distribution in Terrestrial Biota

Some chiral pollutants are very stable and persist in the environment for a long time. Therefore, these pesticides enter into the bodies of different terrestrial animals through the food chain. Reports have also been published indicating the concentrations of various pollutants in terrestrial animals. Möller *et al.* [130] determined α-HCH levels in the liver and brain of sheep (*Ovis ammon*) in Germany. In the fat of the liver, a depletion of the (+)-enantiomer was reported, whereas the situation was reversed in the brain of the same species, with the (+)-enantiomer being dominant. The authors also reported the concentration of α-HCH in sheep and aquatic biota, and determined different values without observing any distribution pattern. The enrichment of α-HCH in the fat, liver and brain tissues of sheep bred in the northern German state of Schleswig-Holstein is shown in Figure 2.3. Pfaffenberger *et al.* [131] analysed liver samples of roe deer for α-HCH, heptachlor *exo*-epoxide and oxychlordane pesticides. In all of the samples, concentrations of α-HCH were reported in the range of 20–140 µg kg^{-1} of fat. The authors have reported a high concentration of α-HCH in liver due to the isomerization of γ-HCH into α-HCH. In addition to this, the concentrations of α-HCH, oxychlordane, heptachlordane and heptachlor *exo*-epoxide have been determined in the hare (*Lepus europaens*) [95, 99]. The concentrations of these pesticides ranged from 20 to 140 µg kg^{-1} of fat. Rimkus and Wolf [132] reported *exo*- and *endo*-isomers of heptachlor in the liver of roe deer. In all of the samples collected, high concentrations of heptachlor *exo*-epoxide were reported. The concentration of oxychlordane was from 10 to 60 µg kg^{-1} of fat, while that of heptachlor *exo*-epoxide was 10 to 100 µg kg^{-1} of fat. Various concentrations of α-HCH, oxychlordane and heptachlor *exo*-epoxide pesticides have also been reported in the hare by many workers [95, 99, 132]. The values of these pesticides in hare liver ranged from 10 to 100 µg kg^{-1} of fat. Portig *et al.* [133] reported the preferential accumulation of (+)-α-HCH. Möller [134], in Germany, analysed the cerebellum and white matter parts of the human brain and reported the different concentrations of α-HCH. Furthermore, high concentrations of α-HCH have been reported in the cerebellum, with a dominant concentration of the (−)-enantiomer compared to the (+)-enantiomer.

Glausche *et al.* [135] reported the presence of PCBs in human milk. Similarly, Ramos *et al.* [136] and Haglund and Wiberg [137] detected PCBs in

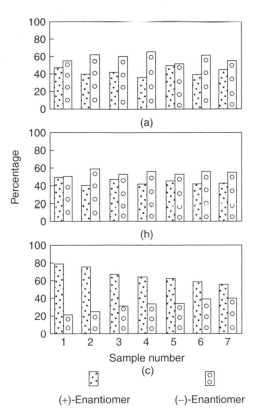

Figure 2.3 The enrichment of α-HCH in (a) fat, (b) liver and (c) brain tissues of sheep bred in the north German state of Schleswig-Holstein [130].

cow's milk. Ellerichmann and coworkers [138, 139] reported the presence of some methyl sulfone PCBs in birds and mammals, including human beings. The methyl sulfone (Me_2SO_2) PCBs found in human liver were 3-Me_2SO_2-2, 5, 6, 2′, 3′, 4′-hexachlorobiphenyl, 3-Me_2SO_2-2, 5, 6, 2′, 4′, 5′-hexachlorobiphenyl and 4-Me_2SO_2-2, 3, 6, 2′, 4′, 5′-hexachlorobiphenyl. These pesticides have also been detected in human lungs [136]. Alder *et al.* [140] found high concentrations of chlordane in the milk of women who consumed a high proportion of seafood in their diet. In the same study, the authors have reported toxaphene in the adipose tissue of monkeys, whose feed contained the same pesticide. Jansen and Jansson [141] reported methyl sulfone PCBs ($MeSO_2$-PCBs) in Baltic seal blubber.

Mattina *et al.* [142] carried out an extensive study on the distribution of *cis*- and *trans*-chlordane pesticides in different plants. Furthermore, the authors reported the presence of these pesticides in the root, stem and leaves of the plants. The studied plants include zucchini (*Capsicum pepo*), pumpkin (*Cucurbita pepo*), cucumber (*Cucumis sativus*), lettuce (*Lactuca sativa*), spinach (*Spinacua oleracea*), pepper (*Capsicum annum*) and tomato (*Lycopersicon esculentum*). The authors reported the enantiomeric ratios of the *cis*- and *trans*-chlordane pesticides, with higher percentages of the (+)- and (−)-enantiomers of the *cis*- and *trans*-chlordanes, respectively. The data are summarized in Table 2.6. Furthermore, the same group [73] have also studied the distribution of *cis*- and *trans*-chlordanes enantiomers in the root, stem, leaves, fruit peel and fruit flesh parts of rhizosphere plants. The authors reported a higher concentration of the (−)-enantiomer of *trans*-chlordane in these plant parts, whereas the (+)-enantiomer of *cis*-chlordane was found in greater concentrations. The occurrence of musk in various human body parts has been reported by a number of workers [126–128]: these pollutants have been reported in adipose tissues and in the milk of women.

Larsson *et al.* [143] described methyl sulfone PCBs (MeSO$_2$-PCBs) in the tissues of animals and humans. The aim of this study was to investigate the presence of atropisomers of MeSO$_2$-PCB congeners in the tissues of rats exposed to a technical PCB product, Clophen A50. The changes in the enantiomer fractions (EFs) of the MeSO$_2$-PCB atropisomers after exposure were also described. Liver, lung, and adipose tissue from rats dosed with Clophen A50 was analysed for the MeSO$_2$-PCB atropisomers of 3-methylsulfonyl-2, 2′, 4′, 5, 6-pentachlorobiphenyl (5-MeSO$_2$-CB91), 4-methylsulfonyl-2, 2′, 3, 4′, 6-pentachlorobiphenyl (4-MeSO$_2$-CB91), 3-methylsulfonyl-2, 2′, 3′, 4′, 5, 6-hexachlorobiphenyl (5′-MeSO$_2$-CB132), 4-methylsulfonyl-2, 2′, 3, 3′, 4′, 6-hexachlorobiphenyl (4′-MeSO$_2$-CB132), 3-methylsulfonyl-2, 2′, 4′, 5, 5′, 6-hexachlorobiphenyl (5-MeSO$_2$-CB149) and 4-methylsulfonyl-2, 2′, 3, 4′, 5′, 6-hexachlorobiphenyl (4-MeSO$_2$-CB149). In all of the tissues analysed, especially the lung, the *p*-MeSO$_2$-PCBs were more abundant than the *m*-MeSO$_2$-PCBs. The concentration ratio was higher for 4-MeSO$_2$-CB149 versus 5-MeSO$_2$-CB149 than for the corresponding 4-/5-MeSO$_2$-CB91 and 4′-/5′-MeSO$_2$-CB132 ratios. The enantioselective MeSO$_2$-PCB analysis of the lung samples showed an excess and a dominance of the second eluting atropisomer of 4-MeSO$_2$-CB149. In both the lung and adipose tissue, small amounts of the first eluting atropisomer of 5-MeSO$_2$-CB149 were present, but this atropisomer was not detected in the liver. No significant time-dependent changes in the EFs of the 4-MeSO$_2$-CB91, 5′-MeSO$_2$-CB132, 4′-MeSO$_2$-CB132, 5-MeSO$_2$-CB149 and 4-MeSO$_2$-CB149 atropisomers

Table 2.6 The enantiomeric ratios of *cis*- and *trans*-chlordane in various plants [142]

Plant	Enantiomeric ratio of *cis*-chlordane	Enantiomeric ratio of *trans*-chlordane
Zucchini		
Roots	0.55	0.42
Stem	0.58	0.36
Leaves	0.58	0.47
Whole fruit	0.60	0.46
Fruit peel	0.59	0.40
Fruit flesh	0.62	0.48
Pumpkin		
Roots	0.51	0.51
Stem	0.46	0.49
Leaves	0.54	0.48
Whole fruit	0.56	0.47
Fruit peel	0.55	0.47
Fruit flesh	0.56	0.47
Cucumber		
Roots	0.54	0.42
Stem	0.52	0.38
Leaves	0.52	0.37
Whole fruit	0.51	0.25
Fruit peel	0.48	0.30
Fruit flesh	0.50	0.22
Lettuce		
Roots	0.55	0.46
Leaves	0.54	0.41
Spinach		
Roots	0.53	0.46
Leaves	0.58	0.45
Pepper		
Roots	0.55	0.48
Stem	0.56	0.54
Leaves	0.52	0.51
Tomato		
Roots	0.55	0.37
Stem	0.59	0.30
Leaves	0.52	0.30

were found for either lung, liver or adipose tissue. The results of this study suggested that enantioselective formation occurred for both *m*- and *p*-MeSO$_2$-PCBs.

Herzke *et al.* [144] reported the content of chlorinated persistent organic pollutants in a total of 44 egg samples from eight different raptor species, collected throughout Norway in the period 1991–7. The contents of eight chlorinated bornanes, nine chlorinated pesticides and 15 polychlorinated biphenyl congeners (PCBs) were determined. The highest average concentrations for the PCBs were found for eggs from the white-tailed sea eagle (*Haliaeetus albicilla*) and the peregrine falcon (*Falco peregrinus*) (average sum PCB concentration: 8.9 and 9.1 μg g^{-1} (w/w), respectively). Merlin (*Falco colombarius*) and sparrowhawk (*Accipiter nisus*) eggs were the most highly contaminated with chlorinated pesticides (average total pesticide concentrations: 3.0 and 4.3 μg g^{-1} (w/w)). For the first time, the chloroborane content was determined in the eggs of Norwegian birds of prey. However, only minor contamination compared to PCBs and conventional chlorinated pesticides was found. The highest total concentration was determined for white-tailed sea eagle eggs (0.09 μg g^{-1}, w/w). No chloroborane contamination was found in osprey (*Pandion haliaetus*) and merlin eggs. No spatial and regional specific trends or pattern distribution was found for the organochlorine contamination in the egg samples analysed. In order to gain information about the enantiomer-specific bioaccumulation and biotransformation capacity of the organism, enantioselective analyses were performed for the chiral contaminants *trans*-chlordane and *oxy*-chlordane and the chloroborane B9-1679. Indications for species-dependent deviation from the racemic distribution (enantiomeric ratio = 1) were found. Peregrine falcon and merlin eggs were characterized by an extremely high enantiomeric excess of (−)-*trans*-chlordane (enantiomeric ratio, ER < 0.01). For the golden eagle (*Aquila chrysaetos*), goshawk (*Accipiter gentilis*) and sparrowhawk eggs, the ERs were between 0.1 and 0.22, demonstrating here also that (−)-*trans*-chlordane was the most abundant enantiomer. For the distribution of *oxy*-chlordane and the B9-1679 enantiomers, no species-dependent differences were found. For all species, ER values between 0.3 and 0.8 were determined. Thus, for *oxy*-chlordane and B9-1679 also, the (−)-enantiomers are the most dominant stereoisomers in the bird of prey eggs analysed. Marucchini *et al.* [145] reported metalaxyl and metalaxyl-M in sunflower plants. The cycling (food chain) of these chiral pollutants in the environment is shown in Figure 2.4, while the distribution of some chiral pollutants in different biota is presented in Table 2.7.

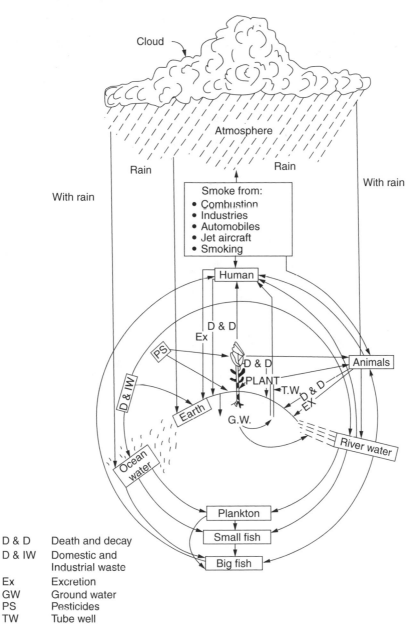

D & D	Death and decay
D & IW	Domestic and Industrial waste
Ex	Excretion
GW	Ground water
PS	Pesticides
TW	Tube well

Figure 2.4 The cycling of chiral pollutants in the environment.

Table 2.7 The distribution of some chiral pollutants in various biota

Chiral pollutant	Ecosystem component	References
α-HCH	Mussel	155, 158
	Eider duck (liver)	153, 158
	Eider duck (kidney)	159
	Eider duck (muscle)	110
	Seal liver	90
	Seal (blubber)	20, 25, 90
	Seal (brain)	90
	Seal (lung)	90
	Female fur seal (milk)	90
	Whale blubber	91
	Flounder (liver)	160
	Cod liver oil	90
β-PCCH	Flounder	25
HHCB	Rudd	43
	Trench liver	43
	Crucian carp liver	43
	Eel	43
	Mussel	43
AHTN	Rudd	43
	Trench liver	43
	Crucian carp liver	43
	Eel	43
	Mussel	43
trans-Chlordane	Baltic herring	161
	Baltic salmon	161
	Baltic seal	161
cis-Chlordane	Baltic herring	161
	Baltic salmon	161
Octachlordane MC4	Baltic herring	161
	Baltic salmon	161
	Baltic seal	161
	Antarctic penguin	161
Octachlordane MC5	Baltic herring	161
	Baltic salmon	161
	Baltic seal	161
	Antarctic penguin	161
Octachlordane MC7	Baltic herring	161
	Baltic salmon	161
	Baltic seal	161
	Antarctic penguin	161
Oxychlordane	Seagull egg	25
Heptachlor *exo*-epoxide	Seagull egg	25

2.3.7 *Distribution in Food Products*

With the above discussion in mind, it may be assumed that chiral xenobiotics are likely to be present in various foodstuffs, as human food comprises the meat of fish, birds and terrestrial animals, and of course vegetables. Many chiral pesticides are used to control insects in vegetables and cereals, and hence the presence of these pesticides in the food products is to be expected. Only a few reports are available dealing the presence of chiral pesticides in various foodstuffs. The cholinesterase inhibition activity of chiral organophosphorous pesticides, as well as that of toxic organophosphorous pesticides nerve gases, are enantioselective in nature. Scientists at the Microbiological and Chemical Exposure Assessment Research Division of NERL in Cincinnati, USA, have found the chiral insecticide permethrin, a pyrethroid insecticide, in spinach. Similarly, malathion has been observed in blackberry extract [146].

In 1965, Lichtenstein *et al.* [147] described the presence of aldrin and dieldrin pesticides in cucumber. In the same year, Thruston [148] reported chlordane in squash tissues. In 1991, Pylypiw *et al.* [149, 150] reported DDE, chlordane and heptachlor epoxide in squash and cantaloupe tissues. Furthermore, the same group [151] reported the presence of chlordane in summer and winter squash and sweet potato from different Connecticut, USA, farms. The authors also examined the market basket data from 1990 to 1997, and suggested that certain crops – cucumber, squash and pumpkin – contain some chiral agrochemicals in their edible portions. Recently, Mattina *et al.* [152] have detected chlordane residues in the edible tissues of carrot, beet, potato, spinach, lettuce, dandelion, zucchini, bush bean and eggplant.

2.4 Conclusions

Chiral pollutants from various sources are contaminating our environment, as discussed above. The presence of chiral pollutants in sea water and sediment indicates the transportation of these pollutants from the point of origin (source) to the sea. There are many pesticides that have already been banned for a long time, but their presence in sea water shows their persistence in the environment. These chiral pollutants, especially pesticides, have been reported in the various body parts of aquatic and terrestrial animals, including human beings, and have entered into the body through the food chain. A detailed search of the literature indicates that only a few reports on the detection of chiral pollutants are available.

Therefore, water, sediment, soil and air samples should be analysed for these pollutants. Moreover, the detection of these notorious pollutants in biological samples – and especially in the blood, urine and other tissues and biological fluids of human beings – should be investigated.

References

1. R. A. Aitken, D. Parker, R. J. Taylor, J. Gopal and R. N. Kilenyi, *Asymmetric Synthesis*, Blackie, New York, 1992.
2. G. Blaschke, H. P. Kraft, K. Fickentscher and F. Köhler, *Drug Res.* **29**, 1640 (1979).
3. B. Knoche and G. Blaschke, *J. Chromatogr. A* **666**, 235 (1994).
4. D. L. Lewis, A. W. Garrison, K. E. Wommack, A. Whittemore, P. Steudler and J. Melillo, *Nature* **401**, 898 (1999).
5. C. K. Jain and I. Ali, *Int. J. Environ. Anal. Chem.* **68**, 83 (1997).
6. R. Kallenborn and H. Hühnerfuss, *Chiral Environmental Pollutants, Trace Analysis and Ecotoxicology*, Springer-Verlag, Berlin, 2001.
7. W. J. M. Hegeman and W. P. J. Laane, *Rev. Environ. Toxicol.* **173**, 85 (2002).
8. D. H. Hutson and T. R. Roberts, *Progress in Pesticide Biochemistry and Toxicology*, Vol. 7, *Environmental Fate of Pesticides*, Wiley, New York, 1990.
9. J. L. Schnoor, *Fate of Pesticides and Chemicals in the Environment*, Wiley, New York, 1991.
10. R. A. Meyers (ed.), *Encyclopedia of Environmental Pollution and Clean-up*, Wiley, New York, 1999.
11. T. F. Bidleman, L. M. Jantunen, K. Wiberg, T. Harner, K. Brice, K. Su, R. L. Falconer, A. D. Lenone, E. J. Aigner and W. Parkhurst, *Environ. Sci. Technol.* **32**, 1546 (1998).
12. R. L. Falconer, A. Leone, C. Bodnar, K. Wiberk, T. F. Bidleman, L. M. Jantunen, T. Harner, W. Parkhurst, H. Alegria, K. Brice and K. Su, *Organohal. Compds* **35**, 331 (1998).
13. T. F. Bidlemann, L. M. Jantunen, T. Harner, K. Wiberg, J. Wideman, R. L. Falconer, E. J. Aigner, A. D. Lenone, J. J. Ridal, B. Kerman, W. Parkhurst, A. Finizio and Y. Szeto, *Organohal. Compds* **31**, 238 (1997).
14. R. L. Falconer, T. F. Bidlemann and D. J. Gregor, *Sci. Total. Environ.* **160/161**, 65 (1995).
15. T. F. Bidlemann, L. M. Jantunen, T. Harner, K. Wiberg, J. Wideman, K. Brice, K. Su, R. L. Falconer, E. J. Aigner, A. D. Leone, J. J. Ridal, B. Kerman, A. Finizio, H. Alegria, W. J. Parkhurst and S. Y. Szeto, *Environ. Pollut.* **101**, 1 (1998).
16. L. M. Jantunen and T. Bidleman, *J. Geophys. Res.* **101**, 837 (1996).

17. T. Harner, H. Kylin, T. F. Bidlemann and W. M. J. Strachan, *Organohal. Compds* **35**, 355 (1998).
18. I. Ali and C. K. Jain, *Curr. Sci.* **75**, 1011 (1998).
19. L. M. Jantunen and T. F. Bidleman, *Arch. Environ. Contam. Toxicol.* **35**, 218 (1998).
20. R. L. Falconer, T. F. Bidleman, D. J. Gregor, R. Semkin and C. Teixeira, *Environ. Sci. Technol.* **29**, 1279 (1995).
21. J. J. Ridal, T. F. Bidleman, B. R. Kerman, M. E. Fox and W. M. J. Strachan, *Environ. Sci. Technol.* **31**, 1940 (1997).
22. L. M. Jantunen and T. Bidleman, *J. Geophys. Res.* **19**, 279 (1996).
23. B. Bethan, W. Dannecker, H. Gerwig, H. Hühnerfuss and M. Schulz, *Chemosphere* **44**, 591 (2001).
24. J. Molsey, A. I. Fisk, K. A. Hobson and R. J. Norstrom, *Environ. Sci. Technol.* **35**, 1920 (2001).
25. H. Hühnerfuss, J. Faller, R. Kallenborn, W. A. König, P. Ludwig, B. Pfaffenberger, M. Oehme and G. Rimkus, *Chirality* **5**, 393 (1993).
26. J. Faller, H. Hühnerfuss, W. A. König and P. Ludwig, *Mar. Pollut. Bull.* **22**, 82 (1991).
27. H. Hühnerfuss, J. Faller, W. A. König and P. Ludwig, *Mar. Chem.* **26**, 2127 (1992).
28. P. Ludwig, Untersuchungen zum enantioselektiven Abbau von polaren und unpolaren chlorieten Kohlenwasserstoffen durch marine Mikroorganismen, Ph.D. thesis, University of Hamburg, 1991.
29. H. Gaul and U. Ziebarth, *Dt. Hydrogr. Z.* **36**, 191 (1983).
30. H. Hühnerfuss, J. Faller, W. A. König and P. Ludwig, *Environ. Sci. Technol.* **26**, 2127 (1992).
31. R. Gatermann, H. Hühnerfuss, G. Rimkus, M. S. Wolf and S. Franke, *Mar. Pollut. Bull.* **30**, 221 (1995).
32. H. D. Eschke, J. Traud and H. J. Dibowski, *Vom. Wasser.* **83**, 373 (1994).
33. C. S. Wong, A. W. Garrison, P. D. Smith and W. T. Foreman, *Environ. Sci. Technol.* **35**, 2448 (2001).
34. B. Pfaffenberger, H. Hühnerfuss, B. Gehrcke, I. Hardt, W. A. König and G. Rimkus, *Chemosphere* **29**, 1385 (1991).
35. B. Bethan, K. Bester, H. Hühnerfuss and G. Rimkus, *Organohal. Compds* **28**, 437 (1996).
36. B. Bethan, K. Bester, H. Hühnerfuss and G. Rimkus, *Chemosphere* **34**, 2271 (1997).
37. S. Franke, C. Meyer, M. Spetch, W. A. König and W. Francke, *J. High Res. Chromatogr.* **21**, 113 (1998).
38. A. Agarwal, ed., *Homicide by Pesticides, What Pollution Does to Our Bodies*, Centre for Science and Environment, New Delhi, 1997.
39. K. P. Singh, R. Takroo and P. K. Ray, *Analysis of Pesticide Residues in Water*, ITRC, Lucknow, 1987.

40. I. Ali and C. K. Jain, *J. Environ. Hydrol.* **9** (Paper 1), 1–7 (2001).
41. H. D. Eschke, J. Traud and H. J. Dibowski, *UWSF-Z Umweltchem. Okotox.* **6**, 183 (1994).
42. H. D. Eschke, H. J. Dibowski and J. Traud, *UWSF-Z Umweltchem. Okotox.* **7**, 131 (1995).
43. S. Franke, C. Meyer, N. Heinzel, R. Gatermann, H. Hühnerfuss, G. Rimkus, W. A. König and W. Francke, *Chirality* **11**, 795 (1999).
44. M. Winkler, J. R. Lawrence and T. R. Neu, *Water Res.* **35**, 3197 (2001).
45. T. Ternes, *Water Res.* **32**, 3245 (1998).
46. K. Wiberg, L. M. Jantunen, T. Harner, J. L. Wideman, T. F. Bidleman, K. Brice, K. Su, R. L. Falconer, A. D. Leone, W. Parkhurst and H. Alegria, *Organohal. Compds* **33**, 2079 (1997).
47. D. W. Kolpin, S. J. Kalkhoff, D. A. Goolsby, D. A. Sneck-Fahrer and E. M. Thurman, *Ground Water* **35**, 679 (1997).
48. D. Gomez de Barreda Jr, M. Gamon Vila, E. Lorenzo Rueda, A. Saez Olmo, D. Gomez de Barreda, J. Garcia de Cuadra and A. Ten C. Peric, *J. Chromatogr. A* **795**, 125 (1998).
49. A. Carrillo, *Geofis. Int.* **37**, 35 (1998).
50. P. Halder, P. Raha and P. Bhattacharya, *Ind. J. Environ. Health* **31**, 156 (1989).
51. N. Thakker and S. P. Pande, *J. Int. Water Works. Assn* **XVIII**, 313 (1986).
52. H. R. Busser, T. Poiger and M. D. Müller, *Environ. Sci. Technol.* **33**, 2529 (1999).
53. K. Kümmerer, *Chemosphere* **45**, 957 (2001).
54. K. Kümmerer, A. Al-Ahmad, B. Bertram and M. Wießler, *Chemosphere* **40**, 767 (2000).
55. H. Hühnerfuss, R. Gatermann, S. Biselli, G. Rimkus, M. Hecker, R. Kallenborn and L. Karbe, *Organohal. Compds* **40**, 401 (1999).
56. R. Gatermann, H. Hühnerfuss, S. Biselli, G. Rimkus, M. Hecker, R. Kallenborn and L. Karbe, *Organohal. Compds* **40**, 595 (1999).
57. S. Biselli, H. Dittmer, R. Gatermann, R. Kallenborn, W. A. König and H. Hühnerfuss, *Organohal. Compds* **40**, 599 (1999).
58. T. Yamagishi, T. Miyazaki, S. Horri and S. Kaneko, *Bull. Environ. Contam. Toxicol.* **26**, 656 (1981).
59. T. Yamagishi, T. Miyazaki, S. Horri and K. Akiyama, *Arch. Environ. Contam. Toxicol.* **12**, 83 (1983).
60. S. Weigel, Entwicklung einer Methode zur Extraktion organischer Spurenstoffe aus großvolumigen Wasserproben mittels Festphasen, Master thesis (Diplomarbeit), University of Hamburg, 1998, p. 86.
61. H. R. Buser and M. D. Müller, *Environ. Sci. Technol.* **29**, 664 (1995).
62. M. D. Müller and H. R. Buser, *Environ. Sci. Technol.* **29**, 2031 (1995).
63. W. Vetter, R. Bartha, G. Stern and G. Tomy, *Organohal. Compds* **35**, 343 (1998).

64. W. Vetter, R. Bartha, G. Stern and G. Tomy, *Environ. Toxicol. Chem.* **18**, 2775 (1999).
65. C. Rappe, P. Haglund, H. R. Buser and M. D. Müller, *Organohal. Compds* **31**, 233 (1997).
66. E. Benicka, R. Novakovski, J. Hrouzek and J. Krupcik, *J. High Res. Chromatogr.* **19**, 95 (1996).
67. C. S. Wong, A. W. Garrison and W. T. Foreman, *Environ. Sci. Technol.* **35**, 33 (2001).
68. E. Aigner, A. Leone and R. Falconer, *Environ. Sci. Technol.* **32**, 1162 (1998).
69. W. Y. Lee, W. Lannucci-Berger, B. D. Eitzer, J. C. White and M. J. Mattina, *J. Environ. Qual.* **32**, 224 (2003).
70. B. D. Eitzer, M. I. Mattina and W. Iannucci-Berger, *Environ. Toxicol. Chem.* **20**, 2198 (2001).
71. G. Fingerling, N. Hertkorn and H. Parlar, *Environ. Sci. Technol.* **30**, 2984 (1996).
72. K. Wiberg, T. Harner, J. L. Wideman and T. F. Bidleman, *Chemosphere* **45**, 843 (2001).
73. J. C. White, M. J. I. Mattina, B. D. Eitzer and W. I. Berger, *Chemosphere* **47**, 639 (2002).
74. S. N. Meijer, C. J. Halsall, T. Harner, A. J. Peters, W. A. Ockenden, A. E. Johnston and K. C. Jones, *Environ. Sci. Technol.* **15**, 1989 (2001).
75. Z. Y. Li, Z. C. Zhang, Q. L. Zhou, R. Y. Gao and Q. S. Wang, *J. Chromatogr. A* **977**, 17 (2002).
76. M. Hutta, I. Rybar and M. Chalanyova, *J. Chromatogr. A* **959**, 143 (2002).
77. H. R. Buser, M. D. Muller, T. Poiger and M. E. Balmer, *Environ. Sci. Technol.* **36**, 221 (2002).
78. A. D. Leone, S. Amato and R. L. Falconer, *Environ. Sci. Technol.* **35**, 4592 (2001).
79. A. Finizio, T. F. Bidleman and S. Y. Szeto, *Chemosphere* **36**, 345 (1998).
80. L. M. Jantunen, H. Kylin and T. F. Bidleman, *Organhal. Compds* **35**, 347 (1998).
81. V. K. Dua, *Bull. Environ. Contam. Toxicol.* **52**, 797 (1994).
82. E. M. Ulrich and R. A. Hites, *Environ. Sci. Technol.* **32**, 1870 (1998).
83. H. R. Buser and M. D. Müller, *Environ. Sci. Technol.* **27**, 1211 (1993).
84. H. Hühnerfuss and W. D. Garrett, *J. Geophys. Res.* **86**, 439 (1981).
85. W. A. König, D. Icheln, T. Runge, B. Pfaffenberger, P. Ludwig and H. Hühnerfuss, *J. High Res. Chromatogr.* **14**, 530 (1991).
86. K. Möller, C. Bretzke, H. Hühnerfuss, R. Kallenborn, J. N. Kinkel, J. Kopf and G. Rimkus, *Angew. Chem. Int. Engl. Ed.* **33**, 882 (1994).
87. R. Kallenborn, H. Hühnerfuss and W. A. Köning, *Angew Chem. Int. Eng. Ed.* **30**, 320 (1991).
88. R. Kallenborn, E. Hartwig and H. Hühnerfuss, *Seevögel* **15**, 31 (1994).

89. H. Hühnerfuss, R. Kallenborn, W. A. König and G. Rimkus, *Organohal. Compds* **10**, 97 (1992).
90. S. Mössner, T. R. Spraker, P. R. Becker and K. Ballschmiter, *Chemosphere* **24**, 1171 (1992).
91. M. D. Müller, M. Schlabach and M. Oehme, *Environ. Sci. Technol.* **26**, 566 (1992).
92. M. D. Müller, W. Vetter, K. Hummert and B. Luckas, *Organohal. Compds* **29**, 118 (1996).
93. K. Hummert, W. Vetter and B. Luckas, *Chemosphere* **31**, 3489 (1995).
94. K. Wiberg, R. Letcher, C. Sandau, R. Norstrom, M. Tysklind and T. Bidleman, *Organohal. Compds* **35**, 371 (1998).
95. H. Hühnerfuss, B. Pfaffenberger and G. Rimkus, *Organohal. Compds* **29**, 88 (1996).
96. K. Wiberg, A. Bergman, M. Olsson, A. Roos, G. Blomkvis and P. Haglund, *Environ. Toxicol Chem.* **21**, 2542 (2002).
97. R. Kallenborn, M. Oehm, W. Vetter and H. Parlar, *Chemosphere* **28**, 89 (1994).
98. H. R. Busser and M. D. Müller, *Environ. Sci. Technol.* **28**, 119 (1994).
99. W. A. König, I. Hardt, B. Gehrcke, D. H. Hochmuch, H. Hühnerfuss, B. Pfaffenberger and G. Rimkus, *Angew Chem. Int. Ed. Engl.* **33**, 2085 (1994).
100. K. L. E. Kaiser, *Environ. Pollut.* **7**, 93 (1974).
101. M. J. Harju and P. Haglund, *Anal. Chem.* **364**, 219 (1994).
102. S. Jansen, *New Scient.* **32**, 612 (1966).
103. J. S. Waid, *PCB and the Environment*, vol. I, CRC Press, Boca Raton FL, 1966.
104. J. De Boer, *Chemosphere* **17**, 1811 (1988).
105. J. C. Duinker, M. T. J. Hillebrand, T. Zeinstra and J. P. Boon, *Aquat. Mamm.* **15**, 95 (1989).
106. R. Kallenborn and H. Hühnerfuss, *Seevögel* **14**, 23 (1993).
107. H. Hühnerfuss, B. Pfaffenberger, B. Gehrcke, L. Karbe, W. A. König and O. Landgraff, *Pollut. Bull.* **30**, 332 (1995).
108. P. Haglund, *Organohal. Compds* **23**, 35 (1995).
109. V. Schurig, A. Glausch and M. Fluck, *Tetrahedron Asymm.* **6**, 2161 (1995).
110. P. Haglund, *J. Chromatogr. A* **724**, 219 (1996).
111. G. P. Blanch, A. Glausch, V. Schurig, R. Serrano and M. J. Gonzalez, *J. High Resolut. Chromatogr.* **19**, 392 (1996).
112. S. Reich, B. Jimenez, L. Marsili, L. M. Hernandez, V. Schurig and M. J. Gonzalez, *Organohal. Compds* **35**, 335 (1998).
113. W. Vetter, U. Klobes, K. Hummert and B. Luckas, *J. High Res. Chromatogr.* **20**, 85 (1997).
114. K. Wiberg, R. Letcher, C. Sandau, J. Duffe, R. Norstrom, P. Haglund and T. F. Bidleman, *Anal. Chem.* **70**, 3845 (1998).

115. C. S. Wong, P. F. Hoekstra, H. Karlsson, S. M. Backus, S. A. Mabury and D. C. Muir, *Chemosphere* **49**, 1339 (2002).
116. C. S. Wong, F. Lau, M. Clark, S. A. Mabury and D. C. Muir, *Environ. Sci. Technol.* **36**, 1257 (2002).
117. H. Hühnerfuss, R. Kallenborn, K. Möller, B. Pfaffenberger and G. Rimkus, in *Proceedings of the 15th International Symposium on GC, Riva*, 1993, p. 576.
118. P. F. Hoekstra, C. S. Wong, T. M. O'Hara, K. R. Solomon, S. A. Mabury and D. C. Muir, *Environ. Sci. Technol.* **36**, 1419 (2002).
119. H. J. de Gues, H. Besselink, A. Brouwer, J. Klungsoyr, B. McHugh, E. Nixon, G. Rimkus, P. G. Wester and J. de Boer, *Environ. Health Prospect.* **1**, 115 (1999).
120. A. T. Fisk, M. Holst, K. A. Hobson, J. Duffe, J. Moisey and R. J. Norstrom, *Arch. Environ. Contam. Toxicol.* **42**, 118 (2002).
121. A. T. Fisk, J. Moisey, K. A. Hobson, N. J. Karnovsky and R. J. Norstrom, *Environ. Pollut.* **113**, 225 (2001).
122. J. Gao, A. W. Garrison, C. Hoehamer, C. S. Mazur and N. L. Wolfe, *J. Agric. Food Chem.* **48**, 6114 (2000).
123. J. Gao, A. W. Garrison, C. Hoehamer, C. S. Mazur and N. L. Wolfe, *J. Agric. Food Chem.* **48**, 6121 (2000).
124. A. W. Garrison, V. A. Nzengung, J. K. Avants, J. Ellington and N. L. Wolf, *Organohal. Compds* **31**, 256 (1997).
125. G. Rimkus and M. Wolf, *Chemosphere* **30**, 641 (1995).
126. G. Rimkus, B. Rimkus and M. Wolf, *Chemosphere* **28**, 421 (1994).
127. B. Liebl and S. Ehrenstorfer, *Chemosphere* **27**, 2253 (1993).
128. S. Müller, P. Schmid and C. Schlatter, *Chemosphere* **33**, 17 (1996).
129. R. Gatermann, S. Biselli, H. Hühnerfuss, G. G. Rimkus, S. Franke, M. Hecker, R. Kallenborn, L. Karbe and W. A. König, *Arch. Environ. Contam. Toxicol.* **42**, 47 (2002).
130. K. Möller, H. Hühnerfuss and G. Rimkus, *J. High Res. Chromatogr.* **16**, 672 (1993).
131. B. Pfaffenberger, I. Hardt, H. Hühnerfuss, W. A. König, G. Rimkus, A. Glausch and V. Schurig, *Chemosphere* **29**, 1543 (1994).
132. G. Rimkus and M. Wolf, *Z. Lebensm. Unters Forsch.* **184**, 308 (1987).
133. J. Portig, K. Stein and H. W. Vohland, *Xenobiotica* **19**, 123 (1989).
134. K. Möller, Untersuchungen zur enantioselektiven Anreicherung von chiralen Schadstoffen im marin und terrestrischen Ökosystem, Master thesis, University of Hamburg, 1993.
135. A. Glausche, G. J. Nicholson, M. Fluck and V. Schurig, *J. High Res. Chromatogr.* **17**, 347 (1994).
136. L. Ramos, B. Jimenez, M. Fernandez, L. Hernandez and M. J. Gonzalez, *Organohal. Compds* **27**, 376 (1996).
137. P. Haglund and K. Wiberg, *J. High Res. Chromatogr.* **19**, 373 (1996).

138. T. Ellerichmann, A. Bergman, S. Franke, H. Hühnerfuss, E. Jakobsson, W. A. König and C. Larsson, *Fresenius Environ. Bull.* **7**, 244 (1998).
139. A. Bergman, T. Ellerichmann, S. Franke, H. Hühnerfuss, E. Jakobsson, W. A. König and C. Larsson, *Organohal. Compds* **35**, 339 (1998).
140. L. Alder, R. Palavinkas and P. Andrews, *Organohal. Compds* **28**, 410 (1996).
141. S. Jansen and B. Jansson, *Ambio* **5**, 275 (1976).
142. M. J. I. Mattina, J. White, B. Eitzer and W. I. Berger, *Environ. Toxicol. Chem.* **21**, 281 (2002).
143. C. Larsson, T. Ellerichmann, H. Hühnerfuss and A. Bergman, *Environ. Sci. Technol.* **36**, 2833 (2002).
144. D. Herzke, R. Kallenborn and T. Nygard, *Sci. Total Environ.* **27**, 291 (2002).
145. C. Marucchini and C. Zadra, *Chirality* **14**, 32 (2002).
146. A. W. Garrison, W. J. Jones, J. J. Ellington and J. Washington, *Selective Transportation and Occurrences of the Enantiomers of Chiral Pesticides in Environmental Matrices and Food Products*, US EPA report APM 127, 12 July 2001.
147. E. P. Lichtenstein, K. R. Schulz, R. F. Skrentny and P. A. Stitt, *J. Econ. Entomol.* **58**, 742 (1965).
148. A. D. Thruston, *J. Assoc. Off. Anal. Chem.* **48**, 952 (1965).
149. H. M. Pylypiw and L. Hankin, *Pesticide Residues in Produce Sold in Connecticut 1990*, Bulletin 886, Connecticut Agricultural Experiment Station, New Haven, 1991.
150. H. M. Pylypiw, E. Naughton and L. Hankin, *J. Dairy Food Environ. Sanit.* **11**, 200 (1991).
151. H. M. Pylypiw, T. Misenti and M. J. I. Mattina, *Pesticide Residues in Produce Sold in Connecticut 1996*, Bulletin 940, Connecticut Agricultural Experiment Station, New Haven, 1997.
152. M. J. I. M. Mattina, W. Iannucci-Berger and L. Dykes, *J. Agric. Food Chem.* **48**, 1909 (2000).
153. H. Hühnerfuss, B. Pfaffenberger, B. Gehrcke, L. Karbe, W. A. König and O. Landgraff, *Organohal. Compds* **21**, 15 (1994).
154. C. Larsson, T. Ellerichmann, S. Franke, M. Athanasiadou, H. Hühnerfuss and A. Bergman, *Organohal. Compds* **40**, 427 (1999).
155. H. Hühnerfuss, *GIT Fachz. Lab.* **36**, 489 (1992).
156. P. Ludwig, H. Hühnerfuss, W. A. König and W. Gunkel, *Mar. Chem.* **38**, 13 (1992).
157. P. Ludwig, W. Gunkel and H. Hühnerfuss, *Chemosphere* **24**, 1423 (1992).
158. H. Hühnerfuss and R. Kallenborn, *J. Chromatogr.* **580**, 191 (1992).
159. H. Raschke, M. Meir, J. G. Burken, R. Hany, M. D. Müller, J. R. van der Meer and H. P. Kohler, *Appl. Environ. Microbiol.* **67**, 3333 (2001).
160. R. Gatermann, H. Hühnerfuss, G. Rimkus, A. Attar and A. Kettrup, *Chemosphere* **35**, 497 (1997).
161. L. M. Jantunen and T. Bidleman, *J. Geophys. Res.* **102**, 279 (1997).

Chapter 3

Chiral Pollutants: Biotransformation, Biodegradation and Metabolism

3.1 Introduction

Various reports have been published indicating that the chemicals in our environment may disrupt normal reproduction and development through inhibition of the androgen receptor [1]. About 25 % of agrochemicals, including pesticides, are chiral in nature and it is known that chiral molecules degrade and metabolize enantioselectively [2, 3]. Also, some of the nonchiral pollutants degrade in the environment, and the degradation products are toxic and chiral in nature. γ-Hexachlorocyclohexane (γ-HCH) is an achiral pesticide, but it degrades into chiral and toxic γ-pentachlorocyclohexene (γ-PCCH) enantiomers (Figure 3.1). Similarly, atrazine is an achiral pesticide and it is transformed into chiral metabolites; that is, 2-chloro-4-ethylamino-6-(1-hydroxy-2-methylethyl-2-amino)-1,3,4,5-triazines (Figure 3.2). In addition to this, many examples are available of the production of chiral metabolites from nonchiral molecules in biological systems. Chiral recognition is an essential issue in pharmacokinetics and pharmacodynamics studies. Therefore, to determine the final fate and toxicities of these chiral pollutants, it is essential to understand their degradation and metabolism mechanisms.

Chiral Pollutants. I. Ali and H. Y. Aboul-Enein
© 2004 John Wiley & Sons, Ltd ISBN: 0-470-86780-9

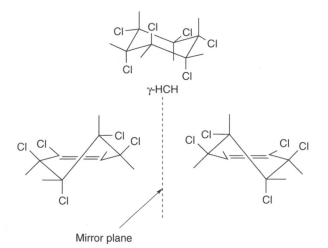

Figure 3.1 The enantioselective biodegradation of γ-HCH into the enantiomers of γ-PCCH.

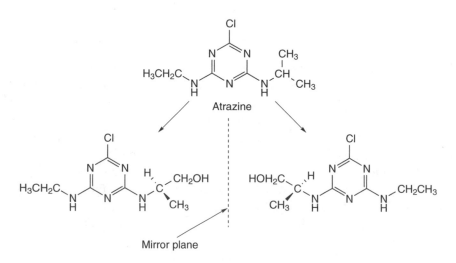

Figure 3.2 The enantioselective biodegradation of atrazine pesticide into the enantiomers of 2-chloro-4-ethylamino-6-(1-hydroxy-2-methylethyl-2-amino)-1,3,5-triazine [117].

The interaction of any xenobiotic, chiral or nonchiral, with biological systems occurs in three steps: penetration, recognition and activation. In the penetration step, the xenobiotic enters the biological system, while the recognition steps involves the binding of the xenobiotic to the receptor or enzyme. The activation step, measured by potency, reaction rate and so on, comprises the molecular effect for example, the activation or blockading of a receptor, or an enzymatic reaction leading to metabolite formation. It is important to mention here that the penetration step is not important when the study is dealing with isolated enzymes or macromolecules. The human body is a highly stereospecific environment, as we have D-sugars and L-amino acids in our systems [4]. Therefore, the enantiomers of xenobiotics react at different rates in the human body, and the toxicities and fates of the enantiomers of xenobiotics are different due to their different biological properties. The differences in the properties and toxicities of chiral xeno-biotics are due to different protein bindings and transport, mechanisms of action, metabolic rates, changes in activity due to metabolism, clearance rates and persistence in the environment [5]. Therefore, to determine the exact fate and toxicities of chiral xenobiotics it is essential to understand the biotransformation, biodegradation and metabolism of chiral pollutants. The present chapter describes these issues in detail.

3.2 The Mechanisms of the Interactions of Chiral Xenobiotics in Biological Systems

Before discussing the enantioselective degradation and metabolism of chiral pollutants, it is essential to describe the mechanisms of the interactions of chiral pollutants in biological systems. Knowledge of chiral interaction mechanisms will make the enantioselective degradation and metabolism topics clear to readers. An understanding of the mechanisms of the action of chiral xenobiotics in biological systems requires three main phases; an initial exposure phase, pharmacokinetics and pharmacodynamic studies. The initial exposure phase is governed by the activity of the chiral pollutants (their affinity for receptors, their agonist or antagonist activities and the activity of metabolites) and also by tissue specificity and its metabolites. This phase is followed by the pharmacokinetic phase (the term is normally used for drugs, but can also be used for pollutants). This phase involves absorption, distribution (e.g. plasma binding), metabolism (routes and rates) and excretion (e.g. clearance rates and routes). The pharmacodynamic studies comprise the interaction of the active enantiomer of the pollutant with the molecular site of action (receptors, enzymes etc.) in the target tissue, which leads to the observed effects and diseases. It is worth mentioning here that enantioselectivity plays an important role not only in pharmacodynamic studies but also in the pharmacokinetic phase. The notorious side effects of the pollutants are influenced by both their pharmacodynamic and pharmacokinetic phases, and enantioselectivity plays an important role in both.

In addition to this qualitative knowledge of the interactions of chiral pollutants in biological systems, the mechanisms at the molecular level are very important in understanding the interactions of chiral pollutants. Basically, the interactions of the enantiomers of chiral pollutants depend on the structure and the active sites of the receptors. The interactions of these enantiomers are based on the well known principle of a 'lock and key' arrangement. The enantiomer that fits to a greater extent, in comparison to others, at the receptor site interacts considerably and its effects are revealed. Moreover, the sites or groups on the receptors also control the binding of the enantiomers to the receptors. This concept of a 'lock and key' arrangement is shown in Figure 3.3. In 1992, Crossley [6] described the interaction sites of a chiral drug with the receptors and, hence, the same model may be used to describe the interactions of chiral pollutants with the biological receptors. A diagrammatic representation of these interactions is presented in Figure 3.4. As the figure shows, the chiral pollutant is assumed to have sites 'a', 'b', 'c' and 'd' that have affinities for specific complementary sites on the receptor, labelled 'A', 'B', 'C' and 'D'. Thus 'Aa' will be a significant interaction, where as 'Ab', 'Ac' and 'Ad' are minimal interactions and are even possibly repulsive. Considering all the possible interactions in Figure 3.4 by keeping each group 'a', 'b', 'c' and 'd' fixed in turn, we find that the *R*-enantiomer will have one four-way and eight one-way significant

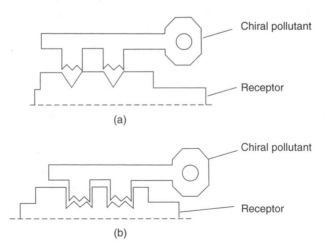

(a)

(b)

Figure 3.3   A schematic representation of the lock and key arrangement of chiral xenobi-otics on receptors. (a) No interaction stage; (b) complete interaction stage.

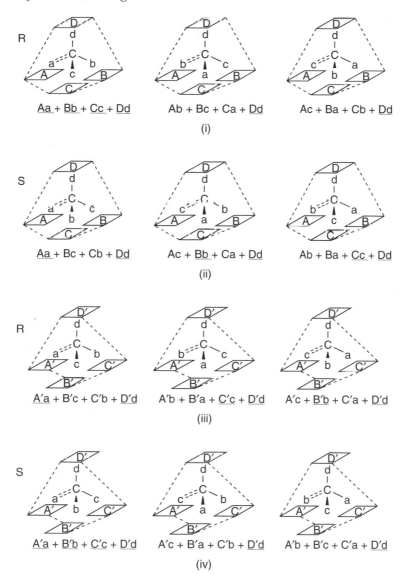

Figure 3.4 Chiral interactions with receptors. The major interactions are indicated as 'Aa'. (i) With d fixed, the *R*-enantiomer can produce one four-way and two one-way major interactions and, if we allow each group to be fixed in turn, there is a total of eight one-way interactions. (ii) The *S*-enantiomer produces three two-way major interactions with d fixed. If we follow a conformational change in the receptor, (iii) the *R*-enantiomer now produces three two-way major interactions and (iv) the *S*-isomer now has one four-way and two one-way major interactions, again with 'd' fixed [20].

interactions. Therefore, the *S*-enantiomer has some affinity for the receptor, and the magnitude of this affinity will depend on the relative importance of the individual interactions and also the degree of cooperation between them. In the enantiomer, such cooperation is at a high level and there is a high degree of molecular fit at the receptor; the *S*-enantiomer will then be essentially inactive at this particular receptor. Whether the *S*-enantiomer produces a response depends on the efficacy at the receptor produced by the two kinds of significant interactions. It is believed that at least three attachment points are required for a chiral interaction of any chiral molecule in a biological system. Moreover, recently, Mesecar and Koshland [7] have described how, sometimes, all four sites of the chiral molecule are required for its chiral interaction with biological receptors. This situation arises when the binding sites of the receptor are in a cleft or on protruding residues.

When the activity of the *R*-enantiomer is minimized, the *S*-enantiomer can still have a significant number of interactions with the receptor. Therefore, its activity may not be zero, and it will be able to antagonize the effect of the *R*-form, at least to some extent. The foregoing treatment assumes that there is no conformational change in the receptor, but this is not necessarily the case, for all conformational changes can produce more than one state in which a pollutant interacts, including inactive states. In this way, the enantiomers of a chiral pollutant form a reversible enantiomer–receptor complex. This triggers a secondary mechanism, such as the opening of anion channels, or catalyses the formation of a second messenger, which is often a cyclic AMP (cAMP). The other kinases are then activated and this chain of reactions finally results in the physiological change that is attributed to the chiral pollutant. The same mechanisms also operate with endogenous agents such as hormones and neurotransmitters. The enormous complexity of living systems and the remoteness of the cause from the effect (i.e. drug administration and pharmacological action) introduce many complications into the study of such relationships.

Molecular pharmacologists and physical scientists have sought to simplify the experimental system as much as possible, by omitting irrelevant factors such as pollutant transport and metabolism and putting the system on a level that is accessible to molecular manipulations and precise physicochemical methods. The recent development of the methodology of quantitative binding experiments on membrane preparations and, subsequently, in isolated conditions has become more sophisticated, precise and simple, and has led to the increasing realization of the pollutant–receptor interaction at the molecular level. This has allowed direct experimental access to receptor binding sites, the recent development of several complementary receptor models and the characterization of the molecular properties of pollutant receptors. Ariëns *et al.* [8], Burt [9] and Hollenberg [10], among others, and their many coworkers, are the molecular pharmacologists who have been at the forefront of the spectacular and explosive progress in receptonology since the 1970s.

3.3 The Fate of Chiral Pollutants in the Ecosystem

An understanding of the fate of xenobiotics in our ecosystem is of great importance, as the health of human beings and other animals is associated with the functioning of our planet. Scientists at the Ecosystem Research Division of NERL/ORD in Athens have been investigating the phenomenon of chirality of pesticides and other chiral pollutants in the environment, and the impact of enantioselectivity on the exposure and fate of chiral pesticides, for the past few years. Chiral xenobiotics transform and degrade enantioselectively, and finally metabolize in human beings and animals, leading to various lethal effects. As discussed above, chiral pollutants exit at every trophic level in the environment and are distributed everywhere in nature. The ultimate fate of chiral pollutants is their degradation and transformation into different metabolites in the ecosystem under varying environmental conditions. However, some of pollutants remain nondegradable for a long time. It is well known that organic pollutants are degradable and transformable under various physical, chemical and biological conditions. Some plants and animals, including bacteria, fungi and algae, have the capacity to transform these chiral pollutants into different toxic and nontoxic products. Different types of bacteria in the environment have the capacity to biodegrade chiral pollutants [11]. The main aim in writing this section is to describe the enantioselective biotransformation and biodegradation of chiral pollutants, for which knowledge of the methods and patterns of biotransformation and biodegradation by animals and microbes are essential. Faller *et al.* [12] raised the question 'Do marine bacteria degrade α-HCH stereoselectively?' and they were the first to reply to this in the affirmative. Therefore, at present, it is known that the biotransformation and biodegradation of chiral pollutants is stereoselective in nature.

3.3.1 Biotransformation

The biotransformation of chiral pollutants includes the conversion of the pollutants into some other products, which may or may not be toxic depending upon the reactions involved in their biotransformation. Normally,

Chiral Pollutants: Distribution, Toxicity and Analysis

the biotransformation takes place by means of biochemical reactions such as halogenation, reduction, oxidation, addition and so on. The rate and fate of the final biotransformation products depends on a number of factors, such as the growing environment of the organisms, the concentration of the xenobiotics and so on. To make this topic clearer, the biotransformation of chiral xenobiotics has been categorized into two classes; biotransformation in animals and in plants (phytotransformation).

Biotransformation in Animals

After the pioneering work of Kallenborn *et al.* [13, 14], who demonstrated the enantioselective degradation of α-HCH in eider duck (*Somateria mollissima*), increasing attention has been paid to this field of study by the world's scientists. Fisk *et al.* [15] studied the enantiomeric ratio of organochlorine pesticides (OCs) – that is, α-HCH, *cis*- and *trans*-chlordane, oxychlordane and heptachlor *exo*-epoxide – in ringed seal blubber. The enantiomer fractions (EFs) in the ringed seals varied considerably from those of other Arctic marine mammals and sea birds, providing additional evidence that the types and characteristics of the enzymes involved in the biotransformation of chiral OCs vary between these organisms. Moisey *et al.* [16] reported the differential biotransformation of α-HCH in invertebrates (four species), pelagic zooplankton (six species), Arctic cod, sea birds (seven species) and the ringed seal. Furthermore, the same authors [17] have also reported differential biotransformation of chlordane pesticides in seven species of sea birds; dovekie (*Alle alle*), thick-billed murre (*Uria lomvia*), black guillemot (*Cepphus grylle*), black-legged kittiwake (*Rissa tridactyla*), ivory gull (*Pagophila eburnea*), glaucous gull (*Larus hyperboreus*) and northern fulmar (*Fulmaris glacialis*). The different concentrations of chlordane in the sea birds indicated different patterns of biotransformation of chlordane. Pfaffenberger *et al.* [18] reported the enantioselective biotransformation of α-HCH in roe deer. The authors made a comparison with the biotransformation in aquatic animals, and observed different patterns of biotransformation of α-HCH in aquatic and terrestrial animals. To summarize, many reports have been published on the enantioselective biotransformation of α-HCH in different animals. Different enzymatic biotransformation pathways have been observed in tissues of mussels, the liver and kidney of eider duck [13, 14] and the liver of flounders [19]. Almost pure (+)-α-HCH was observed in roe deer [18] and the brain of the eider duck [20]. Also, different pathways for the biotransformation of α-HCH in the brains of sheep and humans have been suggested by Möller *et al.* [21, 22]. The different concentrations of HCH due to biotransformation in various biotas are given in Table 3.1.

Table 3.1 The enantiomeric ratios of α-HCH in the various components of the ecosystem

Ecosystem component	Enantiomeric ratio	References
Sea water	0.79–0.94	34, 44, 45 56, 110, 111
Sea water (Great Britain)	1.15	44
Mussel	0.69–0.89	107
Eider duck (liver)	1.4	106
Eider duck (kidney)	1.6	12
Eider duck (muscle)	7.0	12
Seal liver	1.5–1.8	112
Seal (blubber)	1.2–4.5	12, 58
Seal (brain)	7.9	112
Seal (lung)	1.5–1.6	112
Female fur seal (milk)	1.5–1.6	112
Harbour seal (blubber)	2.5–2.9	55, 57, 59
Flounder (liver)	0.80–0.94	13
Cod liver oil	0.98–0.99	112
Rain water	0.96–1.08	46
Air	0.86–1.03	46
Soil	1.10	113
Baltic sea	0.79–0.92	45
North sea	0.81–0.87	14, 50
Blue mussel	0.70–1.04	14
Eider duck (liver)	1.4–2.5	14
Flounder (liver)	0.76–0.98	14
Arctic cod	1.00	114
Ringed seal	1.00	114
Polar bear	1.7	114
Grey seal	1.00	115
Roe deer	0.06–0.40	18
Hare	0.80–1.50	35, 116

Hühnerfuss *et al.* [14, 23] reported the enantioselective biotransformation of PCBs in blue mussels (*Mytillus edulis*). Subsequently, Blanch *et al.* [24] also reported the enantioselective transformation of PCBs in shark (*Centroscymnus coelolepis* B and C) liver. Similarly, various reports have been published on the enantioselective biotransformation of PCBs in different animals [23, 24, 25–34]. Faller *et al.* [12] reported that the different rates of PCB biotransformation were due to their different reaction rates with the biological enzymes. In another study, Hühnerfuss *et al.* [35–37] reported the enantiomeric biotransformation of PCBs in the liver of humans and rats. Wong *et al.* [38] reported the different enantiomeric ratios of PCBs in river biota; that is, fish, bivalves, crayfish, water snakes and barn swallows (*Hirundo rustica*). Nonracemic enantiomeric fractions (EFs) were observed for PCBs 91, 95, 136 and 149 for aquatic and riparian biota from Lake Hartwell. Fish and bivalves at the same location showed marked differences in EFs as compared to sediment, thus suggesting that the PCBs are bioprocessed in biota in a different manner from those found in sediment (e.g. reductive dechlorination). Species-dependent patterns of EFs in PCBs were also observed, which suggested differences in the bioprocesses of PCBs, most likely due to metabolism. One of the major routes for the biotransformation of nitro musks is their conversion into amino musks. In this regard, Rimkus [39] and Gaterman *et al.* [40] identified some amino musks in some aquatic biota in wastewater treatment plants. The same authors have determined the enantiomeric ratios of synthetic polycyclic musks; that is, 1,3,4,6,7,8-hexahydro-4,6,6,7,8,8-hexamethylcyclopenta(g)-2-benzopyrane (HHCB, Galaxolide) and 7-acetyl-1,1,3,4,4,6-hexamethyltetrahydronaphthalene (AHTN, Tonalide).

Phytotransformation

Many chiral xenobiotics are lipophilic and, thus, these pollutants are absorbed into plants and may accumulate to toxic levels. The fate and bioconversion of chiral xenobiotics in plants is, therefore, of importance to the natural world as a whole, and to agriculture and human health. The levels of xenobiotics in plants can be reduced by elimination via roots and leaves, or by metabolism. Again, the rate and the fate of the phytotransformed products depend on various environmental conditions and the concentrations of the xenobiotics.

Gao *et al.* [41] studied the decay of organophosphorus (OP) pesticides (malathion, demeton-S-methyl and crufomate) *in vitro*, using the axenically aquatic cultivated plants parrot feather (*Myriophyllum aquaticum*), duck-weed (*Spirodela oligorrhiza*) and elodea (*Elodea canadensis*). The decay profile of these OP pesticides from the aqueous medium adhered to first order kinetics. However, the extent of the decay and the rate constants were dependent on both the physicochemical properties of the OP compounds and the nature of the plant species. Malathion and demeton-S-methyl exhibited similar transformation patterns in all three plants, with 29–48 % and 83–95 % phytotransformation, respectively, when calculated by mass recovery balance during an 8 day incubation. No significant disappearance and phytotransformation of crufomate occurred in elodea over 14 days, whereas 17–24 % degraded in the other plants over the same incubation period. Using enzyme extracts derived from duckweed, 15–25 % of the three pesticides were transformed within 24 h of incubation, which provided evidence for the degradation of the OP compounds by an organophosphorus hydrolase or by multiple enzyme systems. The results of this study showed that the selected aquatic plants have the potential to accumulate and to metabolize OP compounds. The metabolites of DDT in the extracts of plant tissues are shown in Figure 3.5. Furthermore, the same authors [42] have observed the uptake and phytotransformation of o,p'-DDT and p,p'-DDT *in vitro*, again using the same three axenically cultivated aquatic plants; that is, parrot feather, duckweed and elodea. The decay profile of DDT from the aqueous culture medium again followed first order kinetics for all three plants. During the 6 day incubation period, almost all of the DDT was removed from the medium, and most of it accumulated in or was transformed by these plants. Duckweed demonstrated the greatest potential to transform both DDT isomers; 50–66% was degraded or bound in a nonextractable manner with the plant material after the 6 day incubation. Therefore, duckweed also incorporated less extractable DDT (32–49%) after 6 days than did the other plants. The capacity for phytotransformation/binding by elodea was in between that of duckweed and parrot feather; approximately 31–48% of the spiked DDT was degraded or bound to the elodea plant material. o,p'-DDD and p,p'-DDD are the major metabolites in these plants; small amounts of p,p'-DDE were also found in duckweed (7.9%) and elodea (4.6%) after 6 days. Apparently, reduction of the aliphatic chlorine atoms of DDT is the major pathway for this transformation. This study, which has provided new information on plant biochemistry as related to pollutant accumulation and phytotransformation, has advanced the development of phytoremediation processes. The enantiomeric ratios

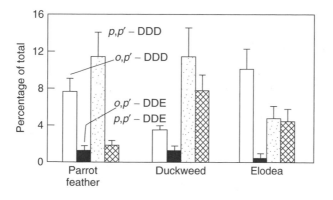

Figure 3.5 The metabolites of DDT in the extracts of plant tissues; that is, parrot feather, duckweed and elodea [42].

of some chiral pollutants resulting from enantioselective biotransformation in different biota are summarized in Table 3.2.

Biodegradation

Many bacteria, fungi and other microbes are present in water, sediments and soils and, therefore, the biodegradation of chiral pollutants is a natural process. Various environmental conditions, such as temperature, moisture (in the case of soil) and the presence of other chemicals (working as catalysts), control the biodegradation of these pollutants. Moreover, the biodegradation of chiral pollutants is also controlled by the enzymatic activities of the microbes. Of course, the biodegradation of chiral pollutants results in chiral degradation products but, sometimes, achiral pollutants also degrade into chiral and toxic metabolites. The biodegradation of some nonchiral pollutants also results in chiral metabolites in the presence of chiral catalysts. The two most important examples are γ-HCH, which degrades into the γ-PCCH enantiomers, and atrazine, which is converted into the 2-chloro-4-ethylamino-6-(1-hydroxy-2-methylethyl-2-amino)-1,3,5-triazine enantiomers by microbial degradation. Such types of achiral pollutant are called prochiral pollutants. Mannervik and Danielson [43] described glutathione transferases as catalysts in the biotransformation of xenobiotics and some drugs. The glutathione transferases are recognized as important catalysts in the biotransformation of xenobiotics, including drugs as well as environmental pollutants. Multiple forms exist, and numerous transferases from mammalian tissues, insects and plants have been isolated and characterized. Enzymatic properties, reactions with antibodies and structural characteristics have been used for classification of the glutathione transferases. The cytosolic mammalian enzymes could be grouped into three distinct classes, α-, μ- and π-; the microsomal glutathione transferase differs greatly from all of the cytosolic enzymes. Members of each enzyme class have been identified in human, rat and mouse tissues. Comparison of known primary structures of representatives of each class suggests a divergent evolution of the enzyme proteins from a common precursor. Products of oxidative metabolism, such as organic hydroperoxides, epoxides, quinones and activated alkenes, are possible 'natural' substrates for the glutathione transferases. Particularly noteworthy are 4-hydroxyalkenals, which are among the best substrates found. Homologous series of substrates give information about the properties of the corresponding binding site. The catalytic mechanism and the active-site topology have also been probed by the use of chiral substrates. Steady state kinetics have provided evidence for

Table 3.2 The enantiomeric ratios of some chiral pollutants in various ecosystem components

Chiral pollutant	Ecosystem component	Enantiomeric ratio	References
HHCB	Rudd	0.66	117
β-PCCH	Sea water	0.97	45, 56, 110
	Baltic sea	0.97	45
	Flounder	1.0	46
γ-PCCH	Sea water	1.15	45, 56, 110
DCPP	Sea water	1.4	110, 111
	Baltic sea	1.15	45
HEPX	Sea water	1.01–1.76	49
trans-Chlordane	Baltic herring	0.42	118
	Baltic salmon	1.19	118
	Baltic seal	0.60	118
	Air (above soil)	0.74	119, 120
	Air (ambient)	0.93	119, 120
	Air (indoor)	0.99	119, 120
cis-Chlordane	Baltic herring	1.35	118
	Baltic salmon	0.38	118
	Air (above soil)	1.11	113, 114
	Air (ambient)	1.04	119, 120
	Air (indoor)	0.98	119, 120
	North Sea	0.94–1.06	49
Octachlordane MC4	Baltic herring	0.70	118
	Baltic salmon	0.70	118
	Baltic seal	2.70	118, 119
	Antarctic penguin	2.2	118
Octachlordane MC5	Baltic herring	0.81	118
	Baltic salmon	0.75	118
	Baltic seal	0.24	118
	Antarctic penguin	0.91	118
Octachlordane MC7	Baltic herring	0.83	118
	Baltic salmon	0.92	118
	Baltic seal	0.86	118
	Antarctic penguin	1.35	118
Oxychlordane	Seagull egg	1.5	46
	Roe deer	7.00–17.00	18
	Hare	1.00–1.5	35, 116
Heptachlor	Air (ambient)	0.99	119, 120
Heptachlor *exo*-epoxide	Seagull egg	1.6	46
	Roe deer	1.00–5.00	18
	Hare	2.5–3.7	35, 116
	Air (ambient)	1.51	119, 120

a sequential mechanism. In general, the biodegradation of chiral pesticides is carried out by different bacteria and, therefore, many reports are available on the degradation of chiral xenobiotics by different microbes. However, only few reports are available on the biodegradation of chiral agrochemicals by fungi and other plants. The biodegradation mechanisms of γ-HCH and atrazine are presented in Figures 3.1 and 3.2. There are many reports on the biodegradation of chiral pollutants, some of which have been discussed herein.

Faller *et al*. [44] investigated the biodegradation of α-HCH in 16 water samples collected from all parts of the North Sea. The authors reported different patterns of biodegradation of this pesticide in the sea water, with preferential degradation of the (+)-enantiomer in the sea area off the east coast of Great Britain. Similarly, Hühnerfuss *et al*. [45] observed the preferential degradation of (+)-α-HCH in the eastern part of the North Sea. The average ER values calculated for 21 seawater samples was 0.87. Furthermore, the authors compared these results with those obtained by laboratory experiments, and the results were found to be similar. Furthermore, In 1993, Hühnerfuss *et al*. [46] studied the decomposition of the two enantiomers, with different velocities, by marine micro-organisms (α-HCH, β-PCCH and γ-PCCH); the enantioselective decomposition of one enantiomer only by marine micro-organisms (DCPP); enantioselective decomposition by enzymatic processes in marine biota (α-HCH, β-PCCH, *trans*-chlordane, *cis*-chlordane, octachlordane MC4, octachlordane MC5, octachlordane MC7, oxychlordane and heptachlor epoxide); enantioselective active transport through the 'blood–brain barrier' (α-HCH); and nonenantioselective photochemical degradation (α-HCH and β-PCCH). Again, this group [35] reported the enantioselective biodegradation of HCH, oxychlordane and heptachlor epoxide in different trophic levels of marine and terrestrial animals.

Jantunen and Bidleman presented the same results in 1996 [47]. The ER values of α-HCH were lower than one in the Bering Chukchi Sea, indicating preferential microbial degradation of one of the enantiomers. Harner *et al*. [48] have studied the microbial degradation of α- and β-HCH in the Arctic Ocean *in situ*, using ER data. The half-lives, due to microbial activity, of (+)-α, (−)-α and γ-HCH were 5.9, 22.8 and 18.8 years, respectively. Bidleman *et al*. [49–54] carried out an extensive study to determine the enantioselective microbial degradation of α-HCH in lakes, coastal bays and their watersheds. The authors reported different rates of degradation for the enantiomers. The biodegradation of some pharmaceuticals, such as antineoplastics, carcinogenics, mutagenics, teratogenics and fetotoxics, in waste water has been studied by Kümmerer *et al.* [55]. Stereochemistry may be crucial for the pharmaceutical activity of the compounds, as well as for their biodegradability in the environment. To determine the enantioselective degradation of chiral pollutants, some workers have carried out the degradation of certain chiral pollutants in the laboratory, under varying experimental conditions. In this direction, the work of Ludwig *et al.* [56] is very important and highly regarded. The authors carried out the microbial degradation of α-HCH and γ-HCH pesticides. α-HCH is the only chiral isomer of the eight available isomers; while γ-HCH is achiral, but is the only active isomer that exhibits insecticidal properties. The biodegradation of these two isomers was reported to be enantioselective. α-HCH converted into the β-PCCH enantiomer, and γ-HCH into the γ-PCCH enantiomer. The biodegradation (ER values) of these isomers is given in Table 3.3. It may be concluded from this table that both of the HCH isomers degraded at different rates (α-HCH at a faster rate), which resulted in different β-PCCH and γ-PCCH enantiomeric ratios. In another experiment, Ludwig *et al.* [57] have studied the biodegradation of 2-(2,4-dihydroxyphenoxy) propionic acid by marine bacteria (Table 3.4 and Figure 3.6). The authors have reported transfer of the *R*-enantiomer, while the *S*-enantiomer persisted in the environment. The degradation product of this pollutant was 2,4-dichlorophenol (2,4-DCP). The trend for the enantioselective degradation of 2-(2,4-dihydroxyphenoxy) propionic acid is shown in Figure 3.6. Buser and Müller [58, 59] studied the biodegradation of α-, β-, γ- and δ-HCHs isomers in sewage sludge in anaerobic conditions. The authors have described significant degradation of the γ- and α-isomers, with half-lives between 20.4 and 99 h respectively. The high enantioselectivity in the transfer of α-HCH was indicated by different rates for the (+)-enantiomer (20.2×10^{-3} h^{-1}) and the (−)-enantiomer (7.26×10^{-3} h^{-1}), resulting in an enrichment of the (−)-enantiomer. Furthermore, it is interesting to note

Table 3.3 The enantiomeric ratios of β-PCCH and γ-PCCH formed by microbial degradation of α-HCH and γ-HCH, respectively [56]

Time (days)	β-PCCH	γ-PCCH
0	–	–
7	1.16	1.20
14	1.32	1.36
21	1.13	1.00
28	1.18	1.01

Table 3.4 The enantiomeric ratios of 2-(2,4-dichlorophenoxy) propionic acid (dichlorprop, or DCPP) formed by microbial degradation [57]

Time (days)	Enantiomeric ratio
0	0.99
3	1.02
7	1.08
10	1.14
14	1.34
17	1.37
21	1.41

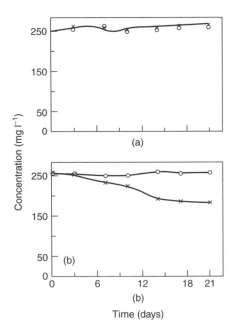

Figure 3.6 The biotransformation of 2-(2,4-dichlorophenoxy) propionic acid insecticide in the presence of marine bacteria [57].

that the authors have reported α- and γ-HCH conversion that is similar to the conversion of γ-HCH in animals described by Ludwig *et al.* [22, 60].

Wong *et al.* [61] described the biodegradation of the enantiomers of eight polychlorinated biphenyls (PCBs) in aquatic sediments from selected sites throughout the United States. Nonracemic ERs for PCBs 91, 95,

132, 136, 149, 174 and 176 were found in sediment cores from Lake Hartwell. Nonracemic ERs for many of the PCBs were also found in bed sediment samples from the Hudson and Housatonic Rivers, thus indicating that some of the PCB biotransformation processes identified at these sites were enantioselective. Patterns in ERs among congeners were consistent with known reductive dechlorination patterns in both river sediment basins. The enantioselectivity of PCB 91 was reversed between the Hudson and Housatonic River sites, which implies that the two sites have different PCB biotransformation processes, with different enantiomer preferences. The different enantiomeric ratios of some chiral pollutants resulting from enantioselective biotransformation and biodegradation in different components of the environment are given in Tables 3.2 and 3.3.

Raschke *et al.* [62] studied the biodegradation of four different classes of aromatic compounds by the *Escherichia coli* strain DH5alpha (pTCB 144), which contains the chlorobenzene dioxygenase (CDO) from *Pseudomonas* sp. strain P51. CDO oxidized biphenyl as well as monochlorobiphenyls to the corresponding *cis*-2,3-dihydro-2,3-dihydroxy derivatives, whereby oxidation occurred on the unsubstituted ring. The authors also reported no oxidation of higher substituted biphenyls. The absolute configurations of several monosubstituted *cis*-benzene dihydrodiols formed by CDO were determined. All had an *S*-configuration at the carbon atom in the *meta*-position to the substituent on the benzene nucleus. With one exception, the enantiomeric excess of several 1,4-disubstituted *cis*-benzene dihydro-diols formed by CDO was higher than that of the products formed by two toluene dioxygenases. Naphthalene was oxidized to enantiomerically pure (+)-*cis*-(1*R*,2*S*)-dihydroxy-1,2-dihydronaphthalene. All absolute con-figurations were identical to those of the products formed by toluene dioxygenases of *Pseudomonas putida* UV4 and *P. putida* F39/D. The formation rate of (+)-*cis*-(1*R*,2*S*)-dihydroxy-1,2-dihydronaphthalene was significantly higher (about 45–200 %) than those of several monosubsti-tuted *cis*-benzene dihydrodiols, and more than four times higher than the formation rate of *cis*-benzene dihydrodiol.

Wiberg *et al.* [63] reported the biodegradation of organochlorine (OC) pesticide residues in 32 agricultural and three cemetery soils from Alabama. The enantiomeric signatures were similar to those obtained from other soils in Canada and the USA. The enantiomeric fractions (EFs) of *o,p'*-DDT showed great variability, ranging from 0.41 to 0.57, while the EFs of the chlordanes and chlordane metabolites were less variable and, in general, differed significantly from racemic mixtures. Garrison *et al.* [64] investigated the microbial and enzymatic degradation of *o,p'*-DDT. The authors reported the degradation of both of the enantiomers of o,p'-DDT through similar pathways but at different rates. o,p'-DDT converted into o,p'- and p,p'-DDD and o,p'- and p,p'-DDE anaerobically, and into o,p'- and p,p'-DDA [2-(2-chlorophenyl)-2-(4-chlorphenyl) ethanoic acid] and 2,2-bis(4-chlorophenyl) ethanoic acid in aerobic conditions. The microbial transformation of these pesticides was reported to be very slow. The authors also reported that chirality in o,p'-DDT in the environment is rare.

Enantioselective depletion of (+)-*trans*-chlordane and (−)-*cis*-chlordane, and enrichment of (+)-heptachlor-*exo*-epoxide and (+)-oxychlordane, has been found in the great majority of the samples with detectable residues. The enantiomeric composition of hexachlorocyclohexane was racemic or close to racemic. Similarly, Garrison *et al.* [65] reported enantioselective biodegradation of 2(2,4-dichlorophenoxy) propionic with 4 and 8 days, as the half-lives of its (−)- and (+)-enantiomers. This is a fortuitous situation, since the (+)-enantiomer is known to be an active antipode while the (−)-enantiomer is a ballast. Schneiderheinze *et al.* [66] have studied the enantioselective biodegradation of phenoxyalkanoic herbicides; that is, 2(2,4-dichlorophenoxy) propionic acid (2,4-DP) and 2-(4-chloro-2-methylphenoxy) propionic acid (MCPP). Racemic mixtures of 2,4-DP and MCPP were applied to three species of turf grass, four species of broadleaf weeds and soil. Preferential degradation of the S-(−)-enantiomer of each herbicide was observed in most species of broadleaf weeds and in soil, while the degradation in all species of grass occurred without enantioselectivity. The biodegradation in all of the systems appeared to follow pseudo first order kinetics, with the fastest degradation rate in broadleaf weeds, followed by the grasses, while the slowest degradation rate was observed in soil. The degradation of 2,4-DP and MCPP in three species of broadleaf weeds is shown in Figures 3.7 and 3.8, respectively, while Figure 3.9 indicates the biodegradation of 2,4-DP and MCPP in soil. Recently, Lewis *et al.* [67] have studied the enantioselective degradation of ruelene, dichlorprop and methyl dichlorprop in soil samples collected from Brazil, Norway and North America. Water slurries of these soil samples were spiked and analysed at different time intervals for these pesticides. The authors have reported the biodegradation of methyl dichlorprop within a day, while the biodegradation of ruelene and dichlorprop was reported after several months. Metalaxyl, with its R-(+)-antipode as an active ingredient, biodegrades through first order kinetics, with disintegration rate constants of 0.063 and 0.011 per day for the two enantiomers [68]. The half-lives of the enantiomers were reported to be 11 and 63 days, respectively. The microbial transformation of metalaxyl in soil-water slurry

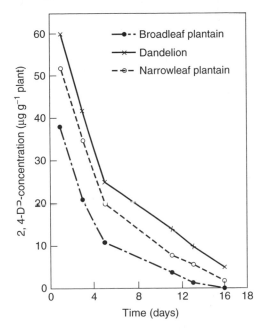

Figure 3.7 The biodegradation of dichlorprop in three species of broadleaf annuals [66].

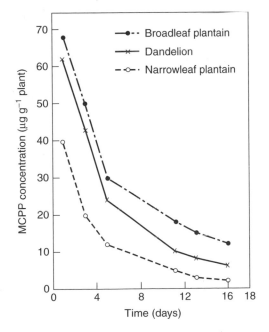

Figure 3.8 The biodegradation of MCCP in three species of broadleaf annuals [66].

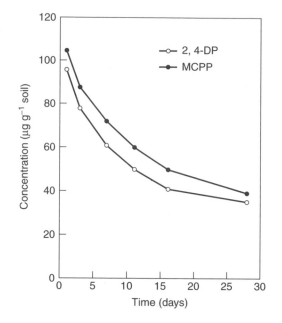

Figure 3.9   The biodegradation of dichlorprop and mecoprop in soils [66].

in aerobic conditions is shown in Figure 3.10. Recently, Buser *et al.* [69] have studied the degradation/dissipation of metalaxyl and its primary carboxylic acid metabolite (MX-acid) in soil under laboratory conditions. The racemic and the enantiopure *R*- and *S*-compounds were incubated in separate experiments. The degradation of metalaxyl was shown to be enantioselective, with the fungicidally active *R*-enantiomer being degraded faster than the inactive *S*-enantiomer, resulting in residues enriched with *S*-metalaxyl when the racemic compound was incubated. The relatively high enantioselectivity suggests that the degradation/dissipation was largely biological. The data indicated conversion of 40–50 % of the metalaxyl to MX-acid, the remaining metalaxyl being degraded via other pathways. The degradation of MX-acid was also enantioselective. Metalaxyl and MX-acid were both configurationally stable in soil, showing no interconversion of *R*- to *S*-enantiomers, or vice versa. Furthermore, the conversion of metalaxyl to MX-acid proceeded with retention of the configuration. The degradation followed approximately first order kinetics, but showed significant lag phases. Marucchini and Zadra [70] investigated the possible stereo- and/or enantioselective degradation in soil and plants (sunflower) of the fungicide metalaxyl (rac-Metalaxyl) and the new compound Metalaxyl-M−*R*(−)-Metalaxyl). The degradation of the two stereoisomers of metalaxyl proved

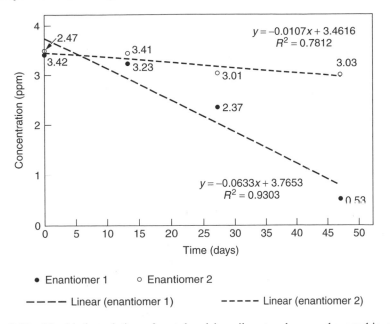

Figure 3.10 The biodegradation of metalaxyl in soil water slurry under aerobic conditions [68].

to be enantioselective and dependent on the media: the (+)-(*S*)-enantiomer showed a faster degradation in plants, while the (−)-(*R*)-enantiomer showed a faster degradation in soil.

3.4 Photochemical Conversion

The photochemical conversion of chiral agrochemicals and xenobiotics is achiral in nature. Several studies have been carried out to determine enantioselective photochemical degradation [45, 71–73], but all of these reports have indicated the nonenantioselective decomposition of chiral pollutants. It can be concluded that the photochemical decomposition of chiral pollutants is nonenantioselective. Busser and Müller [71] and Koske *et al.* [72] studied the photoconversion of cyclodiene, heptachlor, *cis-* and *trans*-chlordane and *cis-* and *trans*-nonachlor, and the authors reported nonenantioselective conversion of these pesticides. However, Hühnerfuss *et al.* [45] reported variation in the enantiomeric ratios of β-PCCH formed by enzymatic action, in the presence of light, provided that the enzymatic process is less effective.

3.5 Metabolism

The metabolism of chiral xenobiotics is also enantioselective in nature and involves two distinct forms; substrate enantioselective and product enantioselective. Recently, a few reviews have discussed this issue [74–76]. The substrate enantioselective form features the differential metabolisms of two enantiomeric substrates under identical conditions. This must be contrasted with product enantioselectivity and, more generally, with product selectivity, which implies the differential formation of stereoisomeric metabolites from a single substrate containing one or more centres of prochirality [77]. Substrate enantioselectivity can originate from chiral recognition either at the binding step or at the catalytic step, or at both. The metabolic interactions between the enantiomers are quite informative in some cases: substrate enantioselectivity involves the metabolic recognition of enantiomeric xenobiotics, and product stereoselectivity arises from the metabolic discrimination of enantiotopic or diastereotopic target groups in a single substrate compound. The observed phenomenon results from different modes of binding of the substrate to the active site of the enzyme. The enzyme substrate complexes are diastereoisomeric and, hence, of different energies, which depend on the nature of the substrate and the active site.

Drug–receptor interactions have been explored by many pharmacologists [78–82], but only a few reports deal with this sort of interaction involving chiral xenobiotics and, therefore, it may be assumed that chiral pollutants interact biologically in a similar way to chiral drugs. Intrinsic activity, affinity and efficacy are very important for the binding of chiral pollutants to receptors. Basically, the binding in chiral xenobiotic–receptor interactions can be quantified by affinity parameters. Moreover, both intrinsic activity and efficacy are essential aspects of the activation step in receptorology, and there is a need for theoretical investigation in this direction. Laduron [83] has emphasized the pitfalls of binding studies – the lack of full receptor specificities for most of the enantiomers. Thus, the xenobiotic metabolism and drug–receptor enantioselective interactions result from chiral recognition at the binding site or at the activation step. This can be understood in molecular terms, since the stereoelectronic conditions for high affinity and high efficiency or velocity may not be identical. The bindings require adequate functional groups and structural features in the chiral molecules. The efficiency of the activation step depends on the relative positioning of the tigger and target groups. The difference between enzymes and receptors loses its significance, and both can be viewed as biological effectors.

The metabolism of xenobiotics in plants occurs in three phases. The first phase comprises transformations of the xenobiotics themselves, by oxidation, reduction and hydrolysis reactions with the primary metabolites. These compounds, along with the parent xenobiotics, are called exocons. In the second phase, the primary metabolites or parent xenobiotics are reacted through hydroxyl, sulfide, primary amine, carboxylic acid, chlorine and nitrate groups with endogenous substances (endocons), resulting in conjugates that are more water soluble compared to the exocons. Depending on the structure of the exocons, the endocons can take the form of glucose or other sugars, amino acids, organic acids or glutathione. Sugar conjugates are frequently processed by the addition of further carbohydrate units or malonic acid, whereas glutathione conjugates are predominantly transformed by a sequence of reactions. Both the first and second phase reactions, including processing, are catalysed by enzymes. After this, the carbohydrates and glutathione conjugates are presumably stored in the vacuole or apoplastic region; these types of reaction come under phase three and may be characterized as internal section [75, 84]. Recently, glutathione conjugates have been shown to be exported into the vacuole by an ATP energized tonoplast carrier [85, 86].

In general, metabolic studies of agrochemicals have been carried out in plant cell suspension cultures. During past few decades, axenic cell suspension and callus cultures of higher plants have been used to carry out metabolic studies of pesticides and xenobiotics. Plant cells have also been used to investigate the phytotoxicity of agrochemicals and the mechanisms of their actions [87–90]. When compared to intact plants grown under environmental conditions, plant cell cultures are artificial systems. Thus, the data obtained from these model experiments is considered to be uncertain in some cases. Plant cell cultures are obtained from aseptic plants, which can be produced from leaves, hypocotyl or roots. These parts are sterilized, cut into pieces and placed on a solidified medium. After few weeks, the callus tissue formed is removed, and the plant cells are grown either as callus culture on a solidified medium or as suspensions in liquid media. Cells may be cultured heterotrophically (chlorophyll deficient in the dark) with an organic carbon source (glucose), mixotrophically or fully phototrophically under light.

Basically, the metabolism of pesticides and xenobiotics in plant cells is considered to be a detoxification process, this also being true for the portion of the applied chemical that is transformed into nonextractable residues. Xenobiotics bound to the plant cell wall are immobile and, thus, are effectively kept away from the sites of primary and secondary plant metabolism. Consequently, phytotoxic effects on the endogenous processes are considerably reduced [75]. In view of all these points, some research work related to the metabolisms of chiral xenobiotics in animals and plants is discussed below.

Glufosinate (GA) is a post-emergence and nonselective herbicide [91, 92]. The L-enantiomer of this herbicide, which is also called phosphinothricin, is a natural microbial phytotoxin, produced by *Streptomyces viridochromogenes* and *Streptomyces hygroscopicus*, and acts as an inhibitor of glutamine synthetase, while the D-enantiomer shows no activity on this enzyme. The herbicidal action involves a rapid accumulation of ammonia, a deficiency in several amino acids, an inhibition of photosynthesis and, finally, the death of the plant cell [91, 93].

Jansen *et al.* [94] studied the metabolism of 14C-glufosinate (GA) in excised shoots and leaves of 20 weed and nonweed species. This herbicide was applied through xylem, and the metabolism was observed after 24–48 h of incubation. The authors reported several metabolized products, such as 2-oxo-4-(hydroxymethylphosphinyl) butanoic acid (PPO), 2-hydroxy-4-(hydroxymethylphosphinyl) butanoic acid (MHB), 3-(hydroxymethylphosphinyl) propionic acid (MPP) and 4-(hydroxymethylphosphinyl) butanoic acid (MPB). However, this pesticide degraded into 4-(hydroxymethylphosphinyl) ethanoic acid (MPA). The structures of these metabolites are given in Figure 3.11, and it may be observed from this figure that MHB is the only optically active metabolite. The metabolism is slow and nonexistent in nature. The first step was the transformation of L-GA to PPO and this intermediate changed rapidly into MPP, which is considered to be the stable metabolite. PPO can also be reduced to the minor transformation product MHB. A possible metabolite of MHB – that is, MPB – was discovered in cell cultures, but was never found to occur *in vivo*. MPB is, therefore, regarded as an artefact produced by cultured cells. In contrast to the L-enantiomer, the D-antipode is apparently not metabolized by plant tissues [95–98]. Komossa *et al.* [97] also reported acetylation of L-GA to nonphytotoxic N-acetyl-glufosinate (NAG), along with trace amounts of nontransgenic metabolites. In carrot (*Daucus carota*), foxglove (*Digitalis purpurea*) and jimson weed (*Datura stramonium*), MMP was the major metabolite, while NAG was the only metabolite found in transgenic sugar beet (*Beta vulgaris*). Also, MPB emerged in trace amounts in foxglove and jimson weed, while MHB and PPO were not detected. Recently, Müller *et al.* [99] have studied the metabolism of glufosinate ammonium in heterotrophic cell suspension and callus cultures of transgenic (*bar-gene*) and

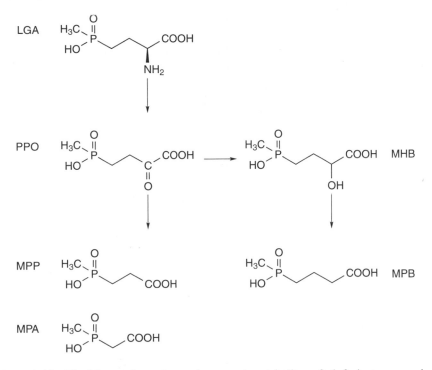

Figure 3.11 The biotransformation pathway and metabolites of glufosinate ammonium salt in plant tissues [94].

nontransgenic sugar beet. Similar studies were also carried out using suspensions of carrot, foxglove and thorn apple (*Datura stramonium*). The authors have reported MPP and *N*-acetyl-L-glucosinate as the major metabolites.

Hühnerfuss *et al.* [46] described the different enantiomeric ratios of eight chiral pollutants and their metabolites in seawater mussels, eider duck, seals and the eggs of some animals. The authors could not give an exact explanation for the appearance of different enantiomeric ratios in the organs of these animals. However, they assumed that the reason lies in differing physiologies and in the differing metabolisms of these chiral pollutants. The metabolisms of many of chiral xenobiotics, such as benzo(a)pyrene and anthracene derivatives, β-naphthoflavone, 3-methylcholanthrene, (+)-BP-7,8-dihydrodiol and (−)-BP-7,8-dihydrodiol, have been studied in various animals. The stereoselective metabolism of benzo(a)pyrene to an ultimate carcinogen has been well known. The metabolic pathway for this is as follows: BP → BP-7,8-oxide → BP-7,8-dihydrodiol → BP-7,8-diol-9,10-epoxide [100]. Similarly, the benzo(a)pyrene analogous polycyclic hydrocarbon chrysene is also metabolized stereospecifically into the corresponding epoxide metabolites [101]. 3-Methylcholanthrene (3MC, Figure 3.12), a potent mutagen and carcinogen, is metabolized to a complex mixture of mutagenic/tumorigenic products, along with some nontoxic products, by mammalian drug metabolizing enzyme systems [102]. 3MC is also used as a cytochrome P-450 isoenzyme inducer in some experimental animals [103]. The C_1 and C_2 positions of 3-MC (Figure 3.12) are major sites of oxidative metabolism, which results in the initial formation of 1-OH-3MC and 2-OH-3MC [104–106]. 1-OH-3MC and 2-OH-3MC are further metabolized to mutagenic and tumorigenic detoxification products [102, 107, 108]. In the metabolism of 3MC by rat liver microsomes, both 3MC *trans*- and *cis*-1,2-diols are formed as metabolites [104, 105]. In rat liver microsomal metabolism of 1-OH-3MC and 2-OH-3MC, 3MC *trans*-1,2-diol is detected, with but very little 3MC *cis*-1,2-diol as a metabolite [105]. In the metabolism of 3MC by rat liver microsome, 1-OH-3MC is enriched in the 1S-enantiomer (53–73 %), whereas 2-OH-3MC is enriched in the 2S-enantiomer (86–98 %). It is interesting to observe that the exact enantiomeric compositions of these metabolites depend on the extent of their further metabolism, and a whether the rats were pretreated with an enzyme inducer.

In 1990, Shou and Yang [109] described the elucidation of the absolute configuration of *trans*- and *cis*-3MC-trans-1,2-diols using circular dichroism (CD) spectra and the hydroxylation of 1R- and 1S-isomers of 3MC in rat liver. The absolute configuration of an enantiomeric 3MC *trans*-1,2-diol was established by the exciton chirality CD method, following conversion to a bis-*p*-*N*,*N'*-dimethylaminobenzoate. The interaction of an enantiomeric 1-OH-3MC with rat liver microsomes resulted in the formation of enantiomeric 3MC *trans*- and *cis*-1,2-diols; the absolute configurations of the enantiomeric 1-OH-3MC and 3MC *cis*-1,2-diol were established on the basis of the absolute configuration of an enantiomeric 3MC *trans*-1,2-diol. The absolute configurations of the enantiomeric 1-OH-3-OHMC were determined by comparing their CD spectra with those of enantiomeric 1-OH-3MC (Figure 3.13). The relative amounts of three aliphatic hydroxylation products formed by rat liver microsomal metabolism of racemic 1-OH-3MC were in the order 1-OH-3-OHMC > 3MC *cis*-1,2-diol > 3MC *trans*-1,2-diol. The enzymatic hydroxylation at C_2 of racemic 1-OH-3MC was enantioselective towards the 1S-enantiomer over the 1R-enantiomer (~3/1), while hydroxylation at the C_3-methyl group was enantioselective towards the 1R-enantiomer over the 1S-enantiomer (~58/42). Rat liver microsomal C_2-hydroxylation of racemic 1-OH-3MC resulted in a 3MC

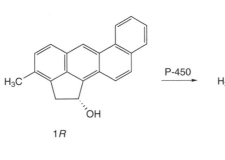

trans-1,2-diol
(24 %)

1S, 2S (63 %)

(t1)

+

cis-1,2-diol
(76 %)

1S, 2R (12 %)

(c1)

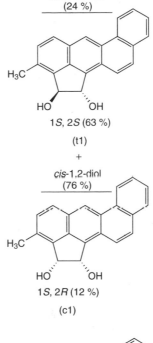

P-450

1R

(a)

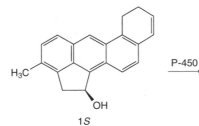

P-450

1S

(b)

1R, 2R (37 %)

(t2)

+

1R, 2S (88 %)

(c2)

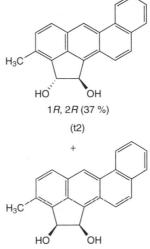

Figure 3.12 The absolute configuration of *trans*- and *cis*-1,2-diols derived from C$_2$-hydroxylation of (1R)-OH-3MC and (1S)- 3MC. Note the change of stereochemical designation (1S → 1R and 1S → 1R respectively) of the hydroxyl group at C$_1$ upon C$_2$-hydroxylation [109].

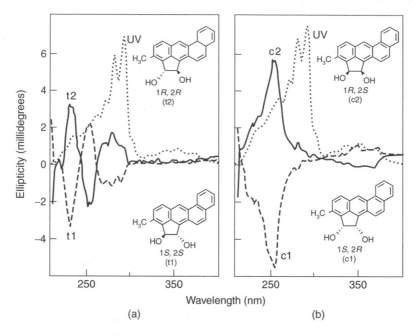

Figure 3.13 UV/Vis and CD spectra of enantiomeric (a) 3MC *trans*-1,2-diols and (b) 3MC *cis*-1,2-diols [109].

trans-1,2-diol with a (1*S*,2*S*)/(1*R*,2*R*) ratio of 63/37 and a 3MC *cis*-1,2-diol with a (1*S*,2*S*)/(1*R*,2*R*) ratio of 12/88, respectively (Figure 3.12).

3.6 Conclusions

Chiral pollutants from various sources are contaminating our environment. These chiral pollutants have been reported in plants and in the various body parts of aquatic and terrestrial animals, including human beings, having entered into the body through the food chain. The biotransformation, biodegradation and metabolism of some chiral pollutants are enantioselective in nature, producing many chiral and nonchiral metabolites. It is also known that some chiral pollutants may interact enantioselectively with an achiral environment in the presence of a chiral catalyst. Knowledge of the biotransformation, biodegradation and metabolism of chiral pollutants is essential in order to understand the exact toxicities and other side effects due to these notorious chiral pollutants. Moreover, prevention and the treatment of the lethal side effects and diseases due to chiral xenobiotics can be carried out scientifically once their metabolisms and their pattern of biodegradation and biotransformation are understood. In spite of these facts, only a few reports are available on this issue – especially on the metabolism of chiral pollutants in animals and plants. Therefore, more studies are required in this field of work to save the planet from the hazardous side effects of chiral xenobiotics.

References

1. S. C. Maness and D. P. McDonnell, K. W. Gaido, *Toxicol. Appl. Pharmacol.* **151**, 135 (1998).
2. N. Kurihara and J. Miyamoto, *Chirality in Agrochemicals*, Wiley, New York, 1998.
3. B. Testa, *TIPS* **7**, 60 (1986).
4. S. Mason, *TIPS* **7**, 20 (1986).
5. D. Stevenson and I. D. Wilson, *Chiral Separations*, Plenum Press, New York, 1988.
6. R. Crossley, *Tetrahedron* **48**, 8155 (1992).
7. A. D. Mesecar and D. E. Koshland Jr, *Nature* **403**, 614 (2000).
8. E. J. Ariëns, A. J. Beld, J. F. Rodrigues de Miranda and A. M. Simonis, in R. D. O'Brien, ed., *The Receptors*, Plenum Press, New York, 1979.
9. D. R. Burt, in H. J. Yamamura, S. J. Enna and M. J. Kuhar, eds, *Neuritransmitter Receptor Binding*, 2nd edn, Raven Press, New York, 1985, p. 41.
10. M. D. Hollenberger, in H. J. Yamamura, S. J. Enna and M. J. Kuhar, eds, *Neuritransmitter Receptor Binding*, 2nd edn, Raven Press, New York, 1985, p. 1.
11. W. Chen and A. Mulchandani, *Trends Biotech.* **16**, 71 (1998).
12. J. Faller, H. Hühnerfuss, W. A. König, R. Krebber and P. Ludwig, *Environ. Sci. Technol.* **25**, 676 (1991).
13. R. Kallenborn, H. Hühnerfuss and W. A. König, *Angew. Chem. Int. Ed. Engl.* **30**, 320 (1991).
14. B. Pfaffenberger, H. Hühnerfuss, R. Kallenborn, A. Köhler-Günther, W. A. König and G. Krüner, *Chemosphere* **25**, 719 (1992).
15. A. T. Fisk, M. Holst, K. A. Hobson, J. Duffe, J. Moisey and R. J. Norstrom, *Arch. Environ. Contam. Toxicol.* **42**, 118 (2002).
16. J. Moisey, A. T. Fisk, K. A. Hobson and R. J. Norstrom, *Environ. Sci. Technol.* **35**, 1920 (2001).
17. A. T. Fisk, J. Moisey, K. A. Hobson, N. J. Karnovsky and R. J. Norstrom, *Environ. Pollut.* **113**, 225 (2001).
18. B. Pfaffenberger, I. Hardt, H. Hühnerfuss, W. A. König, G. Rimkus, A. Glausch and V. Schurig, *Chemosphere* **29**, 1543 (1994).
19. W. A. König, I. H. Hardt, B. Gehrcke, D. H. Hochmuth, H. Hühnerfuss, B. Pfaffenberger and G. Rimkus, *Angew. Chem.* **106**, 2175 (1994).

20. H. Y. Aboul-Enein and I. A. B. Laila, in H. Y. Aboul-Enein and I. W. Wainer, eds, *The Impact of Stereochemistry on Drugs Development and Use*, Chemical Analysis, vol. 142, Wiley, New York, 1997.
21. K. Möller, H. Hühnerfuss and G. Rimkus, *J. High Res. Chromatogr.* **16**, 672 (1993).
22. V. Möller, Schicksal und Toxische Wirkung Chlorierter Organischer Problemstoffe in der Umwelt und Deren Okotoxikologische Bewertung, Thesis, University of Hamburg, 1998, p. 154.
23. H. Hühnerfuss, B. Pfaffenberger, B. Gehrcke, L. Karbe, W. A. König and O. Landgraff, *Pollut. Bull.* **30**, 332 (1995).
24. G. P. Blanch, A. Glausch, V. Schurig, R. Serrano and M. J. Gonzalez, *J. High Res. Chromatogr.* **19**, 392 (1996).
25. S. Jensen, *New Scientist* **32**, 612 (1966).
26. J. S. Waid, *PCB and the environment*, vol. I, CRC Press, Boca Raton FL, 1966, p. 79.
27. J. De Boer, *Chemosphere* **17**, 1811 (1988).
28. J. C. Duinker, M. T. J. Hillebrand, T. Zeinstra and J. P. Boon, *Aquat. Mamm.* **15**, 95 (1989).
29. R. Kallenborn and H. Hühnerfuss, *Seevögel* **14**, 23 (1993).
30. P. Haglund, *Organohal. Compds* **23**, 35 (1995).
31. V. Schurig, A. Glausch and M. Fluck, *Tetrahedron Asymm.* **6**, 2161 (1995).
32. P. Haglund, *J. Chromatogr. A* **724**, 219 (1996).
33. S. Reich, B. Jimenez, L. Marsili, L. M. Hernandez, V. Schurig and M. J. Gonzalez, *Organohal. Compds* **35**, 335 (1998).
34. H. Hühnerfuss, B. Pfaffenberger, B. Gehrcke, L. Karbe, W. A. König and O. Landgraff, *Organohal. Compds* **21**, 15 (1994).
35. H. Hühnerfuss, B. Pfaffenberger and G. Rimkus, *Organohal. Compds* **29**, 88 (1996).
36. T. Ellerichmann, A. Bergman, S. Franke, H. Hühnerfuss, E. Jakobsson, W. A. König and C. Larsson, *Fresenius Environ. Bull.* **7**, 244 (1998).
37. A. Bergman, T. Ellerichmann, S. Franke, H. Hühnerfuss, E. Jakobsson, W. A. König and C. Larsson, *Organohal. Compds* **35**, 339 (1998).
38. C. S. Wong, A. W. Garrison, P. D. Smith and W. T. Foreman, *Environ. Sci. Technol.* **35**, 2448 (2001).
39. G. Rimkus, *Toxicol. Lett.* **111**, 37 (1999).
40. R. Gatermann, H. Hühnerfuss, G. Rimkus, A. Attar and A. Kettrup, *Chemosphere* **36**, 2553 (1997).
41. J. Gao, A. W. Garrison, C. Hoehamer, C. S. Mazur and N. L. Wolfe, *J. Agric. Food Chem.* **48**, 6114 (2000).
42. J. Gao, A. W. Garrison, C. Hoehamer, C. S. Mazur and N. L. Wolfe, *J. Agric. Food Chem.* **48**, 6121 (2000).
43. B. Mannervik and U. H. Danielson, *CRC Crit. Rev. Biochem.* **23**, 283 (1988).

44. J. Faller, H. Hühnerfuss, W. A. König and P. Ludwig, *Mar. Pollut. Bull.* **22**, 82 (1991).
45. H. Hühnerfuss, J. Faller, W. A. König and P. Ludwig, *Mar. Chem.* **26**, 2127 (1992).
46. H. Hühnerfuss, J. Faller, R. Kallenborn, W. A. König, P. Ludwig, B. Pfaffenberger, M. Oehme and G. Rimkus, *Chirality* **5**, 393 (1993).
47. L. M. Jantunen and T. Bidleman, *J. Geophys. Res.* **101**, 837 (1996).
48. T. Harner, L. M. M. Jantunen, T. F. Bidleman and L. A. Barrie, *Geophys. Res. Lett.* **27**, 1155 (2000).
49. L. M. Jantunen and T. F. Bidleman, *Arch. Environ. Contam. Toxicol.* **35**, 218 (1998).
50. J. J. Ridal, T. F. Bidleman, B. R. Kerman, M. E. Fox and W. M. J. Strachan, *Environ. Sci. Technol.* **31**, 1940 (1997).
51. R. L. Falconer, T. F. Bidlemann and D. J. Gregor, *Sci. Total Environ.* **160/161**, 65 (1995).
52. T. F. Bidlemann, L. M. Jantunen, T. Harner, K. Wiberg, J. Wideman, K. Brice, K. Su, R. L. Falconer, E. J. Aigner, A. D. Leone, J. J. Ridal, B. Kerman, A. Finizio, H. Alegria, W. J. Parkhurst and S. Y. Szeto, *Environ. Pollut.* **101**, 1 (1998).
53. R. L. Falconer, T. F. Bidleman, D. J. Gregor, R. Semkin and C. Teixeira, *Environ. Sci. Technol.* **29**, 1297 (1995).
54. L. M. Jantunen and T. Bidleman, *J. Geophys. Res.* **102**, 279 (1997).
55. K. Kümmerer, A. Al-Ahmad, B. Bertram and M. Wiessler, *Chemosphere* **40**, 767 (2000).
56. P. Ludwig, H. Hühnerfuss, W. A. König and W. Gunkel, *Mar. Chem.* **38**, 13 (1992).
57. P. Ludwig, W. Gunkel and H. Hühnerfuss, *Chemosphere* **24**, 1423 (1992).
58. H. R. Buser and M. D. Müller, *Environ. Sci. Technol.* **29**, 664 (1995).
59. M. D. Müller and H. R. Buser, *Environ. Sci. Technol.* **29**, 2031 (1995).
60. P. Ludwig, Untersuchungen zum enantioselektiven Abbau von polaren und unpolaren chlorieten Kohlenwasserstoffen durch marine Mikroorganismen, Ph.D. thesis, University of Hamburg, 1991.
61. C. S. Wong, A. W. Garrison and W. T. Foreman, *Environ. Sci. Technol.* **35**, 33 (2001).
62. H. Raschke, M. Meir, J. G. Burken, R. Hany, M. D. Müller, J. R. van der Meer and H. P. Kohler, *Appl. Environ. Microbiol.* **67**, 3333 (2001).
63. K. Wiberg, T. Harner, J. L. Wideman and T. F. Bidleman, *Chemosphere* **45**, 843 (2001).
64. A. W. Garrison, V. A. Nzengung, J. K. Avants, J. J. Ellington and N. L. Wolfe, *Organohal. Compds* **31**, 256 (1997).
65. A. W. Garrison, P. Schmitt, D. Martens and A. Kettrup, *Environ. Sci. Technol.* **30**, 2449 (1996).

66. J. M. Schneiderheinze, D. W. Armstrong and A. Berthod, *Chirality* **11**, 330 (1999).
67. D. L. Lewis, A. W. Garrison, K. E. Wommack, A. Whittemore, P. Steudler and J. Melillo, *Nature* **401**, 898 (1999).
68. A. W. Garrison, W. J. Jones, J. J. Ellington and J. Washington, *Selective Transportation and Occurrences of the Enantiomers of Chiral Pesticides in Environmental Matrices and Food Products*, US EPA report APM 127, 12 July 2001.
69. H. R. Buser, M. D. Muller, T. Poiger and M. E. Balmer, *Environ. Sci. Technol.* **36**, 221 (2002).
70. C. Marucchini and C. Zadra, *Chirality* **14**, 32 (2002).
71. H. R. Buser and M. D. Müller, *Environ. Sci. Technol.* **27**, 1211 (1993).
72. G. Koske, G. Leupold and H. Parlar, *Fresenius Environ. Bull.* **6**, 489 (1997).
73. H. Hühnerfuss, J. Faller, W. A. König and P. Ludwig, *Environ. Sci. Technol.* **26**, 2127 (1992).
74. B. Testa, *Chirality* **1**, 7 (1989).
75. B. Schmidt, *Recent Res. Agric. Food Chem.* **3**, 329 (1999).
76. E. J. Ariëns, J. J. S. van Resen and W. Welling, *Stereochemistry of Pesticides, Biological and Chemical Problems: Chemicals in Agriculture*, vol. 1, Elsevier, Amsterdam, 1988.
77. B. Testa, *Biochem. Pharmacol.* **37**, 85 (1988).
78. E. J. Ariëns and A. M. Simonis, *J. Pharm. Pharmacol.* **16**, 137 (1964).
79. R. F. Furchgott and P. Bursztyn, *Acad. Sci.* **144**, 882 (1967).
80. R. R. Ruffolo, *J. Auton. Pharmacol.* **2**, 277 (1982).
81. P. Angeli, L. Brasili, M. Gianella, F. Gualtieri and M. Pigini, *Br. J. Pharmacol.* **85**, 783 (1985).
82. D. Mackay, *Pharmacol. Sci.* **9**, 156 (1988).
83. P. M. Laduron, *Biochem. Pharmacol.* **33**, 833 (1984).
84. D. Scheel and H. Sandermann, *Planta* **152**, 253 (1981).
85. J. O. D. Coleman, M. M. A. Blake-Kalff and T. G. E. Davies, *Trends Plant Sci.* **2**, 144 (1997).
86. P. A. Rea, Y. P. Lu and Z. S. Li, *Trends Plant Sci.* **2**, 290 (1997).
87. R. H. Shimabukuro and W. C. Walsh, in G. D. Paulson, D. S. Frear and E. P. Marks, eds, *Xenobiotic Metabolism, In Vitro*, ACS Symposium Series 97, ACS, Washington, DC, 1979, p. 3.
88. R. O. Mumma and R. H. Hamilton, *J. Toxicol. Clin. Toxicol.* **19**, 535 (1982).
89. H. Sandermann, D. Scheel and T. Trenck, *J. Appl. Polym. Sci. Appl. Polym. Symp.* **37**, 407 (1983).
90. N. D. Camper and S. K. McDonald, *Rev. Weed Sci.* **4**, 169 (1989).
91. P. L. Metz, W. J. Iekema and J. P. Nap, *Mol. Breed.* **4**, 335 (1998).
92. F. Schwerdtle, H. Bieringer and M. Finke, *PflSchutz Sonderheft* **9**, 431 (1981).
93. G. Hoerlein, *Rev. Environ. Contam. Toxicol.* **138**, 73 (1994).

94. C. Jansen, I. Schuphan and B. Schmidt, *Weed Sci.* **48**, 319 (2000).
95. W. Dröge, I. Broer and A. Pühler, *Planta* **187**, 142 (1992).
96. W. Dröge, U. Siemeling, A. Pühler and I. Broer, *Plant Physiol.* **105**, 159 (1994).
97. D. Komossa and H. Sandermann, *Pestic. Biochem. Physiol.* **43**, 95 (1992).
98. B. G. Mersey, C. Hall, D. M. Anderson and C. J. Swanton, *Pest. Biochem. Physiol.* **37**, 90 (1990).
99. B. P. Müller, A. Zumdick, I. Schuphan and B. Schmidt, *Pest. Mgmt. Sci.* **57**, 46 (2001).
100. R. L. Chang, A. W. Wood, W. Lewin, H. D. Mah, D. R. Thakker, D. M. Jerina and A. H. Conney, *Proc. Natl Acad. Sci.* **76**, 4280 (1979).
101. M. Nordqvist, D. R. Thakker, K. P. Vyas, H. Yagi, W. Levin, D. E. Ryan, P. E. Thomas, A. H. Conney and D. M. Jerina, *Mol. Pharmacol.* **19**, 168 (1981).
102. P. Sims and P. L. Grover, in H. V. Gelboin and P. O. P. Ts'o, eds, *The Metabolic Activation of Polycyclic Hydrocarbons other than Benzo(a)pyrene, Polycyclic Hydrocarbons and Cancer*, vol. 3, Academic Press, New York, 1981, p. 117.
103. A. Y. H. Lu and S. B. West, *Pharmacol. Rev.* **31**, 277 (1980).
104. T. A. Stoming, W. Bornstein and E. Bresnick, *Biochem. Biophys. Res. Commun.* **79**, 461 (1977).
105. B. Tierney, E. Bresnick, P. Sims and P. L. Grover, *Biochem. Pharmacol.* **28**, 2607 (1979).
106. A. Eastman and E. Bresnick, *Cancer Res.* **39**, 4316 (1979).
107. A. W. Wood, R. L. Chang, W. Levin, P. E. Thomas, D. Ryan, T. A. Stoming, D. R. Thakker, D. M. Jerina and A. H. Conney, *Cancer Res.* **38**, 3398 (1978).
108. W. Levin, M. K. Buening, A. W. Wood, R. L. Chang, D. R. Thakker, D. M. Jerina and A. H. Conney, *Cancer Res.* **39**, 3549 (1979).
109. M. Shou and S. K. Yang, *Chirality* **2**, 141 (1990).
110. H. Hühnerfuss, *GIT Fachz. Lab.* **36**, 489 (1992).
111. H. Hühnerfuss and R. Kallenborn, *J. Chromatogr.* **580**, 191 (1992).
112. S. Mössner, T. R. Spraker, P. R. Becker and K. Ballschmiter, *Chemosphere* **24**, 1171 (1992).
113. M. D. Müller, M. Schlabach and M. Ochme, *Environ. Sci. Technol.* **26**, 566 (1992).
114. K. Wiberg, R. Letcher, C. Sandau, R. Norstrom, M. Tysklind and T. Bidleman, *Organohal. Compds* **35**, 371 (1998).
115. M. D. Müller, W. Vetter, K. Hummert and B. Luckas, *Organohal. Compds* **29**, 118 (1996).
116. W. A. König, I. Hardt, B. Gehrcke, D. H. Hochmuch, H. Hühnerfuss, B. Pfaffenberger and G. Rimkus, *Angew. Chem. Int. Ed. Engl.* **33**, 2085 (1994).

117. R. Kallenborn and H. Hühnerfuss, *Chiral Environmental Pollutants, Trace Analysis and Ecotoxicology*, Springer Verlag, Berlin, 2001.
118. H. R. Buser and M. D. Müller, *Environ. Sci. Technol.* **26**, 1533 (1992).
119. T. F. Bidleman, L. M. Jantunen, K. Wiberg, T. Harner, K. Brice, K. Su, R. L. Falconer, A. D. Lenone, E. J. Aigner and W. Parkhurst, *Environ. Sci. Technol.* **32**, 1546 (1998).
120. R. L. Falconer, A. Leone, C. Bodnar, K. Wiberk, T. F. Bidleman, L. M. Jantunen, T. Harner, W. Parkhurst, H. Alegria, K. Brice and K. Su, *Organohal. Compds* **35**, 331 (1998).

Chapter 4

The Enantioselective Toxicities of Chiral Pollutants

4.1 Introduction

In a nonchiral environment, the enantiomers of a racemate possess the same physical and chemical properties. But in the early 1930s, Easson and Stedman introduced a three-point attachment model that laid the basis for the initial understanding of stereochemical differences in pharmacological activity [1]. The authors described the differences in the bioaffinity of the enantiomers to a common site on an enzyme or receptor surface, with the receptor or enzyme needing to possess three nonequivalent binding sites to discriminate between the enantiomers. Therefore, metabolic and regulatory processes mediated by biological systems are sensitive to stereochemistry, and different responses can often be observed when comparing the activities of a pair of enantiomers. These differences can be expressed in distribution rates, their metabolism and excretion, in antagonistic actions relative to each other, or their toxicological properties. Much work has been carried out on the different biological activities of drugs and other pharmaceuticals [2–6] but, unfortunately, there is little information on the enantioselective toxicities of chiral pollutants. In order to determine the exact toxic load of chiral pollutants, it is essential to know the different concentrations of both enantiomers. The present chapter describes the different toxicities of the two enantiomers of chiral pollutants. About 25 % of

Chiral Pollutants. I. Ali and H. Y. Aboul-Enein
© 2004 John Wiley & Sons, Ltd ISBN: 0-470-86780-9

pesticides are chiral and, therefore, these constitute the greater percentage of chiral pollutants. Mostly pesticides are carcinogenic and toxic, damaging certain body organs such as the liver, kidney, bone marrow and so on, and they also change the enzymatic activities. Some studies have been carried out to establish the enantioselective toxicities of pesticides and other pollutants. The enantioselective toxicities of chiral pollutants are discussed in what follows.

4.2 The Enantioselective Toxicities of PCBs

Polychlorinated biphenyls (PCBs) constitute the greater percentage of the chlorinated chiral pollutants, and before 1930 they were frequently used in industries. Although their use has been banned in many countries since 1970, these pollutants still represent an important class of high priority pollutants due to their long persistence, bioaccumulation and toxicities [7]. PCBs are carcinogenic compounds that are known for their persistence and bioaccumulation in the environment due to their physicochemical properties. Seventy-eight out of the 209 congeners of PCBs have axial chirality in their nonplanar conformation and 19 form stable enantiomers (atropisomers) due to restricted rotation around the central carbon atom (Table 4.1) [8]. Due to these facts, attention has been paid to the toxicity of PCBs in marine and terrestrial biota [9–16]. It has been reported that the non-*ortho*-coplanar PCBs exhibit the highest toxicity, followed by the moderately toxic mono-*ortho* coplanar congeners, while the di-*ortho*-substituted PCBs have turned out to be less toxic [15]. The different toxicities of these PCBs include weight loss, porphyria, teratogenesis, and endocrine and productive malfunctions in various organisms [11]. Ahlborg

Table 4.1 The IUPAC nomenclature for the stable enantiomers (atropisomers) of polychlorinated biphenyls (PCBs)

PCB congener	Chlorine position	PCB congener	Chlorine position
45	2,2′,3,6-	144	2,2′,3,4,5′,6-
84	2,2′,3,3′,6-	149	2,2′,3,4′,5′,6-
88	2,2′,3,4,6-	171	2,2′,3,3′,4,4′,6-
91	2,2′,3,4′,6-	174	2,2′,3,3′,4,5,6′-
95	2,2′,3,5′,6-	175	2,2′,3,3′,4,5,6-
131	2,2′,3,3′,4,6-	176	2,2′,3,3′,4,6,6′-
132	2,2′,3,3′,4,6′-	183	2,2′,3,4,4′,5′,6-
135	2,2′,3,3′,5,6′-	196	2,2′,3,3′,4,4′,5,6-
136	2,2′,3,3′,6,6′-	197	2,2′,3,3′,4,4′,6,6′-

et al. [15] presented a toxic equivalency model, with the help of which the authors described the toxicities of PCBs congeners. The authors calculated the toxic equivalency factors (TEFs) for PCBs. Each PCB was assigned a TEF value based on its toxicity relative to 2,3,7,8-tetrachlorodibenzo-*p*-dioxin (TCDD), which has by definition a TEF of 1.00.

Püttmann *et al.* [17] reported the PCB 139 and PCB 197 congeners as drug metabolizing enzyme (cytochrome P-450, *N*-demethylase and aldrin epoxidase) inducers. The authors reported that the racemic mixtures of PCB 139 are stronger inducers compared to the individual enantiomers. In contrast, the racemic mixture of PCB 197 and its individual enantiomers is only weak a inducer of these enzymes. Furthermore, the same group [18], in 1991, studied the toxicities of the enantiomers of PCBs in chick embryo hepatocyte cultures using 2,2′,3,4,6-pentachlorobiphenyl (PCB 88), 2,2′,3,4,4′,6-hexachlorobiphenyl (PCB 139) and 2,2′,3,3′,4,4′,6,6-octachlorobiphenyl (PCB 197) congeners. The authors studied the induction of drugs metabolizing enzymes and the accumulation of protoporphyrin (PROT) and uroporphyrin (URO). The concentration-related induction activities of cytochrome P-450, ethoxyresorufin-*O*-deethylase (EROD) and benzphetamine-*N*-demethylase (BPDM) were measured. The authors reported that PCB 139 was a stronger potent inducer of EROD than PCB 88 and PCB 197. Furthermore, it was reported that EROD activity was induced to a much greater extent by the (+)-enantiomers of all the congeners studied, with the (−)-enantiomers being inactive for PCB 88 and PCB 197. In regard to the induction activity of cytochrome P-450, the induction was in the order PCB 139 > PCB 197 ≥ PCB 88. The (+)- and (−)-enantiomers of PCB 88 and PCB 197 were of equal potency, whereas the potency of the (+)-enantiomer of PCB 139 was greater than that of its corresponding (−)-enantiomer. The BPDM activity was in the order PCB 197 ≥ PCB 139 > PCB 88. The (−)-enantiomers of PCB 197 and PCB 88 and the (+)-enantiomer of PCB 139 were more potent than their corresponding mirror images. The effects of the enantiomers of the PCB 88, PCB 139 and PCB 197 congeners on the induction of total cytochrome P-450, ethoxyresorufin-*O*-deethylase and benzphetamine-*N*-demethylase are shown in Figures 4.1, 4.2 and 4.3, respectively.

The effects of the enantiomeric concentrations of the PCB 88, PCB 139 and PCB 197 congeners on the accumulation of PROT and URO are summarized in Table 4.2. A perusal of this table indicates that URO accumulation only occurred at higher concentrations (i.e. ≥ 1 μM) of PCB 88 and PCB 197, and at lower concentrations of congener PCB 139 [≥ 0.034 μM for (+)-PCB 139 and ≥0.34 μM for (−)-PCB 139].

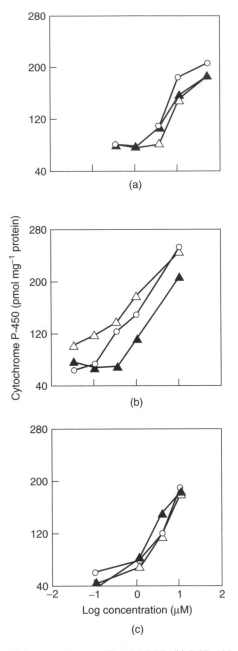

Figure 4.1 The effect of the enantiomers of (a) PCB 88, (b) PCB 139 and (c) PCB 197 congeners on the induction of total cytochrome P-450. △, (+)-enantiomer; ▲, (−)-enantiomer; O, racemic mixture [18].

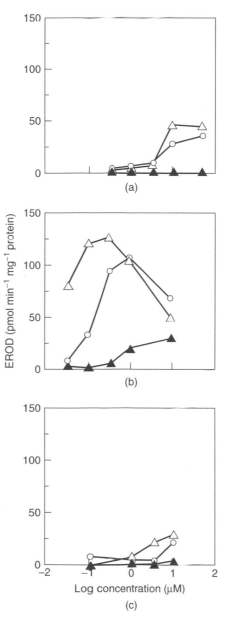

Figure 4.2 The effect of the enantiomers of (a) PCB 88, (b) PCB 139 and (c) PCB 197 congeners on the induction of ethoxyresorufin-*O*-deethylase. $\triangle$, (+)-enantiomer; $\blacktriangle$, (−)-enantiomer; O, racemic mixture [18].

vision) indicated that a PCB and DDE methyl sulfone mixture increased the litter size in these animals [20]. The mixture also influenced some endocrinal parameters in both dam and offspring. A strong respiratory involvement of $MeSO_2$-PCBs has been observed in Japan [21]. Therefore, there are several reports dealing with the toxicities of the $MeSO_2$-PCBs but, unfortunately, no report is available on their enantioselective toxicities. A Swedish–German collaborative project began in 1997, by which the highly enantioselective accumulation of $MeSO_2$-PCBs in the liver of humans and rats has been reported [22, 23], but data on the enantioselective toxicities is still lacking.

4.3 The Enantioselective Toxicities of HCH

HCH is the major pesticide used to control pest hazards all over the world. α-HCH is chiral, while the γ-isomer is achiral but its decomposition results in chiral pentachlorocyclohexene (PCCB) metabolite. Several reports have been published on the enantioselective accumulation of α-HCH pesticides (Chapter 2). Therefore, it is obvious that the biochemistry of α-HCH enantiomers is different. In spite of this, little work has been carried out to determine the enantioselective toxicities of the HCH isomers. Möller *et al.* [24] observed different toxicities of α-HCH enantiomers on cytotoxicity and growth stimulation in primary rat hepatocytes. The cytotoxic effect was determined as a parameter for the acute toxicity of α-HCH, while the growth stimulation may be associated with chronic toxicity; for example, the promotion of tumours. The authors cultured hepatocytes separately in the presence of (+)- and (−)-α-HCH pesticides, and reported 100 % mortality in the presence of 3×10^{-4} M (+)-α-HCH, while at the same concentration of (−)-α-HCH, 75 % mortality was reported. Furthermore, the authors reported no toxic effects of (−)-α-HCH at a lower concentration (1×10^{-4} M), while the mortality rate was 62 % at this concentration of (+)-α-HCH enantiomer. The authors also studied the mimetic rates of α-HCH enantiomers in rats. 5×10^{-5} M concentrations of both of the enantiomers were injected separately into rats, and it was observed that the significant mitosis occurred (factor 2.4) in the presence of the (+)-α-HCH enantiomer, as compared with stimulation by the (−)-α-HCH enantiomer (factor 1.7). Therefore, it may be concluded, from the study carried out by Möller *et al.* [24], that the (+)-enantiomer of α-HCH is more toxic than the (−)-enantiomer. The mortality in primary cultures of rat hepatocytes as a function of the (+)- and (−)-α-HCH concentrations and the stimulation

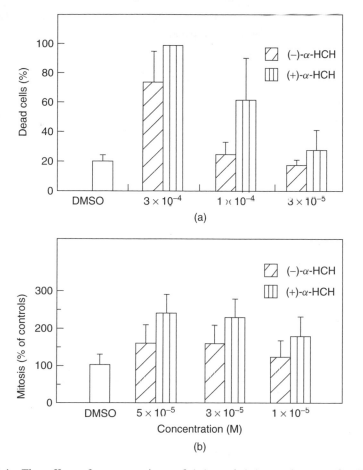

Figure 4.4 The effect of concentrations of (+)- and (−)-enantiomers of α-HCH on (a) mortality in primary cultures of rat hepatocytes and (b) the stimulation of mitosis in primary rat hepatocytes [24].

of mitosis in primary rat hepatocytes by (+)- and (−)-α-HCH are given in Figures 4.4(a) and 4.4(b), respectively.

4.4 The Enantioselective Toxicities of Other Chlorinated Pesticides

Besides PCBs and HCH chlorinated pesticides, some other chiral chlorinated pesticides have been reported to possess different enantioselective toxicities. The most important chiral chlorinated pesticides present in

Table 4.3 The enantioselective toxicities of the enantiomers of chlordene, chlordene epoxide and heptachlor *exo*-epoxide insecticides on German cockroach (*Blattella germanica*) [25]

Insecticide	Dose ($\mu g\,g^{-1}$)					$LD_{28.6}$ ($\mu g\,g^{-1}$)
	300	233	178	126	94.4	
(+)-Chlordene	100%	94.4%	72.2%	33.3%	11.1%	129
(±)-Chlordene	28.6%	–	–	–	–	–
(−)-Chlordene	0.0%	–	–	–	–	–
	200	133.3	88.9	59.3		LD_{50}($\mu g\,g^{-1}$)
(−)-Chlordene epoxide	86.7%	73.3%	66.7%	33.3%		76
(±)-Chlordene epoxide	66.7%	40.0%	13.3%	0.0%		157
(+)-Chlordene epoxide	0.0%	–	–	–		–
	6.0	3.0	1.5	0.75	0.375	LD_{50}($\mu g\,g^{-1}$)
(+)-Heptachlor *exo*-epoxide	100%	86.7%	46.7%	33.3%	6.7%	1.29
(±)-Heptachlor *exo*-epoxide	93.3%	80.0%	33.3%	13.3%	0.0%	1.82
(+)-Heptachlor *exo*-epoxide	93.3%	46.7%	13.3%	0.0%	0.0%	2.98

the environment include chlordene, chlordene epoxide and heptachlor *exo*-epoxide, heptachlor and 2-chloroheptachlor.

Miyazaki *et al.* [25] studied the enantioselective toxicities of cyclodene (chlordene, chlordene epoxide and heptachlor *exo*-epoxide) pesticides on male adults of German cockroach (*Blattella germanica*). A sample of the enantiomers and their racemic mixture was applied to 10–20 insects per dose. The toxicity was expressed as a percentage of the mortality per 24 h after the application of the sample, and the results are given in Table 4.3. It can be concluded that (+)-chlordene, (−)-chlordene epoxide and (+)-heptachlor *exo*-epoxide showed higher toxicities than their corresponding mirror images. The $LD_{28.6}$ value for (+)-chlordene was 129, while the LD_{50} values for (−)-chlordiene epoxide and its racemate were 76 and 157, respectively, which indicates the enantioselective toxicities of these insecticides. Furthermore, the reported LD_{50} values for (−)-, (+)- and racemic heptachlor *exo*-epoxide were 2.98, 1.29 and 1.82, which again indicates the enantioselective toxicities of heptachlor *exo*-epoxide, with the (+)-enantiomer being more toxic. In another study, the same authors [26] reported that only the (−)-enantiomer of chlordene epoxide was insecticidal without any bioactivation, whereas the (+)-enantiomer of chlordene showed toxicity after its biochemical transformation into the corresponding (−)-chlordene epoxide. Again, Miyazaki *et al.* [27] studied the enantioselective toxicities of heptachlor, 2-chlorheptachlor and 3-chlorheptachlor pesticides on the same cockroach (*Blattella germanica*). The LD_{50} values for these pesticides were calculated after 24 h and are summarized in Table 4.4. It has been reported that only heptachlor and 2-chlorheptachlor showed toxicities, while 3-chlorheptachlor was nontoxic. The LD_{50} values for (−)-, (+)- and racemic heptachlor were 5.32, 3.38 and 2.64, respectively. On the other hand, the LD_{50} values calculated for (−)-, (+)- and racemic 2-chlorheptachlor were 100, 50 and 20, respectively. Therefore, it may be concluded that the toxicities of both the (+)-enantiomers of heptachlor and 2-heptachlor are greater than those of their corresponding (−)-enantiomers. On the basis of these results, the theoretical LD_{50} values of the racemic heptachlor and 2-chlorheptachlor should be 4.35 and 75, respectively, but the observed LD_{50} values are somewhat lower (Table 4.4). Therefore, it may be concluded that, in both cases, the toxic potential of one enantiomer is augmented by the other.

Commercial DDT and DDD contain *o,p*-DDT [1,1,1-trichloro-2-(*o*-chlorophenyl)-2-(*p*-chlorophenyl)] ethane and *o,p*-DDD [1,1-dichloro-2-(*o*-chlorophenyl)-2-(*p*-chlorophenyl)] ethane isomers (∼10–20%), which are chiral [28]. Investigations have demonstrated the oestrogenic activity of *o,p*-DDT in fish and mammals [29–34]. It is also known that the (−)-*o,p*-DDT enantiomer is a more active oestrogen mimic than the (+)-enantiomer in rats [35, 36]. Subsequently, Hoekstra *et al.* [37] reported a yeast-based assay to assess the enantiomer-specific transcriptional activity of DDT with

Table 4.4 The enantioselective toxicities of the enantiomers of heptachlor and 2-chlorheptachlor on German cockroach (*Blattella germanica*) [27]

Insecticide	Dose (μg g^{-1})						LD_{50}(μg g^{-1})
	18.0	10.8	6.48	3.88	2.32	1.39	
(+)-Heptachlor	93.3%	66.7%	43.3%	0.0%	0.0%	0.0%	3.38
(±)-Heptachlor	86.7%	93.3%	60.0%	25.0%	0.0%	0.0%	2.64
(−)-Heptachlor	90.0%	46.7%	36.7%	0.0%	0.0%	0.0%	5.32
	200	100	50	25	12.5		LD_{50}(μg g^{-1})
(+)-2-Chloroheptachlor	100%	100%	100%	60%	40%		20
(±)-2-Chloroheptachlor	100%	80%	50%	10%	10%		50
(−)-2-Chloroheptachlor	40%	40%	0.0%	0.0%	0.0%		100

the human oestrogen receptor (hER). (+)-17-β-estradiol, racemic DDT and individual DDT enantiomers were added to yeast cultures and hER activity was measured by quantification of β-galactosidase. The relative activity of *o,p*-DDT was weak as compared to that of estradiol. For *o,p*-DDT, the (−)-enantiomer was the active oestrogen mimic, whereas the hER activity of (+)-*o,p*-DDT was negligible. The presence of the (+)-enantiomer in relatively greater concentrations decreased the transcriptional activity of (−)-*o,p*-DDT. The dose–response curve for the activity of β-galactosidase is shown in Figure 4.5.

Lang *et al.* [38] studied *in vitro* effect of atrazine using liver microsomes of rats, pigs and humans. The authors reported the conversion of atrazine into a racemic mixture of the enantiomers of 2-chloro-4-ethylamino-6-(1-hydroxy-2-methylethyl-2-amino)-1,3,5-triazine. Furthermore, the authors reported the dominance of the *R*-enantiomer in humans, while a higher concentration of the *S*-enantiomer was observed in rats and pigs. The authors also reported the species-dependent enantioselective formation of this metabolite, with *R* : *S* ratios of 76 : 24 in rats, 49 : 51 in pigs and 28 : 72 in human. Therefore, they suggested that the differences in the asymmetric induction of the catalytic reaction on the active centre of the enzyme resulted in different expressions of the stereoselective enzymes which, in turn, led to the observed characteristic changes in the enantiomers. However, this hypothesis has not been verified experimentally. Similarly,

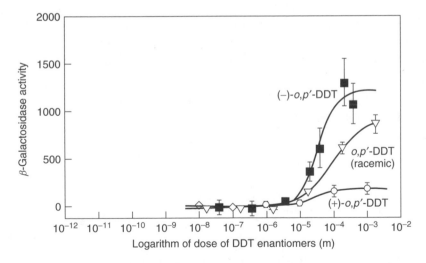

Figure 4.5   The dose–response curve for (+)- and (−)-enantiomers and a racemic mixture of *o,p*-DDT [37].

trans-nonachlor, a major constituent of technical chlordane, is achiral, and the replacement of a chlorine substituent by another atom or group can lead to a chiral pesticide.

4.5 The Enantioselective Toxicities of Phosphorous Pesticides

Besides chlorinated pesticides, some of the phosphorous pesticides are also chiral (see Table 2.1). The phosphorous pesticides also differ in their enantioselective toxicities. These pesticides were introduced in the 1950s to control insects in fruit, vegetables and other crops. Malathion is biotransformed into a racemic malaxon that has anti-acetylcholinesterate (insecticidal) activity. The *R*-enantiomer has a 22 times greater inhibitory potency than the *S*-enantiomer for bovine erythrocyte cholinesterase [39, 40]. The nerve agent, soman, has two chiral centres, and the two (−)-diastereoisomers are more potent inhibitors than their corresponding (+)-counterparts for acetylcholinesterase and α-chymotrypsin.

Miyazaki *et al.* [41, 42] studied the enantioselective toxicities of methamidophose (*O*,*S*-dimethyl phosphoramidothiodate) and acetaphate (*O*,*S*-dimethyl-*N*-acetylphosphoramidothiodate). The authors reported that the (+)-enantiomers were more potent to houseflies than their counterparts (optical isomers). Contrarily, the (−)-enantiomers of these pesticides were reported to be more toxic to German cockroach (*Blattella germanica*). However, the LD_{50} values were very close for both of the enantiomers. Sulfoxidation of propaphos resulted in a racemic mixture of the enantiomers of propaphos sulfoxide, and it has been reported that the (−)-enantiomer was more potent as an inhibitor for German cockroach and bovine erythrocyte acetylcholinesterase as compared with the (+)-enantiomer [42]. Furthermore, the authors studied the toxicities of these two enantiomers on housefly and green leaf hopper and reported a little difference in the toxicities of the two enantiomers on these two insects.

4.6 The Enantioselective Toxicities of Polyaromatic Hydrocarbons (PAHs)

Several chiral polynuclear aromatic hydrocarbons (PAHs) pollute the environment, the most toxic of which include benzo(a)pyrene and anthracene derivatives, β-naphthoflavone and so on. Of course, the toxic nature of PAHs has been known for a long time, but only a few reports are available on the enantioselective toxicities of their transformation products. The benzo(a)pyrenes are wide spread environmental pollutants, with cytotoxic, mutagenic and carcinogenic effects [43]. They are metabolized with high stereospecificity by enzymes in liver microsomes [44–48]. In view of these facts, the determination of the enantioselective toxicities of benzo(a)pyrenes is an important area and hence some reports have been published on this issue.

In 1977, Levin *et al.* [49] studied the carcinogenic toxicities of *trans*-7,8-dihydroxy-7,8-dihydrobenzo(a)pyrene (BP-7,8-dihydrodiol) on mouse skin via a two-stage tumorigenesis system. The applied doses of this toxic hydrocarbon, on the backs of CD-1 mice, were 50, 100 and 200 nmoles of the (+)- and (−)-enantiomers separately. Tumour formation on the skin of the mice was observed after several weeks. The results of the tumour formation due to the application of this hydrocarbon are summarized in Table 4.5. The effects of the dose of BP-7,8-dihydrodiol on the percentage of mice with papilloma formation and the number of papillomas per mouse are shown in Figures 4.6(a) and 4.6(b), respectively. It may be concluded from this table that the (−)-enantiomer is more toxic than the (+)-enantiomers

Table 4.5 The enantioselective tumorigenicity of (+)- and (−)-BP-7,8-dihydrodiol on mice skin: female CD-1 mice (7–8 weeks old) were treated with 100 nmoles of (+)- and (−)-BP-7,8-dihydrodiol [49]

BP-7,8-dihydrodiol	Weeks of promotion	Percentage of mice with tumours	Papillomas per mouse
(−)-	7	13	0.17
(+)-	7	0	0.00
(−)-	9	27	0.51
(+)-	9	7	0.07
(−)-	11	37	0.77
(+)-	11	7	0.07
(−)-	13	50	1.50
(+)-	13	10	0.10
(−)-	15	53	1.70
(+)-	15	17	0.17
(−)-	17	77	3.20
(+)-	17	20	0.30
(−)-	21	77	3.8
(+)-	21	23	0.43

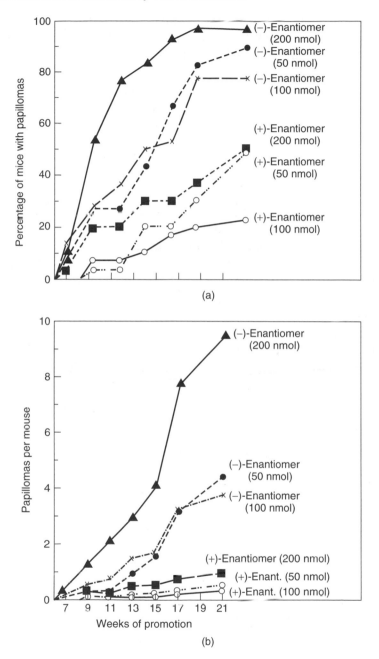

Figure 4.6 The skin tumour initiating activities of the (−)- and (+)-enantiomers of BP-7,8-dihydrodiol in female CD-1 mice: (a) the percentage of mice with papillomas; (b) the number of papillomas per mouse [49].

in all combinations. The maximum tumour formation was observed after 21 weeks. Again, Figures 4.6(a) and 4.6(b) indicate that the effect of the dose was in the order 200 > 50 > 100 nmoles – which is surprising, as a 50 nmoles concentration was more carcinogenic than a 100 nmoles dose. Furthermore, the authors described a 5–10 times higher carcinogenicity of the (−)-enantiomer in comparison to the (+)-enantiomer of BP-7,8-dihydrodiol. The same group has also studied the carcinogenic effect of BP-7,8-dihydrodiol on the skin of newborn mice [50]. The effect of the dose was studied by injecting 20, 40 and 80 nmoles into mice with 1, 8 and 15 day lifespans. The results of this research are given in Table 4.6, and it may be concluded from the table that the (−)-enantiomer of BP-7,8-dihydrodiol is more toxic than its (+)-antipode. Furthermore, the authors reported that the animals treated with (+)-BP-7,8-dihydrodiol and (−)-BP-7,8-dihydrodiol had 0.16 and 9.28 pulmonary adenomas per mouse, respectively. When a five times higher dose was administrated according to the above dosage schedule, (+)-BP-7,8-dihydrodiol caused 2.34 pulmonary adenomas per mouse, while (−)-BP-7,8-dihydrodiol resulted in 32.2 pulmonary adenomas per mouse.

These studies have shown that benzo(a)pyrenes are not carcinogens as such, but that the carcinogenicity is developed due to their bio-transformation into carcinogenic epoxide metabolites [51–56]. The two

Table 4.6 The enantioselective tumorigenicity of (+)- and (−)-BP-7,8-dihydrodiol in newborn mice (8–15 days): Swiss–Webster mice [BLU:Ha(ICR)] were injected with 1400 nmoles of (+)-, (−)- and (±)-BP-7,8-dihydrodiols separately [50]

BP-7,8-dihydrodiol	Sex	Number of mice with pulmonary adenomas	Percentage of mice with pulmonary adenomas	Number of adenomas per mouse	Number of mice with malignant lymphoma	Percentage of mice with malignant lymphoma
(−)-	M	0	0	0	1[a]	100
	F	2	100	5.00	4[b]	80
(+)-	M	27	93	20.5	0	0
	F	17	94	15.3	0	0
(±)-	M	13	100	51.2	13[c]	72
	F	5	100	79.4	8[d]	89

[a]Observed in one animal that died before termination of the study.
[b]Observed in three animals that died between weaning and 17 weeks of age and in one animal that was killed at 17 weeks of age.
[c]Observed in five animals that died between weaning and 17 weeks of age in eight animals that were killed at 17 weeks of age.
[d]Observed in four animals that died between weaning and 17 weeks of age in four animals that were killed at 17 weeks of age.

possible diastereoisomers from *trans*-BP-7,8-dihydrodiol have been characterized [57], and one of them has been shown to be an ultimate carcinogen in newborn mice [58, 59]. The diastereoisomeric BP-7β, 8α-diol-9α, 10α-epoxides are known to be potent mutagens in bacteria and certain mammals [43, 60–62]. Therefore, in 1977, Wood *et al.* [63] studied the differences in mutagenicity of the enantiomers of the diastereoisomeric benzo(a)pyrene-7,8-diol-9,10-epoxides. In Chinese hamster cells, (+)-BP-7β, 8α-diol-9α, 10α-epoxide was four times more toxic than its (−)-antipode. Contrarily, the (−)-7β, 8α-diol-9α, 10α-epoxide was found to be twice as toxic as its (+)-enantiomer in strains TA98 and TA100 of *Salmonella typhimurium*. Slaga *et al.* [64] also studied the enantioselective carcinogenesis of 7β, 8α-diol-9α, 10α-epoxides on mouse skin using a two-stage tumorigenesis system. The authors reported that the toxicity of the (+)-enantiomer of epoxide was 60 % as active as benzo(a)pyrene, whereas the (−)-enantiomer was only 2 % as active as benzo(a)pyrene. The racemic mixture of the epoxide was 25 % as active as benzo(a)pyrene. These findings are summarized in Table 4.7 and Figure 4.7(a) (percentage of papillomas) and Figure 4.7(b) (papillomas per mouse), respectively. It may be concluded that the (+)-enantiomer of 7β, 8α-diol-9α, 10α-epoxide-2 was more toxic than its mirror images.

Table 4.7 The enantioselective tumorigenicity of four optical isomers of BP-7,8-diol-9,10-epoxides [64]

BP-epoxides	Dose (nmoles)	Papillomas per mouse	Percentage of surviving mice with papillomas
On female Charles River CD-1 mice			
(−)-BP-7β, 8α-diol-9β, 10β-epoxide-1	100	0.1	10
	200	0.1	10
(+)-BP-7α, 8β-diol-9α, 10α-epoxide-1	100	0.17	17
	200	0.1	7.0
(−)-BP-7α, 8β-diol-9β, 10β-epoxide-2	100	0.06	6.0
	200	0.1	10
(+)-BP-7β, 8α-diol-9α, 10α-epoxide-2	100	1.1	47
	200	2.6	76
On female Sencar mice			
(−)-BP-7β, 8α-diol-9β, 10β-epoxide-1	100	0.13	13
(+)-BP-7α, 8β-diol-9α, 10α-epoxide-1	100	0.10	10
(−)-BP-7α, 8β-diol-9β, 10β-epoxide-2	100	0.03	3.0
(+)-BP-7β, 8α-diol-9α, 10α-epoxide-2	100	2.0	75

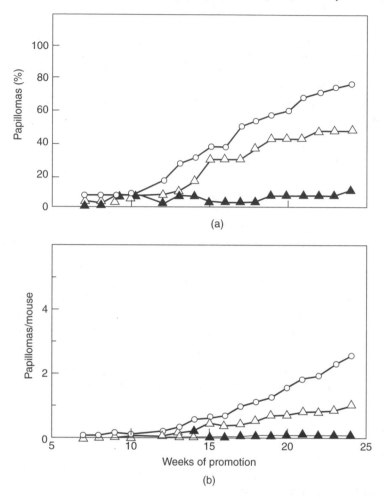

(a)

(b)

Figure 4.7 The skin tumour initiating activities of (+)-BP-7β, 8α-diol-9α, 10α-epoxide-2 (●), (−)-BP-7α, 8β-diol-9β-epoxide-2 (▲) and a racemic mixture of BP-7,8-diol-9,10-epoxide-2 (△) in female Charles River CD-1 mice, with 200 nmole as the dose of each compound separately [64].

The stereoselective metabolism of benzo(a)pyrene to an ultimate carcinogen has been well documented, and the metabolic pathway follows the sequence BP → BP-7,8-oxide → BP-7,8-dihydrodiol → BP-7,8-diol-9,10-epoxide (Figure 4.8). The enantioselective toxicities of BP-7,8-dihydrodiol and BP-7,8-diol-9,10-epoxide have already been discussed, but in order to study the biological effects of (+)- and (−)-benzo(a)pyrene 4,5-oxide, a synthesis of these molecules has been developed based on

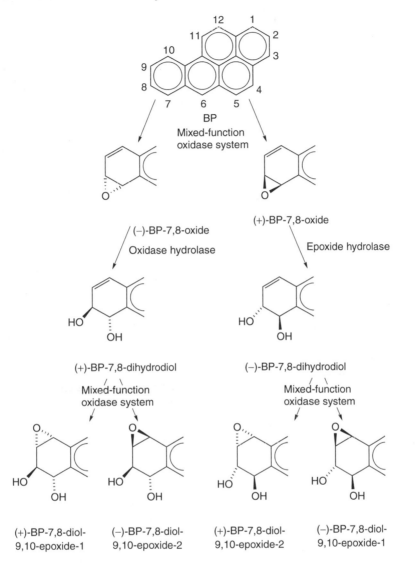

Figure 4.8 The formation and stereochemical structures of the benzo(a)pyrene metabolites responsible for the carcinogenicity of the parent hydrocarbon.

the resolution of (+/−)-cis-4,5-dihydroxy-4,5-dihydrobenzo(a)pyrene by Chang *et al.* [65]. The (−)-enantiomer of benzo(a)pyrene-4,5-oxide was 1.5- to 5.5-fold more mutagenic than the (+)-enantiomer in strains TA98, TA100, TA1537 and TA1538 of *Salmonella typhimurium* and in Chinese hamster V79 cells. In studies with V79 cells, the (−)-enantiomer of benzo(a)pyrene-4,5-oxide was also more cytotoxic than the (+)-enantiomer. When mixtures of the enantiomers were studied in V79 cells, synergistic cytotoxic and mutagenic responses were observed. The greatest cytotoxic and mutagenic effects occurred with a 3 : 1 mixture of the (−)- and (+)-enantiomers of benzo(a)pyrene-4,5-oxide, respectively. Similarly, Levin *et al.* [66] in 1980 also established that BP-7,8-oxide showed enantioselective toxicities on the skin of newborn mice. It was reported that the (+)-enantiomer was more tumorigenic than its corresponding (−)-antipode, and the results are given in Table 4.8. It is interesting to note that the tumour formation potencies of BP-7,8-oxide were in the order racemic mixture > (+)-enantiomer > (−)-enantiomer. The higher toxicity of the racemic mixture may be due to the catalytic tendencies of the enantiomers with respect to one another.

The benzo(a)pyrene analogous polycyclic hydrocarbon chrysene is also metabolized stereospecifically into the corresponding epoxide metabolites [67]. The mutagenicity [68, 69] and tumorigenicity [70, 71] of the chrysene-1,2-dio-3,4-epoxide-1 and -2 diastereoisomers are well known. Wood *et al.* [72] studied the enantioselective toxicities of the four isomers of chrysene-1,2-dio-3,4-epoxide – that is, the -1 and -2 and chrysene-H_4-3,4-epoxides (Figure 4.9) – in bacterial and mammal cells. The authors studied

Table 4.8   The enantioselective tumorigenicity of the optical isomers of BP-7,8-oxide [66]

BP-7,8-oxide	Dose (nmoles)	Percentage of mice with tumours	Number of tumours per mouse
On old mice			
(−)-	100	11	0.11
	400	36	0.43
(+)-	100	38	0.76
	400	55	1.03
(±)-	100	67	1.48
	400	60	1.67
On newborn mice, with 100, 200 and 400 nmole doses for mice with lifespans of 1, 8 and 15 days, respectively			
(−)-		*21	0.33
		+20	0.20
(+)-		*86	1.82
		+83	2.57
(±)-		*93	3.89
		+86	3.39

* and + are female and male mice respectively.

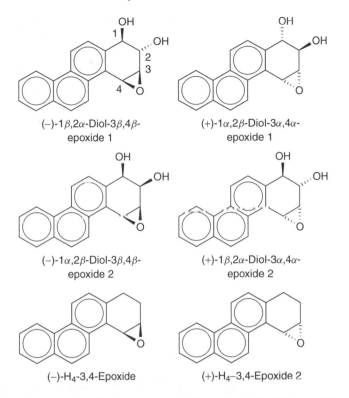

Figure 4.9 The stereochemical structures of chrysene-1,2-diol-3,4-epoxides and tetra-hydroepoxides.

the enantioselective carcinogenicity of chrysene-1,2-diol-3,4-epoxides in histidine-dependent strains of *Salmonella typhimurium* bacteria and in cultured Chinese hamster V79 cells. In strain TA98 of *S. typhimurium*, (−)-chyresene-1,2-diol-3,4-epoxide-2 was found to be 5–10 times toxic than the other three isomers. However, in strain TA100 of *S. typhimurium* and in Chinese hamster V79 cells, (+)-chyresene-1,2-diol-3,4-epoxide-2 was the most mutagenic diol epoxide, and was from 5 to 40 times more active than the other three optical isomers. These studies indicate that the presence and the orientation of the hydroxyl groups in this compound play an important role in the stereochemically based biological activities which result in the different toxicities of the four optical isomers. The mutagenicities of the chrysene-1,2-dio-3,4-epoxide-1 and -2 diastereoisomers on strains TA98 and TA100 of *S. typhimurium* bacteria are given in Table 4.9, while the toxicities of the chrysene-H_4-3,4-epoxide enantiomers on strains TA98 and TA100 of *S. typhimurium* bacteria and

Table 4.9 The enantioselective tumorigenicity of the four optical isomers of chrysene diol epoxide on the TA98 and TA100 strains of of *S. typhimurium* bacteria [72]

Diol-epoxide	Mutagenic activity			
	His$^+$ revertants per nmole		Relative activity	
	TA98	TA100	TA98	TA100
(+)-Chrysene-1,2-diol-3,4-epoxide 1	33	137	9	4
(−)-Chrysene-1,2-diol-3,4-epoxide 1	32	368	8	12
(±)-Chrysene-1,2-diol-3,4-epoxide 1	32	258	−	−
(+)-Chrysene-1,2-diol-3,4-epoxide 2	72	3193	19	100
(−)-Chrysene-1,2-diol-3,4-epoxide 2	377	590	100	18
(±)-Chrysene-1,2-diol-3,4-epoxide 2	252	1768	−	−

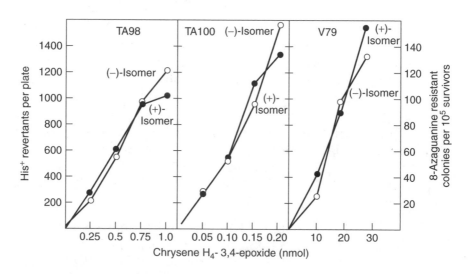

Figure 4.10 The mutagenicity of the (−)- and (+)-enantiomers of chrysene-H$_4$-3,4-epoxide in the TA98 (a) TA100 (b) strains of *Salmonella typhimurium* bacteria and in Chinese hamster cells (c) [72].

Chinese hamster V79 cells are shown in Figure 4.10. It may be concluded from this figure that the (−)-enantiomer of chrysene-H$_4$-3,4-epoxide is more toxic than the (+)-enantiomer in strains TA98 and TA100 of *S. typhimurium* bacteria, while in the case of Chinese hamster V79 cells the (+)-enantiomer is more toxic compared to the (−)-enantiomer. Furthermore, the same group [73] studied the enantioselective toxicities of *trans*-1,2-dihydroxy-1,2-dihydrochrysene (chrysene-1,2-dihydrodiol) and chrysene-1,2-diol-3,4-epoxides in two tumour models. The authors studied the skin, pulmonary and hepatic carcinogenicity that developed as a consequence of these chiral pollutants. The carcinogenicity on the skin is presented in Table 4.10, whereas Table 4.11 represents the extent of the pulmonary and hepatic carcinogenicity in newborn mice. Table 4.10 shows that only 3 % of tumours were found when the (+)-enantiomer was injected, while 67 % carcinogenicity was observed with the (−)-enantiomer of *trans*-1,2-dihydroxy-1,2-dihydrochrysene, which indicates the greater toxicity of the (−)-enantiomer. The authors also reported that the toxicity of the racemic mixture of *trans*-1,2-dihydroxy-1,2-dihydrochrysene was equivalent to that of its (−)-enantiomer. On the other hand, the tumour-initiating percentages of the (+)-, (−)- and (±)-chrysene-1,2-diol-3,4-epoxides (with a 1 ? μmole dose) were 21 %, 13 % and 25 % respectively. Of course, the toxicity of the (+)-enantiomer is higher than the toxicity of the (−)-enantiomers. It is interesting to note that the toxicity of the racemic mixture is greater than the toxicity of either the (+)- or the (−)-enantiomers. This may be due to the fact that the toxicity of the (+)-enantiomer is catalysed by the (−)-enantiomers, or vice versa, or both situations apply. A perusal of Table 4.11 indicates that again (−)-*trans*-1,2-dihydroxy-1,2-dihydrochrysene is more toxic than its (+)-antipode in both male and female mice. However, the (+)-enantiomer of chrysene-1,2-diol-3,4-epoxide-1 was more toxic in male mice, whereas the greater toxicities of its (−)-enantiomer were reported in the female mice. On the other hand, the (+)-enantiomer of chrysene-1,2-diol-3,4-epoxide-2 was more toxic in female mice, while the (−)-enantiomer indicated a greater toxicity in male mice.

Table 4.10 The enantioselective tumorigenicity of the enantiomers of *trans*-1,2-dihydroxy-1,2-dihydrochrysene (chrysene-1,2-dihydrodiol) and chrysene-1,2-diol-3,4-epoxides on mouse skin [73]

Compound	Dose (μmol)	Percentage of mice with tumours	Tumours per mouse
(+)-Chrysene-1,2-dihydrodiol	0.4	3	0.03
	1.2	23	0.40
(−)-Chrysene-1,2-dihydrodiol	0.4	67	1.47
	1.2	83	2.77
(+)-Chrysene-1,2-diol-3,4-epoxide	0.4	13	0.13
	1.2	21	0.24
(−)-Chrysene-1,2-diol-3,4-epoxide	0.4	7	0.07
	1.2	13	0.13
(±)-Chrysene-1,2-diol-3,4-epoxide	1.2	25	0.29

Table 4.11 The enantioselective pulmonary and hepatic carcinogenicity of the enantiomers of *trans*-1,2-dihydroxy-1,2-dihydrochrysene (chrysene-1,2-dihydrodiol) and chrysene-1,2-diol-3,4-epoxides on newborn mice [73]

Compound	Dose (μmol)	Pulmonary tumour		Hepatic tumour	
		Percentage of mice with tumours	Number of tumours per mouse	Percentage of mice with tumours	Number of tumours per mouse
(+)-Chrysene-1,2-dihydrodiol	*1.4	20	0.22	0	0
	+1.4	16	0.54	0.16	0.22
(−)-Chrysene-1,2-dihydrodiol	*1.4	89	10.62	0	0
	+1.4	95	7.41	57	2.3
(±)-Chrysene-1,2-dihydrodiol	*1.4	89	3.78	0	0
	+1.4	84	3.32	26	0.68
(+)-Chrysene-1,2-diol-3,4-epoxide-1	*0.7	21	0.21	0	0
	+0.7	18	0.20	3	0.03
(−)-Chrysene-1,2-diol-3,4-epoxide-1	*0.7	27	0.27	0	0
	+0.7	16	0.20	0	0
(+)-Chrysene-1,2-diol-3,4-epoxide-2	*0.7	88	6.59	0	0
	+0.7	91	4.34	0.23	0.37
(−)-Chrysene-1,2-diol-3,4-epoxide-2	*0.7	8	0.08	0	0
	+0.7	14	0.16	0	0

* and + are female and male mice respectively.

Benz(a)anthracene-3,4-diol-1,2-epoxide (BADE) is a potent carcinogen [74–76] that metabolizes stereochemically and results in enantioselective toxicities. The tumorigenic activities of benz(a)anthracene (BA), the (+)- and (−)-enantiomers of *trans*-3,4-dihydroxy-3,4-dihydrobenz(a)anthracene (BA-3,4-dihydrodiol) and the racemic diastereomers of the BA-3,4-diol-1,2-epoxides – that is, either or both of the diastereomeric 1,2-epoxides derived from BA-3,4-dihydrodiol, in which the epoxide oxygen is *cis*-(diol epoxide-1) or *trans*- (diol epoxide-2) to the benzylic 4-hydroxyl group – were examined in newborn Swiss–Webster mice by Wislocki *et al.* [77]. The mice were administered a total dose of 280 nmoles of compound, in divided doses consisting of 40 nmoles within 24 h of birth, 80 nmoles at 8 days of age, and 160 nmoles at 15 days of age. The experiment was terminated when the animals were 26 weeks of age. BA-3,4-diol-1,2-epoxide-2 was the most potent compound tested. All of the animals treated with BA-3,4-diol-1,2-epoxide-2 developed pulmonary tumours, with an average of 13.3 tumours per mouse. BA-3,4-diol-1,2-epoxide-1 produced pulmonary tumours in 42 % of the mice, with an average of only 0.56 tumours per mouse. The (−)-enantiomer of BA-3,4-dihydrodiol was the second most tumorigenic derivative of BA tested; it produced pulmonary tumours in 71 % of the mice, with an average of 1.88 tumours per mouse. BA and the (+)-enantiomer of BA-3,4-dihydrodiol had little or no tumorigenic activity at the dose tested. A comparison of the average number of pulmonary tumours per mouse revealed that BA-3,4-diol-1,2-epoxide-2 was about 30-fold more tumorigenic than BA-3,4-diol-1,2-epoxide-1, eight-fold more tumorigenic than (−)-BA-3,4-dihydrodiol, and 85-fold more tumorigenic than BA. These data indicate that in newborn mice BA-3,4-dihydrodiol and a BA-3,4-diol-1,2-epoxide are proximate and ultimate carcinogenic metabolites of BA, respectively.

Similarly, Tang *et al.* [78] determined the tumour-initiating activity of (+/−)-*syn*- and (+/−)-*anti*-7,12-dimethylbenz(a)anthracene-3,4-diol-1,2-epoxide (*syn*- and *anti*-DMBADE), the two metabolically formed diol epoxides of DMBA, in the H-ras gene from tumours induced by these compounds. The authors used a two-stage initiation-promotion protocol for tumorigenesis in mouse skin, and found that both *syn*- and *anti*-DMBADE were active tumour initiators, and that the occurrence of papillomas was dependent on the carcinogen dose. All of the papillomas induced by *syn*-DMBADE (a total of 40 mice), 96 % of those induced by *anti*-DMBADE (a total of 25 mice) and 94 % of those induced by DMBA (a total of 16 mice) possessed a –CAA– to –CTA– mutation at codon 61 of H-ras. No mutations in codons 12 or 13 were detected in any tumour. Topical application of *syn*- and *anti*-DMBADE produced stable adducts in mouse epidermal DNA, most of which co-migrated with stable DNA adducts formed after topical application of DMBA. Furthermore, the authors reported that the levels of the major *syn*- and *anti*-DMBADE-deoxyadenosine adducts formed after topical application of DMBA were sufficient to account for the tumour-initiating activity of this carcinogen on mouse skin. Briefly, the authors concluded that the adenine adducts induced by both bay-region diol epoxides of DMBA lead to the mutation at codon 61 of H-ras and, consequently, initiate tumourigenesis in mouse skin. The time course of the papillomas formation and the number of papillomas formed due to DMBA and DMBADE in mice are shown in Figures 4.11 and 4.12, respectively.

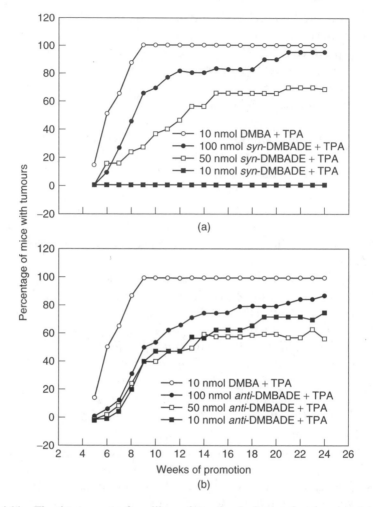

Figure 4.11   The time course of papilloma formation in Sencar female mice: (a) treated with *syn*-DMBA and DMBADE; (b) treated with *anti*-DMBA and DMBADE [78].

4.7 The Enantioselective Toxicities of Other Xenobiotics

In addition to the chiral pollutants discussed above, certain other chiral compounds are found in the environment, and sometimes become harmful to the biota. For example, toxicological information on individual chlorobornanes is scarce, but some reports have recently appeared. The neurotoxic effects of toxaphene exposure on behaviour and learning have

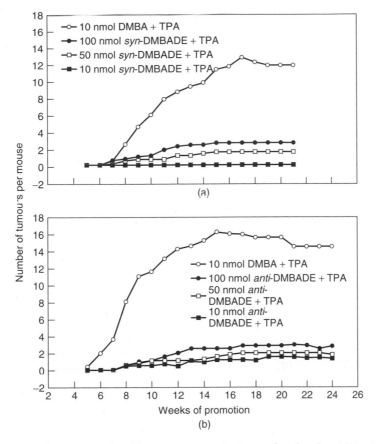

Figure 4.12 The number of papillomas per mouse in Sencar female mice: (a) treated with *syn*-DMBA and DMBADE; (b) treated with *anti*-DMBA and DMBADE [78].

been reported. Technical toxaphene and some individual congeners have been found to be weakly estrogenic in *in vitro* test systems; no evidence for endocrine effects *in vivo* has been reported. The *in vitro* studies have shown technical toxaphene and toxaphene congeners to be mutagenic agents. However, no genotoxicity has appeared in *in vivo* studies, and therefore a nongenotoxic mechanism is proposed. Even then, toxaphene is supposed to possess a potential carcinogenic risk to humans. Up to now, only Germany has established a legal tolerance level for toxaphene, at $0.1 \, \mathrm{mg \, kg^{-1}}$ for fish [79].

In the aquatic environment, a number of other chiral pollutants are available and they differ in their enantioselective toxicities. Tetrodotoxin and saxitoxin (Figure 4.13) are the best examples of this. Tetrodotoxin

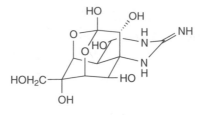

Tetrodotoxin

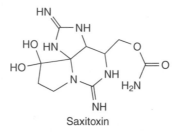

Saxitoxin

Figure 4.13   The chemical structures of tetrodotoxin and saxitoxin pollutants.

is produced by the *Pseudomonas* species of bacteria, and accumulates in fish and other microbes throughout the food chain. Similarly, saxitoxin is produced by the species *Gonayaulax*, and is more harmful to humans than tetrodotoxin, as it accumulates in filter-feeding shellfish, a common human foodstuff all over the world. Therefore, the toxic effects of saxitoxin have been observed in some parts of the world, and hence some filter-feeding shellfish industries have been banned. It has also been reported that the marine environment is rich with the (−)-enantiomer of saxitoxin. Also, various other toxic chemicals such as neurotoxins (anatoxin, homoanatoxin etc.) are produced in the aquatic environment [80]. Therefore, water sometimes becomes toxic due to the presence of these toxins, and several reports have been published on the deaths of cattle and dogs due to these chemicals [81, 82]. Accordingly, the presence of these toxic chiral pollutants may be lethal for human beings. Unfortunately, no report has been published on the enantioselective toxicities of these toxins.

4.8 The Enantioselective Toxicities of Drugs and Pharmaceuticals

Besides chiral pesticides and other chiral pollutants, one of the enantiomers of certain types of drug is also toxic. Therefore, the presence of such a drug enantiomer in the environment may be problematic and hazardous. Weigel [83] reported the presence of several drugs in the aquatic environment, with higher concentrations than reported earlier. Recently, Kümmerer [84] has reported the presence of several drugs in surface, ground and drinking water. Weiss [85] reviewed the toxicities of some drugs and reported that exposure to a variety of drugs can induce haematopoietic damage. These drugs exert their effects through several distinct mechanisms, including the destruction or suppression of haematopoietic stem cells, the cytotoxic destruction of rapidly proliferating precursor cells, immune-mediated haematotoxicity, an altered haematopoietic micro-environment, genetic mutation and microvascular injury. All of these types of injury tend to result in granulocytopenia, and the time course of the changes varies with the injury type. Immune-mediated reactions may occur acutely, or after months or years of treatment with a drug. Drugs that act as mutagens can induce a variety of haematopoietic disorders, including aplastic anaemia, myelodysplasia and leukaemia. Therefore, the time course of the onset of leukopenia after drug or chemical exposure and the rapidity of haematopoietic recovery give clues to the mechanism by which granulopoiesis is suppressed. Furthermore, Atkin *et al.* [86] also reviewed the toxic effects of some drug reactions, revealing the incidence and prevalence of serious adverse effects.

In the late 1950s, thalidomide [(*R*,*S*)-*N*-(2,6-dioxo-3-piperidyl) phthalimide (see Figure 4.14)] was introduced as a sedative in Europe but, unfortunately, a teratogenic effect of this drug was observed on unborn embryos when the drug was administered to pregnant women [87]. In 1960, many babies born in Europe were suffering from malformations and this huge disaster become known as the thalidomide tragedy. Extensive research was subsequently carried out, and it was found that the *S*-enantiomer is responsible for this teratogenic effect. It is very interesting to note that the pure *R*-enantiomer, which possesses tranquillizing properties, racemizes into the *S*-enantiomer, resulting in the teratogenic effects [88–90].

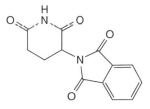

Figure 4.14 The chemical structure of thalidomide.

However, due to the excellent action of thalidomide in erythema nodosum leprosum, a leprosy complication, it is now being marketed in the USA under the trade name of Synovir™. Thalidomide has also been reported to be effective against aphthous ulcers in patients who are infected with HIV. Therefore, thalidomide may become a miracle drug for HIV patients in the future, and that is why it is now available, although very tightly prescribed.

Ifosfamide, a chiral molecule, is a cyclophosphamide analogue, which belongs to a class of anti-cancer drugs called oxazaphosphorine nitrogen mustards. This drug is not an anti-cancer agent as such, but it metabolizes into an alkylating agent (isophosphoramide mustard) which has anti-cancer properties. Accordingly, this drug also possesses toxicity and the toxicity is enantioselective in nature. Masurel *et al.* [91] studied the enantioselective toxicity of this drug on rats. The authors injected the racemic form and the enantiomers separately to nontumour bearing rats at $550-650$ mg kg^{-1} level doses. The results are shown in Table 4.12. The toxicity was observed in terms of mortality and weight loss after 9 days. It may be concluded from this table that the mean weight losses was 30%, 20% and 17% for the (+)-, (−)- and racemic mixtures of ifosfamide (at 650 mg kg^{-1}) respectively. Furthermore, the authors observed signs of acute bladder toxicity, as blood was reported in the urine of a rat when (−)-ifosfamide was injected. Similarly, there are several other drugs whose one enantiomer is toxic. It has been well known for a long time that although L-dopa is used for the treatment of Parkinson's disease, the D-enantiomer of dopa is toxic [92, 93].

Domino [94], in 1986, reported on the enantioselective opioid hallucinogen interactions of N,N-dimethyltriptamine and lysergic acid N,N-diethylamide in rats. In an adult male, trained on a positive reinforced fixed ratio of four (FR4: a reward of 0.01 ml of sugar milk earned on every fourth bar press), N,N-dimethyltriptamine, in concentrations of 3.2

Table 4.12 The survival of nontumour-bearing mice treated with a single dose of ifosfamide [91]

Dose (mg kg^{-1})	Deaths/total		
	(+)-enantiomer	(−)-enantiomer	Racemic mixture
550	0/7 (7)[a]	0/7 (13)	0/7 (12)
600	1/7 (14)	1/7 (13)	0/7 (17)
650	4/7 (30)	3/7 (20)	3/7 (17)

[a]The numbers in parentheses are the percentage of the maximal weight loss for the surviving mice in each treatment group.

and $10.0 \, \mathrm{mg \, kg^{-1}}$, disrupted the established food-rewarded FR4-pressing behaviour in a dose-related way. The low doses of opioid antagonized the effect of both hallucinogens, whereas larger doses enhanced their activities. In contrast to the antagonistic effect of low doses of opioid agonists, (−)-naloxone (17-allyl-4,5-α-epoxy-3,14-dihydroxy-6-morphinanone) enhanced the effect of N,N-dimethyltriptamine and lysergic acid N,N-diethylamide. Furthermore, the author reported an acceleration effect of N,N-dimethyltriptamine due to (−)-naloxone, while the (+)-enantiomer of naloxone was unable to potentiate the effect of N,N-dimethyltriptamine.

4.9 Conclusions

As discussed above, enantiomers of chiral pollutants may have different toxicities to biota, but some reports have been published indicating more toxicities of the racemic mixture in comparison to the pure enantiomers, which may be due to the catalytic properties of one enantiomer with respect to another. The concept of chirality has been explored to a greater extent in drugs and pharmaceuticals but, unfortunately, the enantioselective toxicities of the chiral pollutants have not been fully investigated in detail. Because of the different toxicities of the enantiomers, data on the toxicity of the racemic pollutant are not reliable. Therefore, it is essential to explore the enantioselective hazardous effects and toxicities of chiral pollutants. Moreover, the already existing data on the hazardous effects and toxicities of chiral pollutants should be modified in terms of the enantioselective toxicities.

Chiral pollutants are metabolized stereoselectively in biological systems and that is why the enantiomers have different toxic effects. The stereoselective metabolism of a chiral pollutant is very interesting, and is again a demanding field. Knowledge of the stereoselective metabolism and enantioselective toxicities of chiral pollutants may be useful for the treatment of some diseases caused by the enantiomers of chiral pollutants. For example, the enantioselective carcinogenicity developed by chiral pollutants may be treated effectively by using knowledge about the stereospecific metabolism and enantioselective toxicities of chiral pollutants. Information on the physicochemical degradation of chiral pollutants in the presence of a chiral compound or catalyst is also an important aspect in ascertaining the enantioselective toxicities. Therefore, studies on the stereoselective degradation of chiral pollutants may be used to ascertain the enantioselective toxicities.

References

1. E. H. Easson and E. Stedman, *Biochem. J.* **27**, 1257 (1933).
2. H. Y. Aboul-Enein and I. Ali, *Chiral Separations by Liquid Chromatography and Related Technologies*, Dekker, New York, 2003.
3. S. Allenmark, *Chromatographic Enantioseparation, Methods and Applications*, 2nd edn, Ellis Horwood, New York, 1991.
4. G. Subramanian (ed.), *A Practical Approach to Chiral Separations by Liquid Chromatography*, VCH, Weinheim, 1994.
5. H. Y. Aboul-Enein and I. W. Wainer, eds, *The Impact of Stereochemistry on Drug Development and Use*, Chemical Analysis, vol. 142, Wiley, New York, 1997.
6. T. E. Beesley and R. P. W. Scott, *Chiral Chromatography*, Wiley, New York, 1998.
7. M. J. Gonzalez, M. A. Fernandez and L. M. Hernandez, *Arch. Environ. Contam. Toxicol.* **20**, 343 (1991).
8. W. A. König, B. Gehrcke, T. Runge and C. Wolf, *J. High Res. Chromatogr.* **16**, 376 (1993).
9. S. Tanabe, N. Kannan, A. Subramanian, S. Watanabe and R. Tatsukawa, *Environ. Pollut.* **47**, 147 (1987).
10. S. Safe, *CRC Crit. Rev. Toxicol.* **21**, 51 (1990).
11. S. Safe, *Chemosphere* **25**, 61 (1992).
12. S. Safe, *CRC Crit. Rev. Toxicol.* **24**, 1 (1994).
13. J. C. Duinker, D. E. Schulz and G. Petrick, *Chemosphere* **23**, 1009 (1991).
14. J. Klamer, R. W. P. M. Laane and J. M. Marquenie, *Water Sci. Technol.* **24**, 77 (1991).
15. U. G. Ahlborg, G. C. Becking, L. S. Birnbaum, A. Brouwer, H. J. G. M. Derks, M. Feeley, G. Golor, A. Hanberg, J. C. Larsen, A. K. D. Liem, S. H. Safe, C. Schlatter, F. Waern, M. Younes and E. Yränheikki, *Chemosphere* **28**, 1049 (1994).
16. J. Falandysz, K. Kannan, S. Tanabe and R. Tatsukawa, *Mar. Pollut. Bull.* **28**, 259 (1994).
17. M. Püttmann, A. Mannschreck, F. Oesch and L. W. Robertson, *Biochem. Pharmacol.* **38**, 1345 (1989).
18. E. L. Rodman, S. I. Shedlofsky, A. Mannschreck, M. Püttmann, A. T. Swim and L. W. Robertson, *Biochem. Pharmacol.* **41**, 915 (1991).
19. J. Bakke and J. A. Gustafsson, *Trends Pharmacol. Sci.* **5**, 517 (1984).
20. O. B. Lund, J. Orberg, A. Bergman, C. Larsson, B. M. Bäcklin, H. Hakansson, A. Madej, A. Brouwer and B. Brunström, *Environ. Toxicol. Chem.* **18**, 292 (1998).
21. Y. Nakanishi, N. Shigematsu, Y. Kurita, K. Matsuba, H. Kanegae, S. Ishimaru and Y. Kawazoe, *Environ Health Prospect.* **59**, 31 (1985).

22. T. Ellerichmann, A. Bergman, S. Franke, H. Hühnerfuss, E. Jakobsson, W. A. König and C. Larsson, *Fresenius Environ. Bull.* **7**, 244 (1998).
23. C. S. Wong, A. W. Garrison, P. D. Smith and W. T. Foreman, *Environ. Sci. Technol.* **35**, 2448 (2001).
24. K. Möller, H. Hühnerfuss and D. Wölfle, *Organohal. Compds* **29**, 357 (1996).
25. A. Miyazaki, T. Hotta, S. Marumo and M. Sakai, *J. Agric. Food Chem.* **26**, 975 (1978).
26. A. Miyazaki, M. Sakai and S. Marumo, *J. Agric. Food Chem.* **27**, 1403 (1979).
27. A. Miyazaki, M. Sakai and S. Marumo, *J. Agric. Food Chem.* **27**, 1310 (1979).
28. H. R. Buser and M. D. Muller, *Anal. Chem.* **67**, 2691 (1995).
29. R. M. Donohoe and L. R. Curtis, *Aquat. Toxicol.* **36**, 31 (1996).
30. K. W. Gaido, L. S. Leonard, S. Lovell, J. C. Could, D. Babai, C. J. Portier and D. P. McDonnell, *Toxicol. Appl. Pharmacol.* **143**, 205 (1997).
31. D. M. Klotz, B. L. Ladlie, P. M. Vonier, J. A. McLachlan and S. F. Arnold, *Mol. Cell Endocrinol.* **129**, 63 (1997).
32. S. C. Manness, D. P. McDonnell and K. W. Gaido, *Toxicol. Appl. Pharmacol.* **151**, 135 (1998).
33. L. J. Mills, R. E. Gutjahr-Gobell, R. A. Haebler, D. J. B. Horowitz, S. Jayaraman, R. J. Pruell, R. A. McKinney, G. R. Gardner and G. E. Zaroogian, *Aquat. Toxicol.* **52**, 157 (2001).
34. N. Rajapakse, D. Ong and A. Kortenkamp, *Toxicol. Sci.* **60**, 296 (2001).
35. W. A. McBlain, V. Lewin and F. H. Wolfe, *Can. J. Physiol. Pharmacol.* **54**, 629 (1976).
36. W. A. McBlain, *Life Sci.* **40**, 215 (1987).
37. P. F. Hoekstra, B. K. Burnison, T. Nehel and D. C. Muir, *Toxicol. Lett.* **125**, 75 (2001).
38. D. Lang, D. Griegee, A. Grothusen, R. W. Saalfrank and R. H. Böcker, *Drug Metab. Dispos.* **24**, 859 (1996).
39. O. P. Rodriguez, G. W. Muth, C. E. Berkman, K. Kim and C. M. Thompson, *Bull. Environ. Contam. Toxicol.* **58**, 171 (1997).
40. L. P. A. De Jong and H. P. Benschop, *Proceeding of 3rd International Meeting on Cholinesterases*, American Chemical Society, Washington, DC, 1991, p. 240.
41. A. Miyazaki, T. Nakamura and S. Marumo, *Pestic. Biol. Chem. Physiol.* **33**, 11 (1989).
42. A. Miyazaki, T. Nakamura, M. Kawaradani and S. Marumo, *J. Agric. Food Chem.* **36**, 835 (1987).
43. *Committee on the Biological Effects of Atmospheric Pollutants, Particulate Polycyclic Organic Matter*, National Academy of Science, Washington, DC, (1972).
44. E. Huberman, L. Sachs, S. K. Yang and H. V. Gelboin, *Proc. Natl Acad. Sci.* **73**, 607 (1976).

45. D. R. Thakker, H. Yagi, H. Agaki, M. Koreeda, A. Y. H. Lu and W. Lewin, *Chem. Biol. Interact.* **16**, 281 (1977).
46. D. R. Thakker, H. Yagi, A. Y. H. Lu, W. Lewin, A. H. Conney and D. M. Jerina, *Proc. Natl Acad. Sci.* **73**, 3381 (1976).
47. S. K. Yang and H. V. Gelboin, *Biochem. Pharmacol.* **25**, 2221 (1976).
48. S. K. Yang, D. W. McCourt, J. C. Leutz and H. V. Gelboin, *Science* **196**, 1199 (1977).
49. W. Levin, A. W. Wood, R. L. Chang, T. J. Slaga, H. Yagi, D. M. Jerina and A. H. Conney, *Cancer Res.* **37**, 2721 (1977).
50. J. Kapitulnik, P. G. Wislocki, W. Levin, H. Yagi, D. R. Thakker, H. Akagi, M. Koreeda, D. M. Jerina and A. H. Conney, *Cancer Res.* **38**, 2661 (1978).
51. A. Borgen, H. Darvey, N. Castagnoli, T. T. Crocker, R. Rasmussen and I. W. Wang, *J. Med. Chem.* **16**, 502 (1973).
52. P. Daudel, M. Duquesne, P. Vigny, P. L. Grover and P. Sims, *Fed. Eur. Biochem. Soc. Lett.* **57**, 250 (1975).
53. D. M. Jerina and J. W. Daly, *Science* **185**, 573 (1974).
54. E. C. Miller and J. A. Miller, in H. Busch, ed., *Molecular Biology of Cancer*, Academic Press, New York, 1974.
55. P. Sims and P. L. Grover, *Advn. Cancer Res.* **20**, 165 (1974).
56. P. Sims, P. L. Grover, A. Swaisland, K. Pal and A. Hewer, *Nature* **252**, 326 (1974).
57. H. Yagi, O. Hernandez and D. M. Jerina, *J. Am. Chem. Soc.* **97**, 6881 (1975).
58. J. Kapitulnik, W. Lewin, A. H. Conney, H. Yagi and D. M. Jerina, *Nature* **266**, 378 (1977).
59. J. Kapitulnik, P. G. Wislocki, W. Lewin, H. Yagi, D. M. Jerina and A. H. Conney, *Cancer Res.* **38**, 354 (1978).
60. R. F. Newbold and P. Brookes, *Nature* **261**, 52 (1976).
61. P. G. Wislocki, A. W. Wood, R. L. Chang, W. Lewin and H. Yagi, *Biochem. Biophys. Res. Commun.* **68**, 1006 (1976).
62. A. W. Wood, P. G. Wislocki, R. L. Chang, W. Lewin, A. Y. H. Lu, H. Yagi, O. Hernandez, D. M. Jerina and A. H. Conney, *Cancer Res.* **36**, 3358 (1976).
63. A. W. Wood, R. L. Chang, W. Levin, H. Yagi, D. R. Thakker, D. M. Jerina and A. H. Conney, *Biochem. Biophys. Res. Commun.* **77**, 1389 (1977).
64. T. J. Slaga, W. J. Bracken, G. Gleason, W. Levin, H. Yagi, D. M. Jerina and A. H. Conney, *Cancer Res.* **39**, 67 (1979).
65. R. L. Chang, A. W. Wood, W. Lewin, H. D. Mah, D. R. Thakker, D. M. Jerina and A. H. Conney, *Proc. Natl Acad. Sci.* **76**, 4280 (1979).
66. W. Levin, M. K. Buening, A. W. Wood, R. L. Chang, B. Kedzierski, D. R. Thakker, D. R. Boyd, G. S. Gadaginamath, R. N. Armstrong, H. Yagi, J. M. Karle, T. J. Slaga, D. M. Jerina and A. H. Conney, *J. Biol. Chem.* **255**, 9067 (1980).

67. M. Nordqvist, D. R. Thakker, K. P. Vyas, H. Yagi, W. Levin, D. E. Ryan, P. E. Thomas, A. H. Conney and D. M. Jerina, *Mol. Pharmacol.* **19**, 168 (1981).
68. A. W. Wood, R. L. Chang, W. Levin, D. E. Ryan, P. E. Thomas, H. D. Mah, J. M. Karle, H. Yagi, D. M. Jerina and A. H. Conney, *Cancer Res.* **39**, 4069 (1979).
69. A. W. Wood, W. Levin, D. E. Ryan, P. E. Thomas, H. Yagi, H. D. Mah, D. R. Thakker, D. M. Jerina and A. H. Conney, *Biochem. Biophys. Res. Commun.* **78**, 847 (1977).
70. M. K. Buening, W. Lewin, J. M. Karle, H. Yagi, D. M. Jerina and A. H. Conney, *Cancer Res.* **39**, 5063 (1979).
71. W. Levin, A. W. Wood, R. L. Chang, H. Yagi, H. D. Mah, D. M. Jerina and A. H. Conney, *Cancer Res.* **38**, 1831 (1978).
72. A. W. Wood, R. L. Chang, W. Levin, H. Yagi, M. Tada, K. P. Vyas, D. M. Jerina and A. H. Conney, *Cancer Res.* **42**, 2972 (1982).
73. R. L. Chang, V. V. Levin, A. W. Wood, H. Yagi, M. Tada, K. P. Vyas, D. M. Jerina and A. H. Conney, *Cancer Res.* **43**, 192 (1983).
74. T. J. Slaga, G. H. Gleason, J. DiGiovanni, K. B. Sukumaran and R. G. Harvey, *Cancer Res.* **39**, 1934 (1979).
75. A. C. H. Bigger, J. T. Sawicki, D. M. Blake, L. G. Raymond and A. Dipple, *Cancer Res.* **43**, 5647 (1983).
76. S. Manam, R. D. Store, S. Prahalada, K. R. Leander, A. R. Kraynak, C. L. Hammermeister, D. J. Joslyn, B. J. Ledwith, M. J. Zweiten, M. O. Bradley and W. W. Nichols, *Mol. Carcinogr.* **6**, 68 (1992).
77. P. G. Wislocki, M. K. Buening, W. Levin, R. E. Lehr, D. R. Thakker, D. M. Jerina and A. H. Conney, *J. Natl Cancer Inst.* **63**, 201 (1979).
78. M. S. Tang, S. V. Vulimiri, A. Viaje, J. X. Chen, D. S. Bilolikar, R. J. Morris, R. G. Harvey, T. J. Slaga and J. DiGiovanni, *Cancer Res.* **60**, 5688 (2000).
79. H. J. de Gues, H. Besselink, A. Brouwer, J. Klungsoyr, B. McHugh, E. Nixon, G. G. Rimkus, P. G. Wester and J. de Boer, *Environ. Health Prospect.* **1**, 115 (1999).
80. W. W. Carmichael, in W. W. Carmichael, ed., *The Water Environment – Algal Toxins and Health*, Plenum Press, New York, 1981.
81. O. M. Skulberg, G. A. Codd and W. W. Carmichael, *Ambio* **13**, 244 (1984).
82. J. E. Haugen, O. M. Skulberg, R. A. Andersen, J. Alexander, G. Lilleheil, T. Gallagher and P. A. Brough, *Algae Stud.* **75**, 111 (1994).
83. S. Weigel, Entwicklung einer Methode zur Extraktion organischer Spurenstoffe aus großvolumigen Wasserproben mittels Festphasen, Master thesis (Diplomarbeit), University of Hamburg, 1998, p. 86.
84. K. Kümmerer, *Chemosphere* **45**, 957 (2001).
85. D. J. Weiss, *Toxicol. Pathol.* **21**, 35 (1993).
86. P. A. Atkin, P. C. Veitch, E. M. Veitch and S. J. Ogle, *Drugs & Aging* **14**, 141 (1999).

87. G. J. Annas and S. Elias, *Am. J. Publ. Health* **89**, 98 (1999).
88. R. A. Aitken, D. Parker, R. J. Taylor, J. Gopal and R. N. Kilenyi, *Asymmetric Synthesis*, Blackie, New York, 1992.
89. G. Blaschke, H. P. Kraft, K. Fickentscher and F. Köhler, *Drug Res.* **29**, 1640 (1979).
90. B. Knoche and G. Blaschke, *J. Chromatogr. A* **666**, 235 (1994).
91. D. Masurel, P. J. Houghton, C. L. Young and I. W. Wainer, *Cancer Res.* **50**, 252 (1990).
92. G. C. Cotzias, P. S. Papvasilou and R. Gellene, *New Engl. J. Med.* **280**, 337 (1969).
93. D. C. Poshanzer, *New Engl. J. Med.* **280**, 362 (1969).
94. E. F. Domino, *Pharmacol. Biochem. Behav.* **24**, 401 (1986).

Chapter 5
Sample Preparation

5.1 Introduction

The separation of enantiomers is a difficult and tedious job, due to their identical physicochemical properties. Moreover, the situation becomes more serious and critical when the analysis of the enantiomers is to be performed in the natural samples. Thousands of compounds are present as impurities along with the chiral xenobiotics in environmental and biological samples. Under such circumstances, sample preparation is essential before starting an analysis of the chiral pollutants. Moreover, the chiral pollutants are present at very low concentrations in the unknown sample (at nano- to picogram levels) and are beyond the scope of the detectors used in modern analytical technologies. Therefore, pre-concentration should be carried out to avoid such types of problem. One of the most important trends in simplifying these complications is the generation of simple, rapid and reliable procedures for sample preparation. The development and set-up of the method require the use of materials of known composition; for example, certified reference materials. Therefore, spiking experiments have to be performed for method quality control. In such cases, emphasis has to be placed on the spiking procedures, as they exert an influence on the recovery values. Even if the recoveries of spiked xenobiotics equilibrated in a certain matrix are total, there is no evidence that the incurred xenobiotic will be extracted with the same efficiency. However, it can be concluded that although the present scientific knowledge is not perfect, the use of spiking experiments

Chiral Pollutants.   I. Ali and H. Y. Aboul-Enein
© 2004 John Wiley & Sons, Ltd   ISBN: 0-470-86780-9

helps to minimize the errors. The integration and automation of all of the steps between sample preparation and detection significantly reduces the analysis time, thus increasing the reproducibility and accuracy. For this reason, the development of a new single device for the fully chiral analysis of samples is required. Some review articles have appeared in the literature on this issue [1–5]. Recently, Gomez-Ariza *et al.* [6] reviewed the extraction, concentration and derivatization strategies for environmental samples before they are loaded on to the chromatographic column.

In 1995, the Seventh Symposium on Handling of Environmental and Biological Samples in Chromatography was held on 7–10 May in Lund, Sweden. This symposium continued the series started by the late Dr Roland Frei, one of the early visionaries in sample preparation technologies in the analytical sciences. A survey of the papers presented at this symposium indicates that five points were considered as essential and were highlighted during sample preparation. First, the need for a continuous search for new technologies was realized, so that the high cost due to chemicals and experimental labour may be reduced. The second need realized was for increasing sensitivity, with better and more selective concentration techniques: this has driven scientists to examine affinity and immuno-affinity support that can selectively remove compound classes for further investigation. Thirdly, the development of multidimensional chromatographic techniques was advocated: this would allow online sample clean-up, which provides several advantages, including automation, better reproducibility and a closed system capability. The fourth point considered was the development of better sample preparation techniques, enabling the more effective use of biosensors and other sensors, because exposure to raw matrices can foul many of the sensors. The fifth and last point, which requires considerable attention from scientists, was the quality movement, which has found its way into sample handling [2]. Therefore, the extraction, purification and pre-concentration of the natural samples are very important and essential parts of the chiral analysis of xenobiotics. We have carried out an extensive search of the literature on sample preparation technologies, and we have found only few reports on this subject in chiral analysis studies of pollutants. However, many reports are available on sample preparation in achiral (simple) analysis of pesticides and other agrochemicals. Since sample preparation itself is not enantioselective in nature, the sample preparation methodologies used in achiral analysis may be employed in chiral analysis studies too. A general protocol for sample preparation is given in Scheme 5.1. In view of all these points, the present chapter describes the art of sample preparation, including the sampling, extraction, purification and

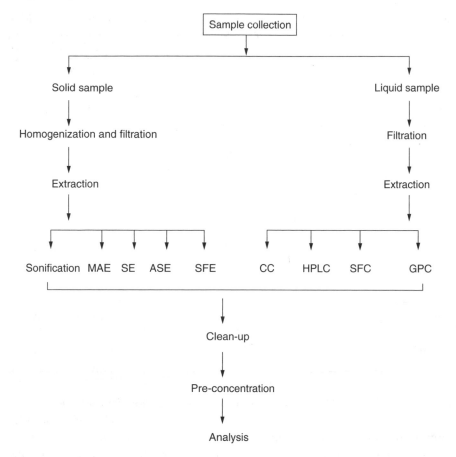

Scheme 5.1 A protocol for sample preparation prior to chiral analysis of the pollu-tants. Note that this is only a brief outline of the sample preparation procedure. MAE, microwave assisted extraction; SE, Soxhlet extraction; ASE, accelerated solvent extrac-tion; SFE, solid phase extraction; CC, column chromatography; HPLC, high performance liquid chromatography; SFC, supercritical fluid chromatography; GPC, gel permeation chromatography.

pre-concentration processes required before analysis of the chiral pollutants in natural samples is undertaken.

5.2 Sampling

Due to the low concentrations of xenobiotics in the environment, sampling requires special attention, and hence it is very important to achieve the correct values of their concentrations in the samples. The sampling strategy includes the selection of sampling sites, the type of sample (grab, mixed or composite), the sample container, volume collection, sample handling, transportation, preservation and storage. The selection of sampling sites and the type of sample are based on the objective and nature of the study. In the case of rivers, up- and downstream samples, and the points at which some tributaries or waste drains join the main stream under study, are normally included. A rigorous clean-up of all components of the sampler and sampling apparatus, the solvent, the adsorbent materials and other chemicals should be carried out very carefully. Possible contamination due to the penetration of the sampler through the surface layers of the water body, which may be highly enriched by pollutants, has to be excluded [1]. Normally, for pesticides and other agrochemicals, samples are collected in dark glass bottles with Teflon® stoppers. Plastic or polyethylene bottles should not be used, as the chiral xenobiotics present in the water may be adsorbed on their walls. The glass bottles should be washed properly with tap water, acids, detergents, tap water again, double distilled water, acetone and finally with working solvents such as hexane, dichloromethane and diethyl ether. The different types of water sampler can be used to collect water samples from rivers, seas, lakes and other water bodies. The Blumer sampler [7], the DHI [8] and a high volume water sampler designed to pump water from a defined depth below the water body surface, outside the wake of the survey vessel, are suitable as samplers [9]. Roinestad *et al.* [10] improved the sampling of indoor air and dust for 23 pesticides using a Tenax ™ TA sampling pump. The collected samples (biological waters and solids and environmental samples) should be transported to the laboratory immediately and kept at 4 °C. Therefore, ice boxes should be used to carry the environmental samples from the site to the laboratory. For many organophosphorous pesticides, the water samples collected should be preserved for up to 3 weeks by adding about 15 ml of chloroform. These collected environmental samples should be handled as soon as possible, and the extract should be kept at 4 °C to avoid any microbial proliferation and decomposition of the chiral pollutants.

5.3 Filtration

The collected environmental samples cannot be analysed directly, and hence they require some pre-processing. Water samples may contain some

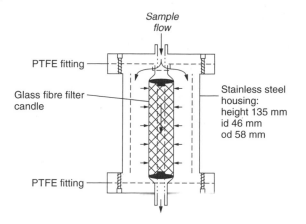

Figure 5.1 A schematic diagram of a filtration unit for use with filter candles [5].

solid impurities, and they will require a filtration process before proceeding to extraction. The filtration of these samples can be carried out by using a vacuum pump and a buccaneer funnel assembly, using Whatman's filter paper. Weigel *et al.* [5] described a PTFE tubing apparatus fitted with a geared pump (Model MCP-Z pump head Z-120 with PTFE gears, magnet 66, Ismatec, Wertheim, Germany) for the filtration of water samples. The pump can be placed behind the extraction unit, to calculate the recovery rates from the purified water. For the handling of environmental samples with a high particulate matter load, the pump should be placed between the filtration and extraction units, because with a rising flow resistance air may be drawn in at the exit of the extraction unit. This may lead to malfunctioning of the pump. Rearrangement of the pump to a position preceding the extraction unit did not cause contamination, as checked by procedural blanks. For online filtration of the sample, glass fibre filter candles (height 82 mm, outer diameter 26 mm, inner diameter 14 mm) were used in a stainless steel housing, as shown in Figure 5.1 [5]. Filtration as such is not required for solid samples, but solid samples can sometimes be manipulated if they contain other solid materials. If dust particles are present in air samples, they should be removed before loading the collected air into the machine. The filtration of air samples is carried out by using perch membranes under pressure.

5.4 Homogenization

Sample preparation using solid environmental and biological samples has been an active area in the analytical sciences for the past few decades.

Composed to liquid or gas samples, the treatment of solid samples is more complicated. Since 1970, acid homogenization methods have been used for the extraction of pollutants from solid environmental and biological samples [11]. The homogenization is carried out in a special apparatus called a homogenizer, which contains a small set of blades and generator probes that cause vigorous mixing and turbulence, to optimize sample contact and preparation. The homogenization of biological and environmental samples with an organic solvent is tedious and time-consuming. Moreover, the consumption of huge volumes of the organic solvent makes this method less popular. The handling of solvents is hazardous from the health viewpoint. The homogenization is assisted by shaking the samples at different speeds. Sometimes, mild heating also affects the extraction of the chiral pollutants during the homogenization procedure. Nowadays, direct or indirect sonication of the samples helps greatly to extract the pollutants from solid samples. Sonication causes a disruption of the cell wall in biological samples – plant cells, bacteria, fungi and yeasts – and it helps to release the pollutants from these biological solid samples. After disruption, the proteins, DNA, RNA and other cellular components can be separated for additional purification or processing.

5.5 Extraction

In the case of water or liquid biological samples, extraction is the next step after filtration, while in case of solid samples it comes after homogenization. Many different devices have been used for extraction procedures. The most important techniques in this field are the classical methods of extraction, liquid–liquid extraction, solid phase extraction and liquid chromatographic techniques. The extraction of solid and liquid samples of environmental and biological origin is described separately.

5.5.1 The Extraction of Solid Samples

The extraction of solid samples starts during the homogenization procedure, which involves the use of a variety of solvents, as discussed above. The best solvent is one in which the solid sample is insoluble while the pollutants of interest are soluble. Several new and improved technologies have also been used for the extraction of organic pollutants from solid matrices. For the best extraction of solid samples, they should be homogenized properly, to give homogenous and finely ground materials that they can pass through a 1 mm sieve. The finely homogenized sample provides a greater surface area to the solvent, which results in maximum extraction. The most important methods for this purpose are Soxhlet extraction, sonication, microwave assisted extraction, accelerated solvent extraction and supercritical fluid extraction (SFE).

Homogenization Extraction

The extraction of pollutants from solid samples can be carried out during the homogenization procedure by using various extracting solvents. In most cases, digestion procedures are used for the release of metal ion species, but no reports are available on digestion procedures for chiral pollutants. However, the chiral pollutants present in sediment, soil and biological samples can be released using homogenization procedures. The solid samples are ground to powdered or paste forms, as per the physical state of the collected solid samples. The extraction of the chiral pollutants can be achieved either by using aqueous or organic solvents or by a mixture of organic solvents. The most important organic solvents used are dichloromethane, chloroform, toluene, hexane, methanol and so on [11–14].

Sonication Extraction

As discussed above, sonication helps in the homogenization of solid samples and, consequently, it can be used for the rapid and easy extraction of pollutants from solid samples. Direct sonication uses a specially designed acoustic tool known as a horn or probe. The horns and probes are available with different cross-sectional areas and lengths, depending on the volume of solution to be processed and the required intensity. The complete assembly, including the horns and tips, is called a sonotrode and is constructed of an inert material such as titanium alloy. The sonotrode is placed directly into the sample that contains the extracting solvent. Ultrasound frequencies, which are beyond the range of human hearing, are used in the sonication process. The ultrasonic energy generated by a piezoelectric transducer at a rate of 20 000–40 000 Hz creates cavitation in liquids; the formation and collapse of countless tiny cavities or vacuum bubbles. The ultrasonic energy causes alternating high and low pressure waves within the liquid and these waves compress and extend the liquid, which promoter the breaking up of solid matrices and extraction of the pollutants into the liquid. The effectiveness of sonication depends on the frequency and amplitude of the ultrasound, the temperature, the surface tension, the vapour pressure, the viscosity and the density of the liquid. It is interesting to note that sonication generates heat in liquid, which augments the extraction process. EPA method SW-846-3550 recommends sonication as the most effective means of extracting nonvolatile and semi-volatile organic pollutants from soils, sludges and other wastes.

Microwave Assisted Extraction

Microwave assisted extraction involves heating solid samples with solvent or solvent mixtures using microwave energy, and the subsequent partitioning of the compounds of interest from the sample to the solvent. It has the potential to be a direct replacement for the conventional Soxhlet extraction technique for solid samples. It is considered to be a suitable method for environmental samples, as it combines rapid operation and low cost chemicals with good efficiency and reproducibility. EPA method SW-846-3550 describes microwave assisted extraction for the acid digestion of inorganic analytes as the best methodology. However, LeBlanc [15] described modifications to this method, which may be used to heat solvents and samples. The heating process helps a lot in the rapid extraction of organic compounds from solid samples [15]. Since 1986, scientists have used microwave heating to extract various compounds from a variety of sample matrices. Recently, LeBlanc [15] reviewed the microwave assisted extraction method for solid samples, listing the benefits of this technology as a shorter extraction time and a reduced solvent consumption. In 1986, Ganzler *et al.* [16] reported the extraction of organic substances from soils, seeds, foodstuffs and animal feeds by using microwave energy. Since then, various modifications have been reported from time to time in microwave assisted extraction technology [17, 18]. In 1992, Bichi *et al.* [19], and in 1993 Onuska and Terry [20], reported the use of this method for the extraction of pharmaceuticals from plant tissues and of pollutants from environmental samples. The authors extracted organochlorine pesticides from soil and sediment samples using a 1 : 1 (v/v) isoctane–acetonitrile mixture. Furthermore, Lopez-Avila *et al.* [21] extended this extraction system for PAHs, phenols and organochlorine and organophosphorous pesticides in soil samples. Mattina *et al.* [22] investigated the use of microwave accelerated extraction to extract taxanes used in ovarian and breast cancer research, from *Taxus* biomass. Recently, microwave assisted steam distillation for the simple determination of polychlorinated biphenyls and organochlorine pesticides in sediments has been reported by Numata *et al.* [23]. This technology has been coupled with liquid chromatography [24] and solid phase extraction methods separately [25], which have resulted in improved analytical techniques for many substances. A schematic diagram of a microwave system is shown in Figure 5.2, indicating a magnetron, an isolator, a wave guide, a cavity and a stirrer. A comparison of the performance of microwave assisted and conventional extraction methods is given in Table 5.1, which indicates the superiority of the microwave assisted method. This technology has good future prospects, as it can be coupled on line with chromatographic techniques.

Soxhlet Extraction

Soxhlet extraction has been used for more than 100 years as an extraction technique for various substances from different matrices. The methodology was developed by the German scientist Franz Von Soxhlet in 1879. Soxhlet extraction is a good alternative for the extraction of organic pollutants from solid samples, but it is not normally used for the extraction of pollutants that may decompose at higher temperatures. Therefore, Soxhlet extraction may be a good method for the extraction of pollutants that have high volatility. In general, organic solvents in which the pollutants are soluble are used in this methodology. EPA method SW-846-3541 has adopted Soxhlet extraction as a suitable method for the extraction of organic pollutants from solid matrices. Soxhlet extraction is a simple and effective, but relatively expensive, method, which requires large amounts of organic solvents. It sometimes requires 18−24 h of operation for the extraction of organic pollutants from environmental samples. Nowadays, automatic Soxhlet assemblies are available which work with a minimum amount of organic solvents (e.g. only

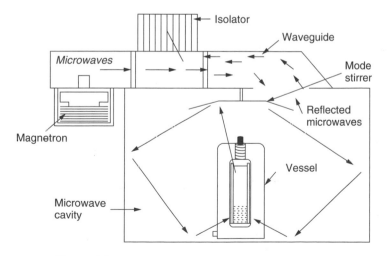

Figure 5.2 A schematic diagram of a microwave [15].

Table 5.1 A comparison of the performance of microwave assisted and conventional extraction methods [15]

Sample	Microwave assisted extraction			Conventional extraction		
	Solvent volume (ml)	Extraction time (min/sample)	Concentration	Solvent volume (ml)	Extraction time (min/sample)	Concentration
Hydrocarbons from soil	30	7	943 mg kg^{-1}	300	60	773 mg kg^{-1}
Organochlorine pesticides from soil	50	7	92.3 %	300	1080	83.4 %
PCBs from soil	25	7	47.7 µg g^{-1}	250	1080	44.0 µg g^{-1}
Pentachloronitrobenzene from radish	25	6	0.42 ppb	300	60	0.36 ppb
Imidazolinone herbicides from soil	20	5	11.2 ppb	400	120	10 ppb

50–100 ml), which makes their operation inexpensive [26]. Guerin [27] compared the extraction recoveries of polynuclear aromatic hydrocarbons (PAHs) from aged clay soil using classical and automatic Soxhlet assemblies, and the author reported comparable results from the two technologies when the extraction was carried out for only 8 h.

With the development of science and technology, various modifications have been made to the Soxhlet extraction assembly, which have resulted in the modern automatic Soxhlet extraction technology. Of these modifications, the procedure of Goldfisch, in which the condensed solvent is allowed to drip through the sample thimble, immediately returning to the boiling solvent, is considered to be the best. In 1996, Ali [28] presented a compact solvent extraction apparatus that is very easy to handle. Recently, Arment [29] has reviewed automatic Soxhlet extraction procedures, with an emphasis on the evolution of the technique from the original design to modern automatic instrumentation. Furthermore, the author discussed several aspects for method optimization and some general considerations. Today, many companies have developed instrumentation for automatic extraction based on Randall's concept of immersing the sample in a boiling solvent. Automatic Soxhlet extraction has gained worldwide acceptance: and EPA method SW-846-3541 described the extraction of analytes from soil, sediment and sludges, and this method is also applicable to many PCBs, as well as semi-volatile organic pollutants. Table 5.2 [29] presents a comparison of the extraction of several pollutants by classical and modern automatic Soxhlet technologies. A perusal of this table indicates that the precision and recovery factors are significantly better with the automatic Soxhlet method than with traditional extraction. Büchi Laboritechnik AG (Flawil, Switzerland) have manufactured a model B-811 with a four-way universal extraction system; that is, a standard Soxhlet, a warm Soxhlet, hot extraction and a continuous drip. Similarly, C. Gerhardt GmbH & Co. Laboratory Instruments of Bonn, Germany, have manufactured the Soxtherm S 306 model, that has a six-way extraction system. Foss-Tecator (Höganäs, Sweden) commercialized and licensed the Randall method in 1975, and marketed an automatic Soxhlet instrument (Figure 5.3).

Accelerated Solvent Extraction

Accelerated solvent extraction (ASE) is a sample preparation technique that works at elevated temperature and pressure, using very small amounts of solvent. The restrictions and limitations of this method are more or less similar to those of Soxhlet extraction technology. The method development is simple, as it involves few operational parameters. The selection of the

Table 5.2 The average percentage recovery and relative standard deviations (RSDs) of some pollutants, using classical and modern automatic Soxhlet extraction methods [29]

Pollutant	Classical Soxhlet		Automatic Soxhlet	
	Average recovery (%)	RSD	Average recovery (%)	RSD
α-BHC	57.4	47.5	94.9	5.5
δ-BHC	65.0	27.1	104.0	9.7
Heptachlor	59.6	34.1	87.1	5.4
Aldrin	69.8	8.8	78.2	5.7
Dieldrin	74.4	20.0	68.8	2.6
Heptachlor epoxide	72.0	20.8	92.4	0.6
trans-Chlordane	75.6	12.5	85.8	2.2
Endosulfane I	76.4	5.5	90.5	2.0
Endosulfane II	78.5	6.7	90.3	10.0
p,p'-DDT	73.6	38.5	61.4	6.5

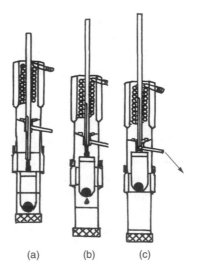

(a) (b) (c)

Figure 5.3 The three-step extraction procedure using the Foss–Tecator Soxhlet Avanti automatic extraction system: (a) boiling, (b) rinsing and (c) the evaporation steps [29].

solvents again depends on the solubility of the pollutant to be extracted. The initial reports on accelerated solvent extraction [30–38] showed that the performance was more or less equal to that of the conventional method, but in a shorter period of time and involving a low amount of solvent. Figure 5.4 shows a schematic representation of this technology, which features solvent

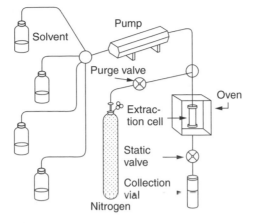

Figure 5.4 A schematic diagram of accelerated solvent extraction [39].

reservoirs, a pump, an extraction cell, an oven, a static valve, a nitrogen gas cylinder and a collection vial. Recently, Richter [39] has reviewed the extraction of solids using the ASE methodology.

ASE method development is very easy, as it requires the selection of the temperature and solvents, which govern the efficiency of the extraction. The other parameters that control the recovery of the extraction are the static time, the flushing volume and the number of static cycles used in the extraction. It is very interesting to note that the first application of this method was reported in the extraction of environmental pollutants [30–38]. Therefore, the ASE method has become very popular in the extraction of environmental samples [39]. Moreover, its use has been expanded for the extraction of many organic and inorganic substances from foodstuffs, polymers, pharmaceuticals and other types of sample. The application of the ASE technique for environmental samples is summarized in Table 5.3.

Supercritical Fluid Extraction

Supercritical fluid extraction (SFE) is a selective technique of sample preparation that enables the preparation of matrices by varying several physical parameters. Nowadays, it is considered to be the best replacement for many extraction technologies, such as accelerated solvent, Soxhlet solvent, microwave assisted extraction and so on. It was originally marketed as a universal extraction tool in 1988 by Isco Inc. (Lincoln, Nebraska, USA), Lee Scientific (Salt Lake City, Utah, USA) and Suprex Corp. (Pittsburgh, Pennsylvania, USA). The basic components of the SFE instrument are a carbon dioxide reservoir, a pump, an extraction vessel, an oven, a restrictor

Table 5.3 The application of accelerated solvent extraction for environmental samples

Samples	Matrix	References
PAHs	Soil, sediment	35, 38
	Dust, sediment	37
	Sediment	165
	Soil	166–169
	Dust, diesel particulate, sediment, muscle tissue, fish tissues	170
PCBs	Air	36
	Soils	169
	Oyster tissue, sewage sludge	37
	Fish tissue	171
	Pine needles, mosses	172
Organochlorine pesticides	Clay, loam, sand	30
	Soil	173
	Sediment	165
	Dust, diesel particulate, sediment, muscle tissue, fish tissues	170
	Pine needles, mosses	172
	Plant	174
	Tomatoes	175
Organophosphorous pesticides	Clay, loam, sand	33
	Plant	174
	Tomatoes	175
	Wheat	176
	Fruit and vegetables	177

(a decompression zone) and a collection vial (Figure 5.5). Depending on the application, some models also contain modifier pumps, which are used to mix carbon dioxide with the organic solvents. Recently, Levy [40] has reviewed the extraction of organic compounds from solid and semi-solid samples using SFE.

In practice, all SFE instruments use carbon dioxide as the primary supercritical fluid. This is due to the fact that at its critical temperature ($31.4\,^\circ$C) and pressure (73 atm) carbon dioxide can be handled easily. Moreover, carbon dioxide is readily and economically available in analytical grade form. Hexane, methylene chloride, acetone and chloroform are suitable solvents, as they can be mixed with carbon dioxide. Several organic

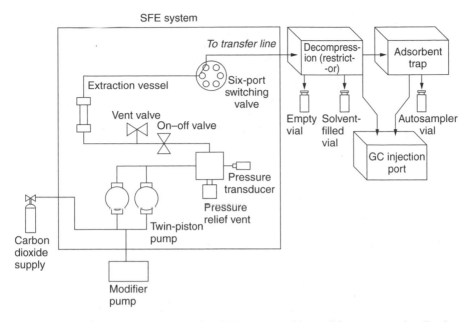

Figure 5.5 A schematic diagram of a SFE system with modifier pump and collection options [40].

compounds are miscible with carbon dioxide, and Table 5.4 [40, 41] lists some of these modifiers and their critical parameters. The values given in this table can be used for the extraction of chiral pollutants. Depending on the application, the use of a third modifier for additional solubility enhancement is sometimes required [42]. The efficiency of this method is increased by variation of the temperature and pressure variables, which have an inverse relationship. Maximizing solvent contact also ensures rapid and efficacious extraction. However, the flow rates of the mobile phase, the SFE time and the SFE mode also control the extraction efficiency. Normally, analysis of the extracted samples is carried out after extraction by SFE, but in 1992 Levy *et al.* [43] coupled SFE with GC and reported the good efficiency of this assembly (Figure 5.6).

SFE has been used for the extraction of many substances from foodstuffs, plants, feeds, and biological and environmental samples [40]. McNally and Wheeler [44] used SFE for the extraction of herbicides and pesticides. France *et al.* [45] extracted organochlorine pesticides from fat samples using this methodology. Similarly, Hopper and King [46] used SFE for the extraction of organochlorine pesticides from food samples. Furthermore, the same group [47] extracted organophosphorous pesticides from

Table 5.4 Critical parameters for carbon dioxide based modifiers [40, 41]

Modifier	$T_c(°C)$	$P_c(atm)$	$V_c(cm^3\ mol^{-1})$
Acetic acid	321.1	57.1	171.1
Acetone	235.2	47.2	216.5
Acetonitrile	247.7	47.7	173.0
Benzene	288.8	48.3	257.8
n-Butanol	288.2	43.6	275.1
tert-Butanol	234.8	41.7	275.0
Carbon tetrachloride	283.3	45.0	275.9
Chloroform	262.3	54.1	241.1
N,N-Dimethylformamide	376.8	54.3	–
Dimethyl sulfoxide	446.8	56.3	237.5
1,4-Dioxane	314.8	51.1	238.8
Ethanol	242.6	62.8	167.1
Ethylacetate	250.9	38.0	285.9
Ethylether	193.55	35.92	280.0
Formic acid	306.8	31.2	435.8
n-Heptanol	358.7	31.2	435.8
Hexane	234.7	29.8	368.4
Hexanol	336.8	34.3	381.0
Isobutanol	274.5	42.4	273.6
Isopropanol	243.4	28.4	219.7
Isopropyl ether	227.2	28.4	386.0
Methanol	239.8	79.1	117.9
Methyl-*tert*-butyl ether	239.7	33.3	–
Methylene chloride	235.1	60.0	180.0
2-Methoxyethanol	302.0	52.2	–
Nitromethane	314.8	62.3	173.0
Pentanol	312.7	38.4	326.0
n-Propanol	363.7	50.5	218.9
Propylene carbonate	–	–	352.0
Sulfur hexafluoride	45.5	37.1	197.0
Tetrahydrofuran	267.3	51.2	224.0
Toluene	319.9	41.5	316.6
Trichlorofluoromethane	198.0	43.2	248.0
Water	374.1	218.2	56.5

T_c, P_c and V_c are the critical temperature, pressure and volume, respectively.

grain by using SFE. Gere *et al.* [48] and Hawthorne *et al.* [49] extracted PCBs from soil and PAHs from fly ash, sediment and dust, respectively. Recently, Zuin *et al.* [50] have described a fast supercritical fluid extraction method coupled with high resolution gas chromatography for multi-residue screening of organochlorine and organophosphorus pesticides in Brazil's medicinal plants.

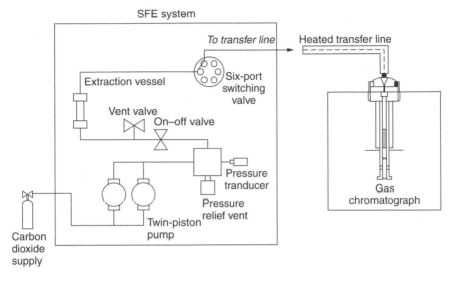

Figure 5.6 A schematic diagram of SFE–GC coupling [40].

5.5.2 The Extraction of Liquid Samples

Normally, liquid samples of environmental and biological origin are of water, blood, serum or urine. The extraction of liquid samples is relatively easy composed with the extraction of solid samples, as the former samples do not require any procedures such as grinding and so on. The most common techniques for the extraction of pollutants from liquid samples are liquid–liquid extraction and solid phase extraction, but with the development of extraction science different kinds of chromatographic modalities have been used as extraction methods. The chromatographic techniques used for this purpose are column chromatography (CC), high performance liquid chromatography (HPLC), supercritical fluid chromatography (SFC), gel permeation chromatography (GPC) and so on. If the chiral pollutants present in solid samples are polar in nature, then the extraction is normally carried out in aqueous solvents. If gas chromatography is required for the analysis of such types of extracted pollutant, further extraction of the already extracted pollutant (in an aqueous solvent or water) is carried out in organic solvents. This is due to the fact that an aqueous solution cannot be injected directly into a gas chromatographic machine, because it may destroy the detector. Contrary to this, if the analysis is carried out by capillary electrophoresis, the pollutants present in organic solvents are extracted into aqueous solvents. This situation arises because the loading of a sample in organic solvent disrupts the continuity of the current in the machine. Recently, the authors have observed this effect when working with capillary electrophoresis in their laboratory.

Liquid–Liquid Extraction

Liquid–liquid extraction is the method used for the extraction of pollutants from liquid samples. The extraction of pesticides and other xenobiotics is normally carried out by simple liquid–liquid extraction (LLE) and solid–liquid extraction (SLE) methods, using organic solvents. According to the distribution law, when brought into contact with an immiscible solvent pair, a solute distributes itself between these solvents on a preferential basis that is determined by many factors, but mainly by the solute and solvent characteristics. In the quantitative LLE of water samples, the right selection of the solvent or solvents must be carried out carefully. Although water quality characteristics, such as ionic strength, turbidity, co-extractive, pH and temperature, may affect the criteria for the solvent selection to some extent, the choice of the solvent or the mixture of solvents should be reviewed in the light of the nature of the solvents, the solubility of the solvents in water, compatibility with the detector and the volatility of the solvents. Given the general principle that like dissolves like, the polarity of the solvent determines its extraction capability to a large extent. The polar xenobiotics may be extracted using polar solvents. For common solvents, the polarity is in the order hexane < benzene < chloroform < methylene chloride < acetonitrile. Solvents that are even partially soluble in water are not good for extraction purposes.

Hexane is not effective as an extraction solvent for all pollutants. In certain cases, when there are suspended solids in the water samples, hexane gives less satisfactory results even though it gives good results (80 % recovery) from spiked distilled water. Therefore, hexane should not be used as a broad spectrum solvent for the extraction of pollutants from water. Methylene chloride is a very effective solvent for multi-residue extraction of pollutants from water, including those samples with considerable turbidity and organic contents [51]. It has low water solubility and it is not too polar to extract excessive co-extractives, as compared with more polar solvents such as diethyl ether and ethyl acetate. However, methylene chloride creates problems in gas chromatographic the electron capture detector. The other organic solvents used are hexane [52], acetone [53], acetic acid [54], benzene [55], toluene [56], methanol [57], acetonitrile [54], petroleum ether [58], ethyl ether [59], iso-octane [60] and pentane [59]. These solvents are used as single, binary or ternary mixtures

Table 5.5 The recovery factors of some organochlorine and organophosphorous pesticides in some solvents

Pesticide	Recovery factor (%)			
	n-Hexane	Benzene	Methylene chloride	Chloroform
Aldrin	92.1	95.0	89.0	–
α-BHC	–	–	91.0	–
β-BHC	–	–	99.0	–
γ-BHC	103.0	96.7	90.0	–
α-Endosulfan	–	–	91.0	–
o,p'-DDT	–	–	94.0	–
p,p'-DDT	96.7	96.1	104.0	–
p,p'-DDD	92.8	97.3	96.0	–
p,p'-DDE	93.4	–	97.0	–
Methoxychlor	–	99.0	97.0	–
Methyl parathion	91.0	95.0	99.0	97.0
Malathion	91.0	99.0	100.0	92.0
Ethion	98.5	103.0	100.0	–
Dimethoate	–	41.0	40.0	96.0

for pollutant extraction. The extraction efficiencies of some pesticides using different solvents are given in Table 5.5.

The general methods for the extraction of pollutants from water samples involve manual or mechanical shaking, stirring with magnetic stirrers, continuous liquid–liquid extraction and the homogenization method [61] using a high efficiency dispenser, which essentially contains high speed blades coupled to a high frequency ultrasonic probe. In brief, the mechanical shaking method is efficient provided that the shaking is sufficient. Different models of shaker can provide different results, particularly with turbid waters, and thorough testing with one's own water samples is needed. The shaking method can cause severe emulsion formation for certain samples (for example, some industrial effluents). Under such conditions, the emulsion is difficult to break up even by high speed centrifugation: this situation may be remedied by the addition of chemicals such as alcohols, salts and so on. The analyst should carefully consider his or her own situation – for example, the type of water and the choice of solvent – before contemplating this approach.

Various workers have used the LLE approach for the extraction and purification of chiral pollutants of musks in river water [62], PCBs in ringed seals [63], HCH in marine water [64], bromocyclen in surface and waste waters and fish [65], and chlordane in salmons, seals and penguins [66].

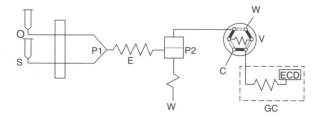

Figure 5.7 A schematic diagram of the online coupling of liquid–liquid extraction with GC. O, 50 ml glass syringe with extraction solvent, flow rate 138 $\mu l\,min^{-1}$; S, 50 ml glass syringe with sample, flow rate 1430 $\mu l\,min^{-1}$; PI, phase segmentor; E, polymer-coated glass extraction coil (5 m × 0.7 mm id); P2, phase separator with Teflon® membrane, 0.2 μm pore size; W, waste; V, rotary valve with variable loop (load position); C, carrier gas inlet; GC, gas chromatography [1].

Online LLE is possible, which reduces the experimental labour, lowers the cost of chemicals and gives better results. Petrick *et al.* [67] have developed a system and coupled it to a GC machine on line. The extraction of the sample is performed in a liquid–liquid segment flow in a glass coil that is internally coated with hydrophobic layer. The authors have compared this system with the classical LLE method and reported 20 % RSD in the classical LLE approach, while RSD was found to be 10 % in their online LLE system. A schematic diagram of this set-up is shown in Figure 5.7. Ezzell *et al.* [33] used the solvent extraction procedure for the extraction of organophosphorus pesticides from soil samples at the $\mu g\,kg^{-1}$ level. An outline of their apparatus is shown in Figure 5.8. The LLE technique is tedious, time-consuming and consumes a large volume of costly solvents.

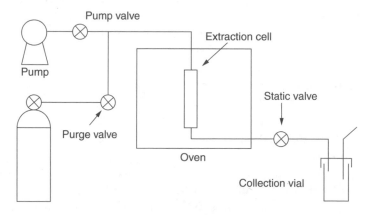

Figure 5.8 A schematic diagram of the accelerated solvent extraction system [33].

The handling of the solvents is hazardous from the health viewpoint. Also, in LLE emulsion formation is another problem in the extraction of the more polar pollutants such as phenoxy acids herbicides and their metabolites [68].

Solid Phase Extraction

The solid phase extraction (SPE) technique is free from the drawbacks of LLE and is very fast and sensitive, the recovery of pollutants ranging from 90 % to 95 % [69]. The use of SPE offers the advantages of convenience, cost savings and minimal consumption of solvents [70], and hence about 50 % of chromatographers use this method for sample preparation [2]. However, during extraction of the environmental samples, both biogenic and anthropogenic compounds are extracted into the organic solvent, and this complex matrix may interfere seriously with the determination of the respective analytes. It has been reported that lipid materials and aliphatic and polynuclear aromatic hydrocarbons are co-extracted with organic compounds into nonpolar solvents such as hexane, which creates great problems for the user [1]. Various columns, discs and cartridges have been used for extraction purposes, but cartridges are the most popular extraction device [49]. Manufacturers have developed new formats for the traditional cartridges. Most SPE cartridges have medical grade polypropylene syringe barrels with a porous PTFE or metal frit that contains 40 μm d_p packings. Cartridges provide certain advantages compared to discs, as liquid flows faster in cartridges than in discs. Most cartridges and discs are silica-based polymeric packings.

The most important cartridges and discs are the SS-401, the XAD-2 (37), the PL RD-S [71], the HDG-C_{18} [72], the RDS-18 [73], the C_{18} cartridge [74], the C_{18} silica-bonded cartridge [75], the SEP-Pak C_{18} [76], extraction fibre [77], ODS-impregnated polymers [78], the C_8 disc [73] and the C_{18} Empore disc [78, 79]. SPE involves the use of nonpolar C_8 and C_{18} phases in the form of cartridges and discs. The discs are preferred to the cartridges, as the discs have a high cross-sectional area that provides several advantages, which are not possible in cartridges. With the disc, the decreased back pressure allows a greater flow rate, while the wide bed decreases the risk of plugging. The embedding of the stationary phase into a disc prevents channelling and improves mass transfer [80]. Offline or online SPE or solid phase micro-extraction on various types of silica-bonded, polymeric or carbon type phase have been used prior to the application of GC [81]. The extraction efficiency in SPE has been optimized and increased by varying a number of parameters, such as the pH, the ionic strength of the sample, the elution solvents, the content of organic modifier in the sample, the elution gravity and so on. The progressive nature of the technique has been established using a polymer membrane containing enmeshed sorbent particles in a web of polymer microfibrils, called a membrane extraction disc. Wells *et al.* [82] have developed a microprocessed optimization of the SPE method for the extraction of various pesticides and environmental pollutants. Barcelo *et al.* [83] has described various aspects of SPE using C_{18} and styrene divinyl benzene Empore extraction discs for the extraction of aqueous samples at low level. The isolation and pre-concentration of most nonpolar or semi-polar pesticides – phenoxy acid herbicides [84], organophosphorus [85], triazines [86], hydroxy triazines [87], organochlorines [81] and fungicides [88] – was achieved by C_{18} or C_8 bonded silica discs or cartridges from water samples. A schematic diagram of a solid phase extraction unit is shown in Figure 5.9 [89]. This figure indicates a housing for a glass cartridge and the bottom is covered with a glass fibre sheet. This cartridge has been used for the extraction of organochlorine pesticides.

The extraction of polar pesticides using nonpolar (C_{18} and C_8) stationary phases is very difficult and, therefore, polar pesticides are extracted successfully on graphitized carbon black (GCB). The porous character of GCB makes it faster in extraction, without any pH adjustment of the environmental water samples. Many workers have recommended styrene divinyl benzene copolymer as the universal SPE system for the extraction of pesticides that are even moderately polar in nature [89–92]. Ion exchangers have also been used for the extraction of semi-polar or polar pesticides [93, 94], but their range of utility is low due to their low capacity (the capacity is decreased by the ions in the water samples) and their inability to extract pesticides that have similar acidic–basic moieties, which

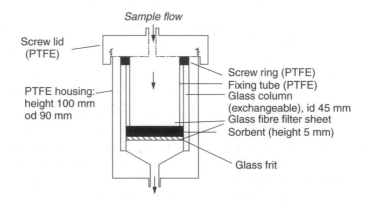

Figure 5.9   A schematic diagram of the solid phase extraction unit [5].

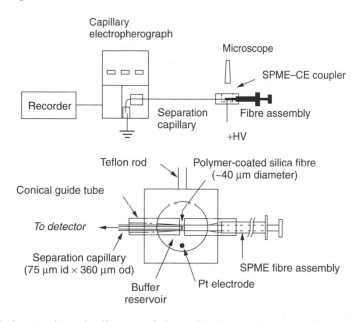

Figure 5.10 A schematic diagram of the solid phase micro-extraction capillary electrophoresis (SPME–CE) system [114].

determine the pH of the sample. However, ion exchangers have been successfully used for the selective extraction of interfering compounds [95, 96]. SPE has been used to enrich pesticides and pollutants from drinking [97, 98], ground [97, 98], waste [99, 100], surface [101–103] estuarine [104, 105], coastal [106] and marine waters [5, 107, 108]. Thus an emphasis has been placed on organochlorine pesticides [109], organophosphorous and nitrogen pesticides [110, 111], chlorophenols [112] and polyaromatic hydrocarbons (PAHs) [113]. Whang and Pawliszyn [114] designed an interface that enables the solid phase micro-extraction (SPME) fibre to be inserted directly into the injection end of a CE capillary. They prepared a partly 'custom-made' polyacrylate fibre to reach the SPME/CE interface (Figure 5.10). The authors tested the developed interface to analyse phenols in water and, therefore, the same technique may be used for chiral resolution of the pollutants. Similarly, Huen and coworkers [115] developed SPE coupling with HPLC for the analysis of 30 pesticides and reported satisfactory results. Brinkman [116] has been a leader in applying column switching to real-world problems, and he has extended his work on the extraction of polar pollutants in river water using SPE coupling with liquid and gas chromatographic machines separately. Van der Hoff *et al.* [117]

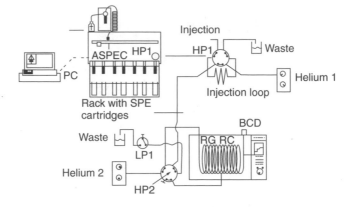

Figure 5.11   A schematic diagram of the ASPEC–GC–ECD equipment. HP1 and HP2, high pressure six-way valves; LP11, low pressure three-way valves; RG, retention gap; RC, retention column; T1, press fit connection between RG and RC; T2, three-way press fit connection for solvent vapour exit [117].

has described an automatic sample preparation with extraction column (ASPEC) coupled with gas chromatography (GC) system for the extraction of organochlorine pesticides and pyrethroid insecticides. The assembly developed by these authors is shown in Figure 5.11. They used a loop type interface to couple ASPEC to the GC machine, and the system developed was found to be suitable for the analysis of the reported pesticides. Furthermore, the authors also studied the elution pattern of the reported pesticides on a silica-based solid phase cartridge (Figure 5.12). This figure indicates differing elution patterns for the pesticides. α-HCH eluted first, while cypermethrin and deltamethrin eluted later. Recently, Sosa *et al.* [118] has described an extraction method for the screening of pesticide residues in water. The developed method was used for the extraction of a variety of pollutants, such as organochlorine and organophosphate pesticides, in water samples. The analytical technique employed for the extraction and clean-up steps involved solid phase extraction with C_{18} cartridges.

Many modifications have been made to SPE over time, and a new version of SPE has emerged as solid phase micro-extraction (SPME): several papers have been published on this issue in the literature [115]. In automatic SPME, a polymer-coated fibre, which is used instead of a GC autosampler needle, is dipped into a solution of water or held in the headspace of the sample to enable the analytes of interest to diffuse into the coating. After establishing equilibrium, the system withdraws the needle and places it in the heated GC injection port, and the analytes are thermally desorbed on the GC

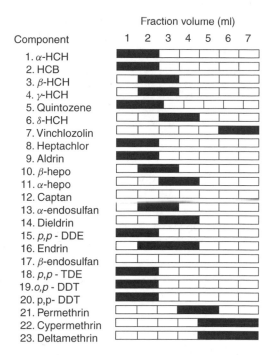

Figure 5.12   The pattern of elution of pesticides from a silica solid phase cartridge by hexane–isopropanol (99.9 : 0.1, v/v) mixture [117].

column. Applications of SPME include the extraction of herbicides in water [119], pesticides in soils [120] and organohalogen compounds in water and air [121]. Barcelo [122] reviewed the status of SPE as a suitable methodology for the extraction of the pollutants from water samples. Other papers presented at the 7th Symposium on Handling of Environmental and Biological Samples in Chromatography, held in Lund, Sweden, in 1995, describe the extraction of herbicides [123–125], organochlorine and organophosphorous pollutants [126, 127] and of pesticides in water [128, 129], soil [130] and leachates [123] using SPME.

Liquid Chromatographic Extraction

In recent years, liquid chromatographic techniques – column chromatography (CC), high performance liquid chromatography (HPLC), supercritical fluid chromatography (SFC) and gel permeation chromatography (GPC) – have received widespread attention for the extraction of organic pollutants from environmental samples [3]. The popularity of chromatographic techniques in the extraction of pollutants is due to the manifold developments of this technology, which can be used for the extraction of a wide range of pollutants. Moreover, high speed and a low consumption of costly solvents are the other assets of chromatographic methodologies, together with the coupling of the chromatographic instrument with other analytical techniques.

Column Chromatography

Column chromatography is a classical modality of liquid chromatography, which still exists due to its ease of operation and its inexpensive nature. It has been used extensively for the extraction and purification of environmental and biological samples. Various types of adsorbents have been used in column chromatography: the most important are the florisil column, used for organochlorine, PCBs, PCNBs, strobane, trifluralin [111, 131] and organophosphorus [132], and the silica gel column (10 %), used for nitrogen–phosphorous pesticides [133]. This modality has been replaced by high pressure chromatographic technologies, and hence its use in sample preparation is gradually decreasing.

High Performance Liquid Chromatography (HPLC)

HPLC has been used in the extraction and purification of pollutants in environmental and biological samples. Its manifold applications are due to the availability of a number of stationary phases, particularly reversed phase columns of various dimensions. HPLC instruments used for this purpose are similar to the analytical machine but with the use of a different column, which depends on the nature of the pollutants to be extracted and purified. Many of the reports available in the literature deal with the use of HPLC as an extraction and purification technique for environmental and biological samples. For determinations of polar pesticides, HPLC appears to be the most appropriate technique for purification purposes [134, 135]. Smith *et al.* [136] used low pressure HPLC for the extraction of polychlorinated dibenzofurans and dioxins. Furthermore, Blanch *et al.* [137] used low pressure HPLC for the purification and extraction of PCBs from shark liver. Similarly, Ramos *et al.* [138] used the Smith *et al.* [136] method for the extraction of PCBs in dolphin liver. Bethan *et al.* [139, 140] used a LiChrosorb 100-Si HPLC column for the extraction of bromocyclen collected from river water and of α-HCH in marine water. However, HPLC cannot achieve a reputation as the universal and highly applicable extraction technique, due to its low range with regard to preparative chromatography. Therefore, further advancement of the technique is required, especially with regard to the development of preparative columns.

Supercritical Fluid Chromatography (SFC)

Supercritical fluid chromatography (SFC) is the latest development in liquid chromatographic technology. SFC is a viable sample preparation strategy for use with liquid samples, especially for selective applications. Fortunately, analysts have several new sample preparation strategies to choose from to accomplish an analytical objective. The technical and instrumental components of this methodology have already been discussed in Section 5.5.2.2. Hawthorne [141] reviewed the status of this technology in sample preparation. Hawthorne [141] reported the extraction of PCBs using SFC. Similarly, Karlsson *et al.* [142] reported a wide variety of applications of SFC for environmental, food and biological samples. The authors described the extraction and purification of PCBs in sediment samples. Lawrence [143] used SFC for the extraction and purification of polar pesticides and other contaminants from food matrices. Notar and Leskovsek [144] also used SFC for the extraction of PAHs in soil and sediment samples. Other applications of SEC for the extraction and purification of pesticides have been published [68, 145–147]. Barnabas *et al.* [148] described the main components of Carlo Erba SFC technology (Figure 5.13), which features a carbon dioxide gas cylinder, a syringe pump, an oven, a cartridge and a collection vial. The authors used this set-up for the extraction of organochlorine and organophosphorous pesticides.

Despite the good efficiency and recovery factors of SFC, it cannot achieve a respectable status in separation science. This is due to its high cost and its inability to handle simultaneous and parallel extraction. Moreover, the high solubility of water in supercritical carbon dioxide makes it unfit for routine applications [92]. Another problem in SFC

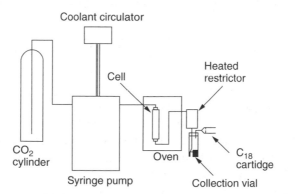

Figure 5.13   The main components of the Carlo Erba SFE system [148].

extraction is optimization of the experimental conditions by optimizing the temperature, the pressure, the amount and type of modifier, the extraction time and the cell volume. Also, SFC columns are not currently available as disposable cartridges. Moreover, SFC has low range of application, with poor reproducibility [149]. However, the coupling of SFC with SPE has resulted in the rapid and sensitive extraction of pesticides, with good recovery rates. Alzaga *et al.* [150] compared SFC and LLE for some pesticides. Briefly, this technology requires further advancement to reach its horizons.

Gel Permeation Chromatography (GPC)

The working principle of gel permeation chromatography is based on the size of the pollutants. The working column is made of polymer beads and the pollutants are separated by a filtration mechanism. The large-sized pollutants are eluted earlier than the small-sized ones. The smaller pollutants enter into the beads, while the larger-sized pollutants passes through the intra-particle spaces between the beads and hence are eluted first. The use of gel permeation chromatography (GPC) in clean-up processes is limited, as it is not capable of cleaning-up pesticides or any other contaminants of a similar size. However, a very few reports [111, 151] are available on the clean-up of pesticides by GPC. A Bio-Beads S-X3 column was used for the clean-up of organophosphorous pesticides [111]. A comparison of GPC, sweep co-distillation and florisil column chromatography for organochlorine has been discussed [79]. Furthermore, a comparison of various types of GPC columns for the clean-up of 12 selected pesticides from soil samples has been reported [149]. Dabek-Zlotorzynska *et al.* [152] also reviewed the sample pre-treatment methodologies for environmental analysis, including the above-mentioned methods. Some other reviews have also appeared in the literature in the past few years on this issue [153–156]. Puig and Barcelo [149] used size exclusion chromatography for the extraction of organochlorine pesticides.

5.6 Membrane Methods in Sample Preparation

Membrane filtration is also one of the most important and newly developed methods for sample preparation. Many papers on this issue were presented at the 7th Symposium on Handling of Environmental and Biological Samples in Chromatography, held on 7–10 May 1995 in Lund, Sweden. Knutsson *et al.* [157] described this technique as a combination of dialysis and LLE, and applied it to the extraction of acidic herbicides and chlorophenols in water samples. Shen and coworkers [158] extracted organic acids in soil samples. The authors used a supported liquid membrane impregnated with 10 % tri-*n*-octylphosphine oxide in di-*n*-hexylether. Microdialysis is also a sample preparation technique in which the user can insert small probes, coated with membranes, into the sample and make *in vivo* observations without disturbing the sample [156]. Buttler *et al.* [159] studied various microdialysis probes used for bioprocess monitoring. The dialysis factor of the probe was determined by several parameters; that is, the membrane area, the physicochemical properties of the membrane, the flow rate of the perfusion liquid, the diffusion coefficients of the analytes and various interactions between the analytes and the membrane. Membrane technologies have been used for the extraction and purification of pharmaceuticals in biological samples, but no reports are available on sample preparation in environmental samples. However, this method has good prospects in the extraction and purification of pollutants in biological and environmental samples. On the other hand, Majors [160] described sample preparation for environmental analysis using solid phase extraction, supercritical fluid extraction, microsolvent extraction, microwave assisted and accelerated solvent extraction, selective extraction based on molecular recognition, supported liquid membranes, microdialysis and chromatography.

5.7 Clean-up

Natural environmental and biological samples contain thousands of other substances as impurities and these get co-extracted with the pollutants of interest. Due to the similar properties of the co-extractives, in any analytical technology they usually interfere with the analysis of the pollutants. Therefore, in general, a clean-up procedure is required for those samples that are not clear; mostly samples of blood, serum, food, plant extracts and industrial and municipal effluents. For higher recoveries of analytes, clean-up is required to be at least 85 % [131].

The clean-up of the extracted pollutants may be carried out by column chromatography, gel permeation chromatography, sweep co-distillation, liquid–liquid partition, cartridges and discs. The columns used for this purpose are nuchar carbon, silica, XAD-2, alumina and florisil, while the cartridges and discs used are C_8, C_{18} and so on. The florisil, alumina and silica columns are suitable for the cleaning up of organochlorine pesticides, but care must be taken during the clean-up of organophosphorous pesticides using these columns, as recovery may be poor due to adsorption. The cleaning up of organophosphorus pesticides in water samples was achieved by a C_{18} Empore disc, disposable C_{18} cartridges and a C_{18} Empore disc plus florisil [111]. Various nonpolar cartridges and discs have been used for the SPE technique, but reports are only available on the clean-up of pesticides, using two SPE cartridges (C_{18} and SCX) installed vertically in series for efficient operation [69]. The use of membrane discs for clean-up has increased significantly in the past few years. SPE membrane discs may also be used in conjunction with supercritical fluid extraction in the pre-concentration of metal ions. Online SPE may be coupled with membrane discs for the clean-up of pollutant species in water and other liquid biological samples. Using an SFC methodology, greater recovery rates of the pollutants may be achieved. The use of gel permeation chromatography (GPC) in clean-up process is limited, as it is not capable of cleaning up the molecules of the pollutants due to their small size.

5.8 Pre-concentration

In general, the pollutant levels in environmental and biological samples are below the detection limits of modern detectors and, therefore, pre-concentration of the extracted samples is required. Normally, the volume of the extracted solvent containing pollutants is reduced to 5–10 ml. The classical approach of evaporation of the solvents is used to concentrate the extracted pollutants. However, some techniques such as purge and tape devices (cryogenically cooled capillary tapes) have been used for pre-concentration purposes [161]. Solvents with low boiling points are considered to be the best from the pre-concentration point of view, but the selection of the solvent depends on several factors, as already discussed. The large volume of solvent obtained after extraction is evaporated by the distillation unit. Open evaporation of extracts on a water bath may cause severe losses of the pollutants and, therefore, evaporation should be carried out at low temperature under reduced pressure. Normally, a rotary flash evaporator or a Kuderna–Danish assembly is used for this purpose. Sometimes, pure nitrogen is also allowed to pass through the evaporator assembly, so that oxidation of the pollutants may be avoided. Sometimes, the loss of pollutants takes place during evaporation and to check this some keeper solvent (toluene or iso-octane, which have high boiling points) is added to the extract before the evaporation is started. Besides this conventional method, other techniques such as chromatography, SPE and so on have also been used for pre-concentration purposes. SPE membrane discs have also been used in conjunction with SFC in the pre-concentration of pesticides [162]. Online SPE has been coupled with membrane discs for the concentration of pesticides in river water and ground water [163]. Using a SFC methodology, >90 % recovery of the pesticides has been reported [164]. A comparative study for the pre-concentration of pesticides has been discussed [162]. The clean-up methods for some organic pollutants are given in Table 5.6.

During extraction procedures, especially in LLE methodology, some invisible moisture may obviously remain in the samples along with the organic solvents, which may again create problems in the detection of the pollutants. Therefore, moisture should be removed from the sample prior to its loading on to the analytical machine, especially on GC. Normally, the moisture is removed from the organic solvent by the addition of anhydrous sodium sulfate of high purity. Sometimes, the extracted

Table 5.6 Some clean-up methods for pollutants

Method	Pollutants	Co-extractive	Nature
1. Partitioning			
(a) Liquid–liquid partitioning	OC and OP	Fats, waxes, lipids, pigments and some polar compounds	General
(b) Partitioning adsorption column	OC and OP	Fats, phospholipids, waxes and other polar substances	General
2. Gel permeation chromatography	OP	Fats and lipids	General
3. Liquid–liquid adsorption chromatography (adsorption column)	OC and OP	Fats, phospholipids, pigments and polar impurities	General
(a) Flirosil			
(b) Alumina			
(c) Silica gel			
(d) Nuchar carbon			
4. Chemical clean-up Acidic Alkaline Acid base	OC and OP	Fats and neutral molecules	Specific
5. Sweep co-distillation	OP	Comparatively nonvolatile impurities	–

OC and OP denote organochlorine and organophosphorous pesticides respectively.

and concentrated solvent is passed through a glass column that contains anhydrous sodium sulfate.

5.9 Conclusions

Sample preparation in the analysis of pollutants is very important, as thousands of impurities are present in natural samples. The quantitative values of the pollutants depend on the strategy adopted in sample preparation. It is not possible to get 100 % recovery of any pollutant from the sample by any method of sample preparation and, therefore, it is important to carry out spiking experiments are in sample preparation technologies. Briefly, to get maximum recovery of the pollutants, sample preparation should be handled very carefully. Nowadays, online micro-SPE methods are also available for the extraction and purification of environmental and biological samples. The online coupling of extraction and purification avoids the possibility of contamination. Moreover, time and costly chemicals can be saved by using online extraction technologies. However, the recovery problem still exists and more advanced modifications to these technologies are required.

References

1. H. Hühnerfuss and R. Kallenborn R., *J. Chromatogr. A* **580**, 191 (1992).
2. R. E. Majors, *LC–GC* **13**, 555 (1995).
3. C. K. Jain and I. Ali, *Int. J. Environ. Anal. Chem.* **68**, 83 (1997).
4. D. Martinez, M. J. Cugat, F. Borrull and M. Calull, *J. Chromatogr. A* **902**, 65 (2000).
5. S. Weigel, K. Bester and H. Hühnerfuss, *J. Chromatogr. A* **912**, 151 (2001).
6. J. L. Gomez-Ariza, E. Morales, I. Giraldez, D. Sanchez-Rodas and A. J. Velasco, *J. Chromatogr. A* **938**, 211 (2001).
7. R. C. Clark, M. Blumer and O. S. Raymond, *Deep Sea Res.* **14**, 125 (1967).
8. H. Gaul and U. Ziebarth, *Dtsch. Hydrogr. Z.* **36**, 191 (1983).
9. T. C. Sauer, G. S. Durrell, J. S. Brown, D. Redford and P. D. Boehm, *Mar. Chem.* **27**, 235 (1989).
10. K. S. Roinestad, J. B. Louis and J. D. Rosen, *J. AOAC Int.* **76**, 1121 (1993).
11. Ph. Quevauviller, *J. Chromatogr. A* **750**, 25 (1996).
12. R. Rubio, A. Padro, J. Alberti and G. Rauret, *Mikrochim. Acta* **109**, 39 (1992).
13. E. H. Larsen and S. H. Hansen, *Mikrochim. Acta* **109**, 47 (1992).
14. I. Tolosa, J. Dach and J. M. Bayon, *Mikrochim. Acta* **109**, 87 (1992).
15. C. LeBlanc, *LC–GC* **17**, S30 (1999).
16. K. Gangzler, A. Salgo and K. Valko, *J. Chromatogr. A* **371**, 299 (1986).

17. W. Freitag and O. Johan, *Die Angew. Makromol. Chem.* **175**, 181 (1990).
18. *US Pat.* 5,002,784 (1991), J. Pare, J. Lapointe and M. Sigouin.
19. C. Bichi, F. Beliarab and P. Rubiolo, *Lab. 2000* **6**, 36 (1992).
20. F. Onuska and K. Terry, *Chromatographia* **36**, 191 (1993).
21. V. Lopez-Avila, R. Young and W. Beckert, *Anal. Chem.* **66**, 1097 (1994).
22. M. I. Mattina, W. I. Berger and C. Denson, *J. Agric. Food Chem.* **45**, 4691 (1997).
23. M. Numata, T. Yarita, Y. Aoyagi and A. Takatsu, *Anal. Chem.* **75**, 1450 (2003).
24. S. Stout, A. Dacunha and M. Safarpour, *J. AOAC Int.* **80**, 426 (1997).
25. M. McNair, Y. Wang and M. Bonilla, *J. High Res. Chromatogr.* **20**, 213 (1997).
26. R. K. Smith, ed., *Handbook of Environmental Analysis*, 3rd edn, Genium, Schenectady, New York, 1997, p. 319.
27. T. F. Guerin, *J. Environ. Monit.* **1**, 63 (1999).
28. I. Ali, *J. Chem. Ed.* **73**, 285 (1996).
29. S. Arment, *LC–GC* **17**, S38 (1999).
30. B. E. Richter, J. L. Ezzel, D. Felix, K. A. Roberts and D. W. Later, *Am. Lab.* **27**, 24 (1995).
31. F. Höfler, J. Ezzel and B. E. Richter, *Labor Praxis* **19**, 58 (1995).
32. F. Höfler, J. Ezzel and B. E. Richter, *Labor Praxis* **19**, 62 (1995).
33. J. L. Ezzel, B. E. Richter, W. D. Felix, S. R. Black and J. E. Meikle, *LC–GC* **13**, 390 (1995).
34. B. E. Richter, J. L. Ezzel, D. Felix, K. A. Roberts and D. W. Later, *Int. Lab.* **25**, 18 (1995).
35. F. Höfler, D. Jensen, J. Ezzel and B. E. Richter, *Chromatographie* **1**, 68 (1995).
36. J. Ezzel and B. E. Richter, *Am. Environ. Lab.* **8**, 16 (1996).
37. B. E. Richter, B. A. Jones, J. L. Ezzell, N. L. Porter, N. Avdalovic and C. Pohl, *Anal. Chem.* **68**, 1033 (1996).
38. D. Jensen, F. Hoefler, J. J. Ezzell and B. E. Richter, *Polyarom. Compds* **9**, 233 (1996).
39. B. E. Richter, *LC–GC* **17**, S22 (1999).
40. J. M. Levy, *LC–GC* **17**, S14 (1999).
41. S. H. Page and M. L. Lee, *J. Microcol. Sepn* **4**, 261 (1992).
42. J. M. Levy, E. Storozynsky and M. Ashraf-Khorassani, in F. V. Bright an M. E. P. McNally, eds, *Supercritical Fluid Technology*, American Chemical Society, Washington, DC, 1992, p. 336.
43. J. M. Levy, A. C. Rosselli, E. Storozynsky, R. Ravey and M. Ashraf-Khorassani, *LC–GC* **10**, 386 (1992).
44. M. E. P. McNally and J. R. Wheeler, *J. Chromatogr.* **447**, 53 (1988).
45. J. E. France, J. W. King and J. M. Snyder, *J. Agric. Food Chem.* **39**, 1871 (1991).

46. M. L. Hopper and J. W. King, *J. Assoc. Off. Anal. Chem.* **74**, 661 (1991).
47. J. W. King, M. L. Hopper, R. G. Luchtefeld, S. L. Taylor and W. L. Orton, *J. AOAC Int.* **76**, 857 (1993).
48. D. R. Gere, C. R. Knipe, P. Castelli, J. Hedrick, L. G. Randall-Frank, H. Schulenberg-Schell, R. Schuster, L. Doherty, J. Orolin and H. B. Lee, *J. Chromatogr. Sci.* **31**, 246 (1993).
49. S. B. Hawthorne, D. J. Miller and M. S. Krieger, *J. Chromatogr. Sci.* **27**, 347 (1989).
50. V. G. Zuin, J. H. Yariwake and C. Bicchi, *J. Chromatogr. A* **985**, 159 (2003).
51. C. Sanchez-Brunete, L. Martinez and J. L. Tadeo, *J. Agric. Food Chem.* **42**, 2210 (1994).
52. G. H. Tan, *Analyst* **117**, 1129 (1992).
53. G. H. Tan and K. Vijayaletchumy, *Pestic. Sci.* **40**, 121 (1994).
54. M. J. Waliszcwski and G. A. Szymczynski, *J. Chromatogr.* **321**, 480 (1985).
55. J. L. Sericano and A. F. Pucci, *Bull. Environ. Contam. Toxicol.* **33**, 138 (1983).
56. F. Mangani, G. Crescentini and F. Bruner, *Anal. Chem.* **53**, 1627 (1981).
57. G. S. Raju, J. A. Millette and S. U. Khan, *Chemosphere* **26**, 1429 (1993).
58. C. G. Gomez, M. I. Arufe Martinez, J. L. Romeo Palanco, J. J. Gamero Lucas and M. A. Vizcaya Rojas, *Bull. Environ. Contam. Toxicol.* **55**, 431 (1995).
59. H. Faeber, S. Peldszus and H. F. Schoeler, *Vam. Wasser.* **76**, 13 (1991).
60. E. C. Goosen, R. G. Bunschoten, V. Engelen, D. de Jong and J. H. M. van den Berg, *J. High Res. Chromatogr.* **13**, 438 (1990).
61. L. J. Bernal, M. J. Del Nozal and J. J. Jiminez, *Chromatographia* **34**, 468 (1993).
62. N. L. Olson, R. Carrell, R. K. Cumming and R. Rieck, *LC–GC* **7**, 12 (1994).
63. L. M. Devi, M. Baldi, L. Pennazi and M. Liboni, *Pestic. Sci.* **35**, 63 (1992).
64. G. Durand, V. Bouvot and D. Barcelo, *J. Chromatogr. A* **607**, 319 (1992).
65. G. E. Miliadis, *Bull. Environ. Contam. Toxicol.* **54**, 837 (1995).
66. A. de Kok, M. van Opstal, T. de Jong, B. Hoogcarspel, R. B. Geerdink, R. W. Frei and U. A. Th. Brinkmann, *J. Environ. Anal. Chem.* **18**, 101 (1984).
67. G. Petrick, D. E. Schulz and J. C. Duinker, *J. Chromatogr.* **435**, 241 (1988).
68. S. Hartik and J. Tekel, *J. Chromatogr. A* **733**, 217 (1996).
69. H. Sabik, S. Cooper, P. Lafrance and J. Fournier, *Talanta* **42**, 717 (1995).
70. S. Chiron, S. Dupas, P. Scribe and D. Barcelo, *J. Chromatogr. A* **665**, 295 (1994).
71. T. H. M. Noij and M. M. E. van der Kooi, *J. High Res. Chromatogr. A* **18**, 535 (1995).
72. M. Shi, X. Zhu, G. Wang and Z. Hu, *Fenxi Huaxue* **17**, 299 (1989).
73. H. G. Stan, *J. Chromatogr. A* **643**, 227 (1993).
74. C. de la Colina, F. Sanchez-Raseor, G. D. Cancella, E. R. Toboada and A. Pena, *Analyst* **120**, 1723 (1995).

75. J. W. Readman, L. L. W. Kwong, D. Grondin, J. Bartocci, I. P. Villencuve and L. D. Mee, *Environ. Sci. Technol.* **27**, 1940 (1993).

76. G. E. Miliadis, *Bull. Environ. Contam. Toxicol.* **52**, 25 (1994).

77. R. Eisert and K. Levsen, *Fresenius J. Anal. Chem.* **351**, 555 (1995).

78. T. S. Thompson and L. Morphy, *J. Chromatogr. Sci.* **33**, 393 (1995).

79. P. Armishaw and R. G. Millar, *J. AOAC Int.* **76**, 1317 (1993).

80. J. Triska, *Chromatographia* **40**, 712 (1995).

81. C. Crespo, R. M. Marce and F. Borrull, *J. Chromatogr. A* **670**, 135 (1994).

82. M. S. M. Wells, D. D. Reimer and M. C. Wells-Knecht, *J. Chromatogr. A* **659**, 337 (1993).

83. D. Barcelo, S. Chiron, S. Lacorte, E. Martinez, J. S. Salau and M. C. Hennion, *Trends Anal. Chem.* **13**, 352 (1994).

84. T. Suzuki and S. Watanabe, *J. Chromatogr. A* **541**, 359 (1991).

85. M. Psathaki, E. Manoussaridou and E. G. Stephanou, *J. Chromatogr. A* **667**, 241 (1994).

86. M. Stahl, M. Luhrmann, H. G. Kicinski and A. Kettrup, *Z. Wasser. Abwasser. Forsch.* **22**, 124 (1989).

87. S. Hartik, J. Lehotany and J. Tekel, *J. High Res. Chromatogr. A* **17**, 756 (1994).

88. J. S. Salau, R. Alonso, G. Batllo and D. Barcelo, *Anal. Chim. Acta* **293**, 109 (1994).

89. S. Guenu and M. C. Hennion, *J. Chromatogr. A* **665**, 243 (1994).

90. M. Hiemstra and A. de Kok, *J. Chromatogr. A* **667**, 155 (1994).

91. S. Lacorte and D. Barcelo, *Anal. Chim. Acta* **296**, 223 (1994).

92. V. Pichon and M. C. Hennion, *J. Chromatogr. A* **665**, 269 (1994).

93. C. C. J. Land, *LC–GC* **7**, 205 (1994).

94. M. Adolfsson-Erici and L. Renberg, *Chemosphere* **23**, 845 (1991).

95. C. E. Goewie, P. Kwakman, R. W. Frei, U. A. Th. Brinkman, W. Maasfeld, T. Seshadri and A. Kettrup, *J. Chromatogr.* **284**, 73 (1984).

96. A. di Corcia and M. Marchetti, *J. Chromatogr. A* **541**, 365 (1991).

97. J. Beltram, F. J. Lopez and F. Hernandez, *Anal. Chim. Acta* **283**, 297 (1993).

98. M. W. Brooks, D. Tessier, D. Soderstrom, J. Jemkins and J. M. Clark, *J. Chromatogr. Sci.* **28**, 487 (1990).

99. O. Fiehn and M. Jekel, *J. Chromatogr. A* **725**, 85 (1996).

100. S. Lacorte and D. Barcelo, *J. Chromatogr. A* **725**, 85 (1996).

101. A. J. H. Louter, C. A. van Beekvelt, P. Cid, P. C. Montanes, J. Slobodnik, J. J. Vreuls and U. A. Th. Brinkman, *J. Chromatogr. A* **725**, 67 (1996).

102. M. Fernandez, M. Ibanez, Y. Pico and J. Manes, *Arch. Environ. Contam. Toxicol.* **35**, 377 (1998).

103. C. Aguilar, I. Ferrer, F. Borrul, R. M. Marce and D. Barcelo, *Anal. Chim. Acta* **386**, 237 (1999).

104. S. Chiron, E. Martinez and D. Barcelo, *J. Chromatogr. A* **665**, 283 (1994).

105. M. A. Blackburn, S. J. Kirby and M. J. Waldock, *Mar. Pollut. Bull.* **38**, 109 (1999).
106. J. L. Zhou, T. W. Fileman, W. A. House, J. L. A. Long, R. F. C. Mantoura, A. A. Meharg, D. Osborne and J. Wright, *Mar. Pollut. Bull.* **37**, 330 (1998).
107. T. I. R. Utvik, G. S. Durell and S. Johnsen, *Mar. Pollut. Bull.* **38**, 977 (1999).
108. M. T. Galceran and O. Jauregui, *Anal. Chim. Acta* **304**, 75 (1995).
109. J. C. Molto, C. Albelda, G. Font and J. Manes, *Int. J. Environ. Anal. Chem.* **41**, 21 (1990).
110. I. Tolosa, J. W. Readman and L. D. Mee, *J. Chromatogr. A* **725**, 93 (1996).
111. S. Lacorte, C. Molina and D. Barcelo, *Anal. Chim. Acta* **281**, 71 (1993).
112. L. Schmidt, J. J. Sun, J. S. Fritz, D. F. Hagen, C. G. Markell and E. F. Wisted, *J. Chromatogr. A* **641**, 57 (1993).
113. S. Kira, M. Sakano and Y. Nogami, *Bull. Environ. Contam. Toxicol.* **58**, 878 (1997).
114. C. Whang and J. Pawliszyn, *Anal. Commun.* **35**, 353 (1998).
115. J. M. Huen, R. Gillard, A. G. Mayer. B. Baltensperger and H. Kern, Abstracts, 7th Symposium on Handling of Environmental and Biological Samples in Chromatography, 7–10 May 1995, Lund, Sweden.
116. U. A. Th. Brinkmann and J. J. Vreuls, Abstracts, 7th Symposium on Handling of Environmental and Biological Samples in Chromatography, 7–10 May 1995, Lund, Sweden.
117. G. R. van der Hoff, S. M. Gort, R. A. Baumann, P. van Zoon and U. A. Th. Brinkmann, *J. High Res. Chromatogr.* **14**, 465 (1991).
118. A. Sosa, A. Lenardon and M. L. Mattassini, *J. Chromatogr. Sci.* **41**, 92 (2003).
119. G. R. B. Webster, K. N. Graham, L. P. Sarna and H. Ng, Abstracts, 7th Symposium on Handling of Environmental and Biological Samples in Chromatography, 7–10 May, 1995, Lund, Sweden.
120. S. Bengtsson and T. Berglöf, Abstracts, 7th Symposium on Handling of Environmental and Biological Samples in Chromatography, 7–10 May 1995, Lund, Sweden.
121. J. Czerwinski and J. Namiesnik, Abstracts, 7th Symposium on Handling of Environmental and Biological Samples in Chromatography, 7–10 May 1995, Lund, Sweden.
122. D. Barcelo, Abstracts, 7th Symposium on Handling of Environmental and Biological Samples in Chromatography, 7–10 May 1995, Lund, Sweden.
123. S. Emelianov and A. Tsyganov, Abstracts, 7th Symposium on Handling of Environmental and Biological samples in Chromatography, 7–10 May 1995, Lund, Sweden.
124. C. LeDoare and G. Durank, Abstracts, 7th Symposium on Handling of Environmental and Biological Samples in Chromatography, 7–10 May 1995, Lund, Sweden.

125. B. Grass and J. Nolte, Abstracts, 7th Symposium on Handling of Environmental and Biological Samples in Chromatography, 7–10 May 1995, Lund, Sweden.
126. I. Tolosa, J. W. Readman and L. D. Me, Abstracts, 7th Symposium on Handling of Environmental and Biological Samples in Chromatography, 7–10 May 1995, Lund, Sweden.
127. C. Molina, P. Grasso, E. Benfenati, A. Provini and D. Barcelo, Abstracts, 7th Symposium on Handling of Environmental and Biological Samples in Chromatography, 7–10 May 1995, Lund, Sweden.
128. A. Martin-Esteban, P. Fernandez and C. Camara, Abstracts, 7th Symposium on Handling of Environmental and Biological Samples in Chromatography, 7–10 May 1995, Lund, Sweden.
129. S. Guenu and M. C. Hennion, Abstracts, 7th Symposium on Handling of Environmental and Biological Samples in Chromatography, 7–10 May 1995, Lund, Sweden.
130. T. Renner, D. Baumgarten, K. H. Jensen and K. K. Unger, Abstracts, 7th Symposium on Handling of Environmental and Biological Samples in Chromatography, 7–10 May 1995, Lund, Sweden.
131. T. A. Pressley and J. E. Longbottom, *US Environmental Protection Agency, Off. Res. Dev. (Rep.)* **EPA-600/4-82-006**, 31 (1982).
132. S. Galassi, E. Gosso and G. Tartari, *Chemosphere* **27**, 2287 (1993).
133. J. Gandraß, G. Bormann and R. D. Wilken, *Fresenius J. Anal. Chem.* **353**, 70 (1995).
134. D. Barcelo, *Chromatographia* **25**, 928 (1988).
135. G. Durand, S. Chiron, V. Bouvot and D. Barcelo, *Int. J. Environ. Anal. Chem.* **49**, 31 (1992).
136. L. M. Smith, D. L. Stalling and J. L. Johnson, *Anal. Chem.* **56**, 1830 (1984).
137. G. P. Blanch, A. Glausch, V. Schurig, R. Serrano and M. J. Gonzalez, *J. High Res. Chromatogr.* **19**, 392 (1996).
138. L. Ramos, L. M. Hernandez and M. J. Gonzalez, *Anal. Chem.* **71**, 70 (1999).
139. B. Bethan, K. Bester, H. Hühnerfuss and G. Rimkus, *Chemosphere* **34**, 2271 (1997).
140. B. Bethan, W. Dannecker, H. Gerwig, H. Hühnerfuss and M. Schulz, *Chemosphere* **44**, 591 (2001).
141. S. B. Hawthorne, Abstracts, 7th Symposium on Handling of Environmental and Biological Samples in Chromatography, 7–10 May 1995, Lund, Sweden.
142. L. Karlsson, E. Björklund, M. Järemo, J. A. Jönsson and L. Mathiasson, Abstracts, 7th Symposium on Handling of Environmental and Biological Samples in Chromatography, 7–10 May 1995, Lund, Sweden.
143. J. E. Lawrence, Abstracts, 7th Symposium on Handling of Environmental and Biological Samples in Chromatography, 7–10 May 1995, Lund, Sweden.

144. M. Notar and H. Leskovsek, Abstracts, 7th Symposium on Handling of Environmental and Biological Samples in Chromatography, 7–10 May 1995, Lund, Sweden.
145. J. S. Ho, P. H. Tan, J. W. Eichelberger and W. L. Budde, *Natl. Meet. Am. Chem. Soc., Div. Environ. Chem.* **33**, 313 (1993).
146. I. J. Barnabas, J. R. Dean, S. M. Hitchen and S. P. Owen, *J. Chromatogr. A* **665**, 307 (1994).
147. J. Hedrick and L. T. Taylor, *Anal. Chem.* **61**, 1986 (1989).
148. I. J. Barnabas, J. R. Dean, S. M. Hitchen and S. P. Owen, *Anal. Chim. Acta* **291**, 261 (1994).
149. D. Puig and D. Barcelo, *J. Chromatogr. A* **673**, 55 (1994).
150. R. Alzaga, G. Durand, D. Barcelo and M. J. Bayona, *Chromatographia* **38**, 502 (1994).
151. S. Bengtsson and A. Ramberg, *J. Chromatogr. Sci.* **33**, 554 (1995).
152. E. Dabek-Zlotorzynska, R. Aranda-Rodriguez and K. Keppel-Jones, *Electrophoresis* **22**, 4262 (2001).
153. P. R. Haddad, P. Doble and M. Macka, *J. Chromatogr. A* **856**, 145 (1999).
154. J. S. Fritz and M. Macka, *J. Chromatogr. A* **902**, 137 (2000).
155. S. Pedersen-Bjegaard, K. E. Rasmussen and T. G. Halvorsen, *J. Chromatogr. A* **896**, 95 (2000).
156. R. E. Major, *LC–GC* **13**, 364 (1995).
157. M. Knutsson, L. Mathiasson and J. A. Jönnson, Abstracts, 7th Symposium on Handling of Environmental and Biological Samples in Chromatography, 7–10 May 1995, Lund, Sweden.
158. Y. Shen, V. Obuseng, L. Grönberg and J. A. Jönsson, Abstracts, 7th Symposium on Handling of Environmental and Biological Samples in Chromatography, 7–10 May 1995, Lund, Sweden.
159. T. Buttler, Abstracts, 7th Symposium on Handling of Environmental and Biological Samples in Chromatography, 7–10 May 1995, Lund, Sweden.
160. R. E. Majors, *LC–GC* **13**, 542 (1995).
161. J. Faller, H. Hühnerfuss, W. A. König, R. Krebber and P. Ludwig, *Environ. Sci. Technol.* **25**, 676 (1991).
162. S. Bengtsson, T. Bergloef, S. Grant and G. Jonsaell, *Pestic. Sci.* **41**, 55 (1994).
163. S. Chiron, A. Fernandez-Alba and D. Barcelo, *Environ. Sci. Technol.* **27**, 2352 (1993).
164. L. J. Ezzell and B. E. Richter, *J. Microcol. Sepn.* **4**, 319 (1993).
165. O. Heemken, N. Theobald and B. Wenclawiak, *Anal. Chem.* **69**, 2171 (1997).
166. J. R. Dean, *Anal. Commun.* **33**, 191 (1996).
167. J. R. Dean, A. Santamaria-Rekondo and E. Ludkin, *Anal. Commun.* **33**, 413–416 (1996).
168. N. Saim, J. R. Dean, M. P. Abdullah and Z. Zakaria, *J. Chromatogr. A* **791**, 361 (1997).

169. K. Li, M. Landriault, M. Fingas and M. Llompart, in *Proceedings of the 15th Technical Seminar on Chemical Spills*, Environment Canada, Ottawa, Ontario, 1998, pp. 115–128.
170. M. M. Schantz, J. J. Nichols and S. A. Wise, *Anal. Chem.* **69**, 4210 (1997).
171. J. L. Ezzell, B. E. Richter and E. Francis, *Am. Environ. Lab.* **8**, 12 (1996).
172. K. D. Wenzel, A. Hubert, M. Manz, L. Weissflog, W. Engewald and G. Schü-ürmann, *Anal. Chem.* **70**, 4827 (1998).
173. J. Fischer, M. Scarlett and A. Stoit, *Environ. Sci. Technol.* **31**, 1120 (1997).
174. B. Debrunner, *Chimia* **50**, 305 (1996).
175. S. J. Lehotay and C. H. Lee, *J. Chromatogr. A* **785**, 313 (1997).
176. J. Ezzell, *Am. Environ. Lab.* **10**, 24 (1998).
177. H. Obana, K. Kikuchi, M. Okihashi and S. Hori, *Analyst* **122**, 217 (1997).

Chapter 6

The Analysis of Chiral Pollutants by Gas Chromatography

6.1 Introduction

Nowadays, there are many methods for the determination of the enantiomeric ratios of chiral pollutants. The most important techniques used for this purpose have been discussed in Chapter 1. Among these, gas chromatography has also been used for chiral resolution of the environmental pollutants. Chiral recognition in GC was first utilized for the configurational analysis of amino acids in 1966, by Gil-Av *et al.* [1], and since then GC has been used for chiral resolution of various drugs, pharmaceuticals, agrochemicals and environmental pollutants. In spite of the development of several techniques (LC, CE etc.) for enantiomeric resolution of chiral pollutants, GC still maintains its status in this field. This is due to the fact that most of the environmental pollutants and their derivatives are volatile at the working temperature of GC. Moreover, some of the pesticides are transparent to UV radiation and, therefore, electron capture detector (ECD), which is used only in GC, is the best choice for the detection of such types of pesticides. The most notable advantages of GC are that the relative peak areas of the enantiomers or the enantiomeric ratios can be determined with high precision (relative standard deviations generally better than $\pm 2\%$), whereas the precision of quantification in organic trace analysis is typically 15–20%. The variations in the recovery rates do not affect the

Chiral Pollutants. I. Ali and H. Y. Aboul-Enein
© 2004 John Wiley & Sons, Ltd ISBN: 0-470-86780-9

determination of the ER values [2]. Chiral resolution of the environmental pollutants can be carried out using chiral stationary phases (CSPs). Various CSPs have been used for this purpose and these phases will be discussed in detail in Section 6.2. Since GC can analyse only volatile compounds, the derivatization of some chiral pollutants is required before GC analysis takes place. Because of the large number of publications available in this field, it is not possible to comment on all of them. However, the intention of this chapter is to discuss the art of chiral resolution of the environmental pollutants by GC and its current status.

6.2 Chiral Selectors

Chiral resolution in GC is carried out using chiral stationary phases (CSPs); that is, GC columns containing chiral compounds. Many chiral compounds are used for the preparation of chiral stationary phases. The most commonly used chiral compounds are cyclodextrins (CD) and their derivatives. The merit of CD derivatives for chiral resolution in GC is the great spectrum of resolvable classes of compounds. The modified CDs are capable of resolving chiral pollutants over a high range of GC temperatures.

As pointed out earlier, CDs are the most important and most suitable chiral selectors in GC. In 1983, Koscielski *et al.* [3] reported chiral resolution by GC on cyclic glucose oligomers, so-called cyclodextrins. Cyclodextrins (CDs) are cyclic and nonreducing oligosaccharides, obtained from starch. Villiers [4] described CDs in 1891, and called them cellulosine. Subsequently, Schardinger [5] identified three different forms of naturally occurring CDs – that is, α-, β- and γ-CDs – and referred to them as Schardinger's sugars. These are also known as cyclohexamylose (α-CD), cycloheptamylose (β-CD), cycloctamaylose (γ-CD), cycloglucans, glucopyranose and Schardinger dextrins. From 1911 to 1935, Pringsheim, in Germany, carried out an extensive research work and reported that these sugars formed stable complexes with many compounds [6]. By the mid-1970s, each of the naturally occurring CDs had been structurally and chemically characterized, and many more complexes had been studied [7]. Therefore, the ability of CDs to form complexes with a wide variety of molecules has been documented [8–13]. The formation of complexes and their binding constants are determined and controlled by several different factors; namely, hydrophobic interactions, hydrogen bondings and van der Waals interactions. Therefore, CDs and their derivatives have been widely used in separation science since the early 1980s [7, 14]. The evolution of

CDs as chiral selectors in the gas and liquid chromatographic separations of enantiomers has become a subject of interest over the past two decades. The presence of the chiral hollow basket/cavity of these molecules makes them suitable for chiral resolution of a wide range of racemic compounds. At present, the use of CDs as chiral selectors for enantiomeric resolution by gas chromatography is very common. As chiral selectors, CDs have been used in the form of chiral stationary phases (CSPs).

6.2.1 Structures and Properties

CDs are cyclic oligosaccharides comprising 6–12 D-(+)-glucopyranose units in α-(1,4) linkage with chair conformation. These are three types that have different numbers of glucopyranose units; that is, α-, β- and γ-CDs containing six, seven and eight glucopyranose units, respectively. The secondary 2- and 3-hydroxyl groups line the mouth of the CD cavity, while the primary 6-hydroxyl groups are on the opposite end of the molecules. This arrangement makes CD cavity hydrophilic in nature, which results in the formation of inclusion complexes with a variety of molecules. However, the interior of the cavity is slightly hydrophobic in nature. Many derivatives of CDs have been synthesized with various derivatizing groups such as acetyl, hydroxypropyl, naphthyl, ethylcarbamoyl, 3,5-dimethylphenyl carbamoyl, tosyl, trimethyl, dimethyl, phenyl carbamoyl, propyl carbamoyl and aminomethylbenzyl. These derivatives modify the hydroxyl groups of CDs without destroying the CD cavity. Derivatized CDs exhibit significantly different behaviour as compared to native CDs. The derivatization results in partial blocking of the mouth of the CD cavity and eliminates some or all of the hydrogen bondings, leading to a significant difference in enantioselectivity. The internal diameter of these CDs varies from α to γ due the presence of a different number of glucopyranose units. The most stable three-dimensional molecular configuration for these nonreducing cyclic oligosaccharides takes the form of a toroid, with the upper (larger) and lower (smaller) opening of the toroid presenting secondary and primary hydroxyl groups, respectively. The interior of the toroid is hydrophobic as a result of the electron-rich environment provided in large part by the glycosidic oxygen atoms. CDs are highly stable in acidic and basic conditions. The chemical structures of CDs and their derivatives are shown in Figure 6.1.

As discussed above, CDs are capable of forming inclusion complexes with a variety of molecules. The stoichiometries (guest : CD) of CDs are 1 : 1, 1 : 2 and 2 : 1, depending on the structures of the molecules [15].

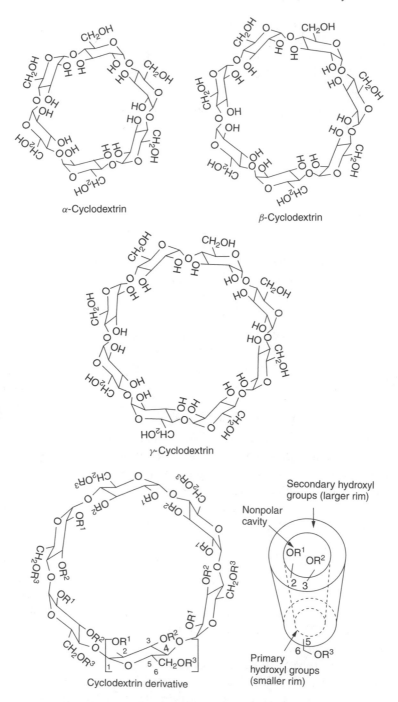

Figure 6.1   The chemical structures of cyclodextrins and their derivatives.

Table 6.1 The Properties of CDs

Property	α-CD	β-CD	γ-CD
Number of glucopyranose units	6	7	8
Molecular weight	972	1135	1297
pKa	12.33	12.20	12.08
Inner diameter (nm)	0.45–0.57	0.62–0.78	0.79–0.95
Outer diameter (nm)	1.37	1.53	1.69
Height (nm)	79	79	79
Cavity volume (nm^3)	0.174	0.262	0.472
Cavity volumes (per ml)	104	157	256
(per g)	0.10	0.14	0.20
Solubility in water (g ml^{-1})	14.5	1 85	23.2
Chiral centres	30	35	40
$[\alpha]_D^{25}$	+150.5	+162.0	+177.4

Complexation in CDs is molecule specific and generally ternary complexes are favoured. However, α-CD includes single phenyl and naphthyl groups (small molecules), β-CD accepts naphthyl and heavily substituted phenyl groups, and γ-CD attracts bulky steroid type molecules. The stability constants of complex formation vary from 10^{-2} to 10^{-5} M^{-1} with a formation time $T_{1/2} \sim 0.001 - 1.0$ ms. These complex formation properties of CDs have been utilized for chiral resolution of enantiomers, as the two enantiomers form inclusion diastereoisomeric complexes. Other properties of CDs are given in Table 6.1.

6.2.2 Preparation and Commercialization

CDs are produced from starch by the action of *Bacillus macerans amylase* or the enzyme cyclodextrin transglycosylate (CTG) [16–19]. The latter enzyme can be used to produce CDs of specific sizes by controlling the reaction conditions. In the past few years, enantiomers have been resolved using peralkylated α-, β- and γ-CDs dissolved in polysiloxanes and coated within glass or fused silica capillary tubing [20, 21]. Subsequently, the CDs were linked to the solid supports. In 1979, Harada *et al.* [22] polymerized and crosslinked a CD with a gel support, and the CSP developed was tested for the chiral resolution of mandelic acid and its derivatives. Various workers have subsequently bonded all three CDs with different solid supports [23–33]. Of course, these CDs bonded to gel support have been used for the chiral resolution of different racemates, but they suffer from certain drawbacks because of their poor mechanical strength and efficiency in both GC and HPLC. An improvement to these

CSPs was introduced through covalent bonding between CDs and silica gel via propylamine, ethylene diamine and other related linkages [34–39]. Again, these developed phases have not been able achieve the status of the best CSPs due to some associated problems. The major drawbacks related to these CSPs were the tedious synthesis, involving the formation of nitroxide in the reaction mixture, their poor loading capacities, their hydrolytic instability and the effect of amine on enantioseparation. These drawbacks were overcome by preparing the more hydrolytically stable CSPs [40–55] on which most chiral resolutions have been carried out. The development of derivatized CD-based CSPs was a major step forward in the field [56–64]. The biggest advantage of these CSPs is the versatility that they offer, being capable of use over a high range of temperatures.

The three CDs and their derivatives can be bonded to silica gel using silane reagents. The most commonly used silane reagents are 3-glycidoxypropyl trimethoxysilane, 3-glycidoxypropyldimethylchlorosilane and 3-glycidoxypropyltriethoxysilane. The preparation of CD-based CSPs is a two-step process. The silane reagent is attached to silica gel (spherical with 5 μm as particle diameter) by refluxing in dry toluene at 95 °C for 3 h. The mixture is then cooled, filtered, washed several times with toluene and methanol, and then dried at 60 °C over phosphorus pentaoxide under vacuum. In the second step, the treated silica gel is allowed to react with CD in pyridine at 80 °C [41, 65]. The chemical pathways for these reactions are shown in Figure 6.2. The synthetic procedure varies slightly depending upon the use of a silane reagent and the type of CD. From the theoretical point of view, the linkage can occur through either the primary or secondary hydroxyl groups on the CDs, but in practice primary hydroxyl groups are more reactive since they are less sterically hindered. If more than one chloro- or methoxy- group is present in the silane reagent, it is possible that more than one bonding link may occur between the silane reagent and the silica gel. A CD has many free hydroxyl groups and, therefore, it may bond with silica gel through more than one silane bond. The high stability of CD bonded phases supports the binding of the CD through more than one bond, but no experimental support in this regard has been presented thus far. At present, CD-based CSPs are one of the most inexpensive tools for chiral resolution purposes.

6.2.3 Other Chiral GC CSPs

König *et al.* [66–68] separated acylated-α-hydroxy acid esters on (*S*)-α-hydroxyalkanoic acid-(*S*)-α-phenylethylamide based CSPs. However, the

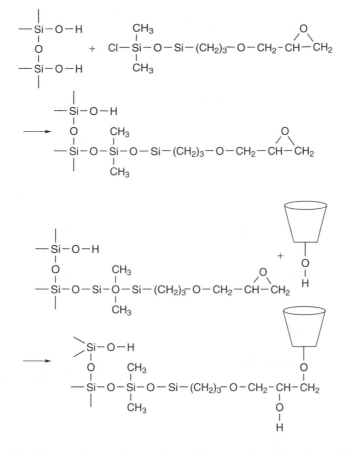

Figure 6.2 The chemical pathway of immobilization of cyclodextrins on silica gel.

thermal instability of these phases was poor and hence they could not be used on a commercial level. Furthermore, Oi *et al.* [69–71] used low molecular weight stationary phases for the separation of hydroxy acids and other compounds. Frank *et al.* [72, 73] copolymerized dimethylsiloxane with (2-carboxypropyl)methylsiloxane and, further, coupled the carboxy-group with the amino group of L-valine-*tert*-butylamide, a polysiloxane with a chiral side chain. This chiral polymer was marketed under the trade name of Chirasil-Val[®] and showed a wide range of applications at high temperature (up to 200 °C) [74–76]. On this concept, however, a different approach was adopted by Saeed *et al.* [77, 78]: after hydrolysis of the cyanoalkyl chain of the polysiloxanes OV-225 or Silar-10C, they obtained chiral polymers by coupling L-valine-*tert*-butylamide to the carboxylic groups. However, according to König [79], the temperature stability and selectivity towards amino acid enantiomers were not quite as good as in the case of Chirasil-Val®. Similarly, König and Benecke [80] functionalized cyanoalkylpolysiloxanes by reducing the nitrile group to aminomethyl groups and coupling *N*-acetylated amino acids. More versatility in applications was achieved by preparing chiral polymers by modifying the polysiloxane XE-60 [68, 81, 82]. Fused silica capillary columns with Chirasil-Val® and XE-60-L-valine-(*S*)-α-phenylethylamide have become commercially available. In 1985, Schomburg *et al.* [83] crosslinked Carbowax 20M and acryloyl-L-valine-(*S*)-α-phenylethylamide, which was immobilized inside the capillary columns. The CSP developed was suitable for the chiral separation of many compounds but, again according to König [79], the separation factors (for a variety of racemates) were poor for this phase. Schurig *et al.* [84, 85] utilized the concept of the association of chiral molecules with chiral transition metal ion complexes, and hence developed CSPs based on diamide phases for the chiral resolution of various pollutants.

In 1980, König *et al.* [67] developed chiral stationary phases from mandelic acid and α-phenylethylamine. The Z-derivative of mandelic acid was prepared by the Thamm reaction [86] and coupled with α-phenylethylamine after the formation of mixed anhydride with chloro-carbonic acid ethyl ester. Analogously, Z-mandelic acid was coupled with *tert*-butylamine, L-valine cyclohexyl ester, L-valine *tert*-butylamide and *R*-phenylglycinol. The structures of these chiral compounds are given in Figure 6.3. The compound XI was prepared by esterification of *S*-mandelic acid, with cyclohexanol and *p*-toluenesulfonic acid as catalysts. Compound VII was obtained from *S*-mandelic acid by reaction with dodecanoyl chloride and coupling with *tert*-butylamine. Compounds X and I were obtained from Z-*S*-mandelic acid with *N'*,*N*-carbonyldiimidazole, respectively. The authors resolved the enantiomers of α-substituted carboxylic acids. Furthermore, the same group [68] developed chiral stationary phases from *S*-2-hydroxyisopentanoic acid by coupling with *S*-α-phenylethylamine. The structures of the CSPs developed are given in Figure 6.4. The authors reported the chiral resolution of hydroxy acids. It is worth mentioning here that these chiral selectors are not stable over a high range of temperatures and hence are not in common use.

In 1988, Macherey-Nagel, of Düren, Germany, introduced the commercial GC CSPs under the trade name Lipodex [87]. Subsequently, in 1990, Armstrong and coworkers [88–91] coated polar CD derivatives on a fused silica column: this technique has been commercialized by Advanced

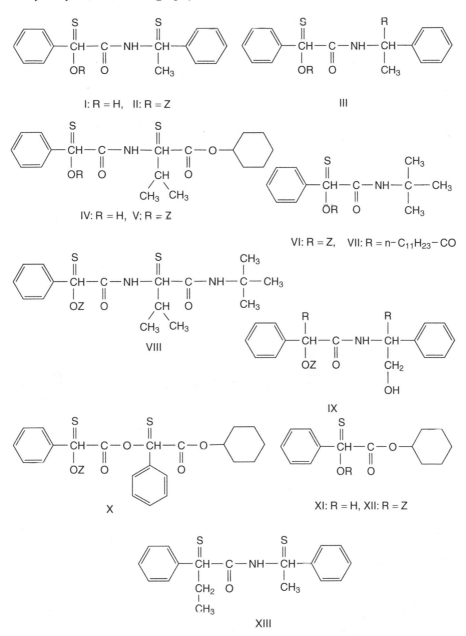

I: R = H,   II: R = Z

III

IV: R = H, V; R = Z

VI: R = Z,   VII: R = n–C₁₁H₂₃–CO

VIII

IX

X

XI: R = H, XII: R = Z

XIII

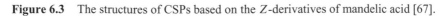

Figure 6.3   The structures of CSPs based on the *Z*-derivatives of mandelic acid [67].

No.	Name	Formula

I   *O*-Benzyloxycarbonyl-*S*-3-ph-enyllactic acid-*tert*-butylamide

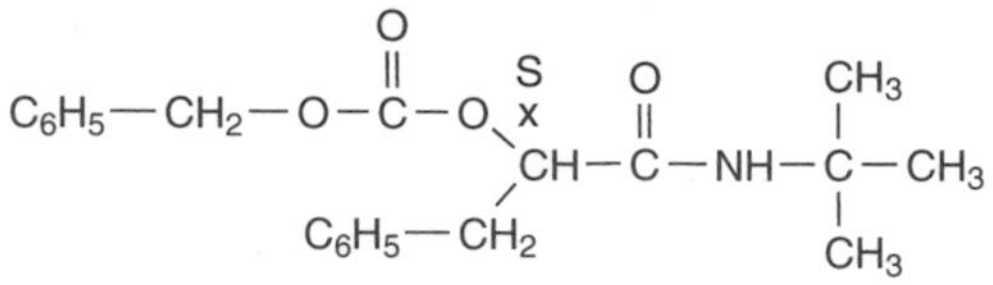

$$C_6H_5-CH_2-O-\overset{\overset{\text{O}}{\|}}{C}-O\underset{CH-\overset{\overset{\text{O}}{\|}}{C}-NH-\overset{CH_3}{\underset{CH_3}{|}}{C}-CH_3}{\overset{\text{S}}{\underset{C_6H_5-CH_2}{\text{x}}}}$$

II   *S*-2-Hydroxyisopentanoic acid-*S*-α-phenylethylamide

$$CH_3-CH-\overset{\overset{\text{S}}{\text{x}}}{\underset{OH}{CH}}-\overset{\overset{\text{O}}{\|}}{C}-NH-\overset{\overset{\text{S}}{\text{x}}}{\underset{CH_3}{CH}}-C_6H_5$$
$$\underset{CH_3}{|}$$

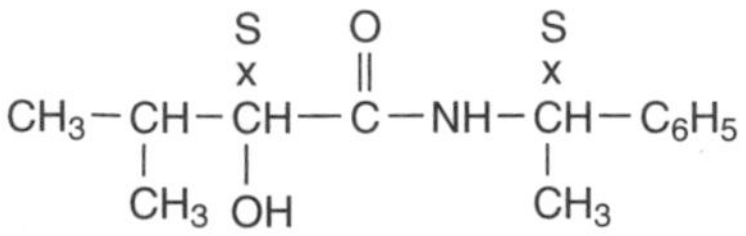

III   *S*-2-Hydroxyoctanoic acid-*S*-α-phenylethylamide

$$CH_3-(CH_2)_5-\overset{\overset{\text{S}}{\text{x}}}{\underset{OH}{CH}}-\overset{\overset{\text{O}}{\|}}{C}-NH-\overset{\overset{\text{S}}{\text{x}}}{\underset{CH_3}{CH}}-C_6H_5$$

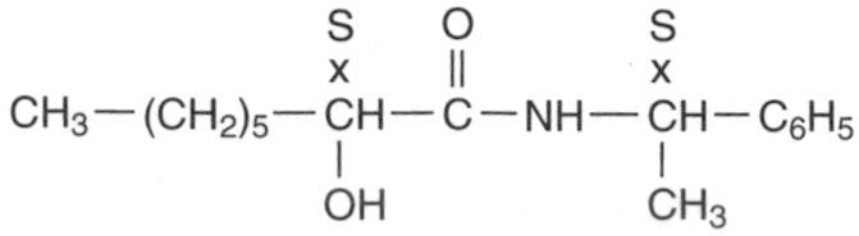

IV   XE-60-*S*-valine-*S*-α-phenylethylamide

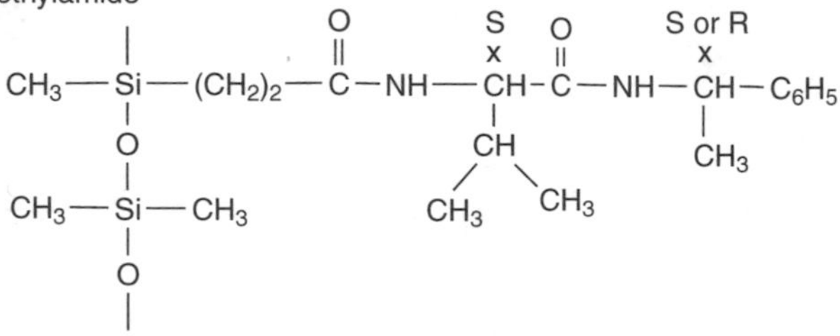

$$CH_3-\overset{\overset{|}{Si}}{\underset{O}{|}}-(CH_2)_2-\overset{\overset{\text{O}}{\|}}{C}-NH-\overset{\overset{\text{S}}{\text{x}}}{\underset{CH}{CH}}-\overset{\overset{\text{O}}{\|}}{C}-NH-\overset{\overset{\text{S or R}}{\text{x}}}{\underset{CH_3}{CH}}-C_6H_5$$
$$CH_3-\underset{O}{\overset{|}{Si}}-CH_3 \qquad \overset{}{\underset{CH_3 \quad CH_3}{CH}}$$

V   XE-60-*S*-valine-*R*-α-phenylethylamide

Figure 6.4   The structures of CSPs based on *S*-α-phenylethylamide [68].

Separation Technologies Inc., of Whippany, NJ, USA. In 1991, the same company received two global patents for the use of two specific cyclodextrin derivatives for the separation of enantiomers by capillary GC. This technology is known by the trade name of Chiraldex. Various Chiraldex columns have been developed, using the prefixes A, B and G for the α-, β- and γ-CDs, respectively [92]. Similarly, Supelco have also developed various chiral stationary phases under the trade name of Dex: α-Dex, β-Dex and γ-Dex CSPs on fused silica capillary columns containing premethylated α-, β- and γ-CDs coupled to a phenyl that contains a stationary polysiloxane

Table 6.2   Various cyclodextrin-based commercial GC CSPs

Commercialized CSP	Company
Lipodex A [heptakis(2,3,6-tri-O-n-pentyl)-α-CD] Lipodex B [heptakis(3-O-acetyl-2,6-di-O-n-pentyl)-α-CD] Lipodex C [heptakis(2,3,6-tri-O-n-pentyl)-β-CD] Lipodex D [heptakis(3-O-acetyl-2,6-di-O-n-pentyl)-β-CD] Lipodex E [octakis(3-O-butyryl-2,6-di-O-n-pentyl)- γ-CD]	Macherey-Nagel, Düren, Germany
PMHP-α-CD [heptakis(O-(S)-2-hydroxypropyl)-per-O- methyl)-α-CD] Dipentyl-α-CD [heptakis(2,6-di-O-n-pentyl)-α-CD] PMHP- β-CD [heptakis(O-(S)-2-hydroxypropyl)-per-O-methyl)-β-CD Dipentyl-β-CD [heptakis(2,6-di-O-n-pentyl)-β-CD] DPTFA-β-CD [heptakis(2,6-di-O-n-pentyl-3-O-trifluoroacetyl)-β-CD]	Advanced Separation Technologies, Whippany, NJ, USA
Chiraldex G-TA (γ-CD, trifluoroacetyl) Chiraldex A-PH (β-CD, hydroxypropyl) Chiraldex B-PH (β-CD, hydroxypropyl) Chiraldex G-PH (β-CD, hydroxypropyl) Chiraldex A-DA (β-CD, dialkyl) Chiraldex B-DA (β-CD, dialkyl) Chiraldex G-DA (β-CD, dialkyl) Chiraldex G-NP (β-CD, propionyl) Chiraldex G-BP (β-CD, butyryl) Chiraldex B-PM (β-CD, permethylated) Chiraldex B-DM (β-CD, dimethyl)	Advanced Separation Technologies, Whippany, NJ, USA
α-Dex 120 (α-CD) β-Dex 110 (β-CD) β-Dex 120 (β-CD) γ-Dex 120 (γ-CD)	Supelco, Bellefonte, PA, USA

co-phase [93]. Some other phases have also been commercialized, such as Chirasil-Dex columns [94–96]. Some of the commercialized CD-based columns produced by various companies are listed in Table 6.2.

6.3 Applications

As discussed earlier, Gil-Av *et al.* [1, 97] in 1966 were the first to use GC for the configurational analysis of N-trifluoro-acetylated (N-TFA) amino acid esters, applying acetylated amino acids and dipeptides as chiral selectors in glass capillary columns. The GC technique was subsequently used for the chiral resolution of drugs, pharmaceuticals, agrochemicals and environmental pollutants [79, 87, 92, 98–100]. The chiral resolution of environmental pollutants was achieved on bonded or coated CSPs in the form of capillary columns. Cyclodextrins have frequently been used for this purpose. However, some other chiral compounds were also used as chiral selectors in GC. The contribution of the groups of Kallenborn and Hühnerfuss to the field of chiral resolution of environmental pollutants by GC is significant. These groups analysed the chiral ratios of the various environmental pollutants in different components of the marine and aqueous ecosystems and in organisms.

Kallenborn *et al.* [101] resolved the enantiomers of α-HCH on heptakis(3-O-butyryl-2,6-di-O-pentyl)-β-CD as the CSP, using helium as the mobile phase at 100 kPa. Furthermore, the same group [102] resolved the enantiomers of α-HCH in North Sea water samples. The authors used heptakis(3-O-butyryl-2,6-di-O-pentyl)-β-CD as the CSP and helium (100 kPa) as the mobile phase. Faller *et al.* [103] separated the enantiomers of α-HCH in water samples representing different parts of the North Sea on heptakis(3-O-butyryl-2,6-di-O-pentyl)-β-CD as CSP and helium as the mobile phase (100 kPa). The correlation between the enantiomeric ratios and the concentrations of α-HCH and γ-HCH allowed a tentative characterization of the different microbiological degradation pathways in the North Sea. In the plume of the River Elbe, Skagerrak Water and the Norwegian coastal current, higher γ-HCH concentrations correlated with an excess of $(-)$-α-HCH, which might in part be indicative of a transformation of γ-HCH into α-HCH. Ludwig *et al.* [104] resolved the enantiomers of γ-PCCH and β-PCCH that were formed during the partial degradation of the prochiral pesticide γ-HCH and chiral α-HCH, respectively, by microbial degradation. The authors used CD and helium as chiral stationary and mobile phases, respectively. Hühnerfuss *et al.* [105] reported the chiral resolution of some xenobiotics and their derivatives in the marine ecosystem. The authors reported the chiral separation of α-HCH, β-PCCH and γ-PCCH on cyclodextrin-based CSP (coated on 30 m DB608 and 25 m NB-54 HNU capillary columns) in water samples collected from 14 sampling stations in the Baltic and North Seas, using the mobile phase of helium (150 kPa). Furthermore, the same group [106] investigated the chiral ratio of α-HCH in various regions of the marine ecosystem, using CD and helium as the CSP and mobile phase, respectively. The authors used a silica column (25 m $\times$ 0.25 mm id) coated with a 50 : 50 mixture of Lipodex C [heptakis(2,3,6-tri-*O*-n-pentyl)-*β*-CD] and OV-1701. In another study, Hühnerfuss *et al.* [107] determined the enantiomeric ratios of *α*-HCH, *β*-PCCH and *γ*-PCCH, *trans*-chlordane, *cis*-chlordane, octachlordane MC4, octachlordane MC5, octachlordane MC7, octachlordane and heptachlor epoxide in various marine ecosystem components. The authors used a fused silica capillary column coated with 50 % heptakis(2,3,6-tri-*O*-n-pentyl)-*β*-CD and 50 % OV-1701, with helium (45 kPa) as the carrier gas. Moisey *et al.* [108] determined the concentrations of HCH isomers (*α*-, *β*- and *γ*-) and the enantiomeric ratios of *α*-HCH in the Northwater Polynya Arctic marine food web. The isomeric and enantiomeric concentrations of HCH were determined in water, sediment, benthic invertebrates, pelagic zooplankton, Arctic cod (*Boreogulus salda*), sea birds and ringed seals (*Phoca hispida*). The authors used *β*-DEX 120 (20 % nonbonded permethylated *β*-CD) as the CSP. Furthermore, the same group [109] determined the chiral ratios of various organochlorines in the ringed seal. The chiral pollutants determined were *α*-HCH, PCBs, *cis*- and *trans*-chlordane, oxychlordane and heptachlor epoxide. The authors used a fused silica column (0.25 mm id) of *β*-DEX 120 (20 % nonbonded permethylated *β*-CD). Bethan *et al.* [110] investigated chiral concentrations of *α*-HCH in seawater, air and bulk deposition samples from the North Sea, using a CD-based CSP.

Buser and Müller [111] resolved the enantiomers of chlordane by chiral high resolution chromatography. Octa- and nonachlordane, oxychlordane, heptachlor *exo*- and *endo*-epoxide were analysed using two *β*-CD derivatives, dissolved in polysiloxanes, as the CSPs. Silylated *β*-CD showed an increased enantiomer resolution as compared to permethylated *β*-CD. However, silylated *β*-CD was less suitable for the analysis of a technical chlordane mixture, because of co-elution of the enantiomers of different major octachlordanes. Furthermore, the same authors [112] reported a chiral resolution method for chlordane and its metabolites using high resolution gas chromatography with air and water samples. *α*- and *β*-CDs on silicon columns were used as CSPs. Eitzer *et al.* [113] analysed the chiral and compositional profiles of *cis*- and *trans*-chlordane in soil samples collected from New Haven, Connecticut, USA, using chiral gas chromatography. The authors used a commercial *γ*-DEX 120 column for the chiral resolution. Furthermore, the same group [114] studied the compositional and chiral profiles of weathered chlordane residues in soil and vegetative components using the *γ*-DEX 120 column. Fisk and coworkers [115] determined the chiral concentrations of chlordane and its metabolites in seven species of Arctic sea birds from the Northwater Polynya (NOW), which is a large area of year-round open water, found in the High Arctic between Ellesmere

Island and Greenland. The NOW has high biological productivity compared with other Arctic marine areas, and supports large populations of several seabird species. Seven species of sea birds, dovekie (*Alle alle*), thick-billed murre (*Uria lomvia*), black guillemot (*Cepphus grylle*), black-legged kittiwake (*Rissa tridactyla*), ivory gull (*Pagophila eburnea*), glaucous gull (*Larus hyperboreus*) and northern fulmar (*Fulmaris glacialis*), were collected in May and June 1998 to determine chlordane concentrations in the liver and fat, and to examine species differences, relationships with stable isotopes of nitrogen and the enantiomeric fractions (EFs) of the chiral components. The CSP used for these studies was β-DEX 120.

Schurig and Glausch [94] resolved the enantiomers of several PCBs (84, 91, 95, 132, 136 and 149) on a Chirasil-Dex GC column (Figure 6.5). Furthermore, the same groups [95] resolved PCBs 45, 95 and 139 on octakis(2,6-di-O-methyl-3-O-n-pentyl)-γ-CD diluted in a 1 : 1 ratio with OV-1701 and PCB congeners 91, 95, 132, 136 and 149 on an immobilized Chirasil-Dex column. Hydrogen was used as the mobile phase, with 1.0 and 0.4 bar as the pressure for the first and second columns, respectively. Wong and Garrison [116] investigated seven commercially available chiral capillary gas chromatography columns containing modified cyclodextrins for their ability to separate the enantiomers of 19 stable chiral polychlorinated biphenyl (PCB) atropisomers, and for their ability to separate these enantiomers from achiral congeners, which is necessary in order to trace the environmental analysis of chiral PCBs. The enantiomers of each of the 19 chiral PCBs were at least partially separated on one or more of these columns. The enantiomeric ratios of 11 atropisomers could also be quantified on six columns, as they did not co-elute with any other congener containing the same number of chlorine atoms, and thus could be

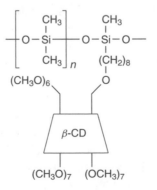

Figure 6.5   The structure of Chirasil-Dex CSP [94].

quantified using gas chromatography – mass spectrometry. Analysis of the lake sediment, which was heavily contaminated with PCBs, showed enantioselective occurrence of PCB 91, proof positive of enantioselective *in situ* reductive dechlorination at the sampling site. Blanch *et al*. [96] resolved the enantiomers of PCBs 132 and 149 in Atlantic Ocean shark (*Centroscymnus coelolepis*) liver using a Chirasil-Dex column. Ramos *et al*. [117] resolved the enantiomers of PCBs 45, 84, 88, 91, 95, 131, 132, 135, 136, 139, 144, 149, 171, 174, 175, 176, 183, 196 and 197 on a Chirasil-Dex column (30 m × 0.25 mm id). The authors reported that their method was reliable and reproducible, without the PCB co-elution problem. The method developed was used for the determination of PCB atropisomers in dolphin liver samples. The authors also claimed the chiral resolution of nine PCBs in a single run for the first time.

Buser and Müller [118] reported the chiral resolution of 2,4-DDT, 2,4-DDD, 4,4-DDT, 4,4-DDD and related compounds using chiral high resolution chromatography, using *tert*-butyldimethylsilyl β-CD as the chiral selector. Bethan *et al*. [119] utilized a modified CD as the CSP for the chiral resolution of bromocyclen in water and in muscle tissue samples of trout and beam from the River Stör, Germany. The authors used 50 : 50 OV-1701 and heptakis(6-*O*-*tert*-butyldimethylsilyl-2,3-di-*O*-methyl)-β-CD as the CSP. Franke *et al*. [120] studied the enantiomeric ratios of polycyclic musks HHCB (Galaxolide™) and AHTN (Tonalide™). Gas chromatography using modified cyclodextrins as chiral stationary phases, coupled to high resolution mass spectrometry, enabled enantioselective analysis even under unfavourable matrix conditions. The gas chromatographic elution order of (4*S*,7*RS*)- and (4*R*,7*RS*)-HHCB was assigned using synthetic (4*S*,7*RS*)-HHCB. The authors used 50 : 50 OV-1701 and heptakis(6-*O*-*tert*-butyldimethylsilyl-2,3-di-*O*-methyl)-β-CD as the CSP. Armstrong *et al*. [121] resolved 19 racemic environmental pollutants on a fused silica capillary column containing heptakis(*O*-(*S*)-2-hydroxypropyl)-β-CD. The 19 environmental pollutants included rodenticides (coumachlor and coumafuryl), insecticides (crufomate, bulan, fonofos and mitotane), insect repellent (ethohexadiol), herbicides and fungicides (ancymidol, silvex, napropamide, phenyl mercuric lactate, 2-[3-chlorphenoxy]propionamide and 2-chloropropionic acid) and halocarbons (1,2-dichloropopane, 2-bromo-1-chloropropane, 1,2-dibromo-3-chloropropane, 2,3-dichlorobutane and α-1,2,3,4,5,6-hexachlorocyclohexane). König and Sievers [67] resolved the enantiomers of various toxic hydroxy acids on *Z*-derivatives of (*R*)-mandelic acids as CSPs. Furthermore, the same group [68] resolved the enantiomers of hydroxy acids on a chiral stationary phase prepared from

S-2-hydroxyisopentanoic acid and *S*-2-hydroxyoctanoic acid by coupling with *S*-α-phenylethylamine.

The separation of the isomers of some environmental pollutants (chlordanes, toxaphenes etc.) is a challenging job, while the chiral resolution of these pollutants is extremely difficult. A multidimensional gas chromatographic (MDcGC) approach for such types of chiral resolution of environmental pollutants has been proposed as the best choice. The MDcGC technique involves the use of two chiral columns of different polarities in series, each with a separate temperature control. The remarkable advantage of this technique, in both a qualitative and a quantitative sense, is a consequence of the fact that a valveless pneumatic system, which involves a live T-piece, allows a preselected small fraction to be cut from the eluate of the first column and transferred quantitatively and reproducibly to the second column. This technique may be used for the complete separation of chiral pollutants, with increased sensitivity and selectivity. Examples of this technique for the chiral resolution of environmental pollutants can be found in several publications [93–96, 122, 123].

Another approach to the chiral resolution of chlordanes, toxaphenes and so on is tandem gas chromatography; that is, the direct coupling of two capillary columns (chiral and achiral columns and vice versa) with different stationary phases in series. However, the likelihood of finding an optimum tandem system is rather small if the column properties (e.g. phase, length and film thickness) are selected by a trial and error method. Jones and Purnell [124] developed a general theory that allows the optimization of separation columns coupled in series; in particular, of the calculation of the length of combination necessary for the baseline separation of a compound mixture. The theory was verified experimentally using complex mixtures that could not be separated by single columns of any length [125]. Oehem *et al.* [126] reported the coupling of achiral followed by chiral and chiral followed by achiral columns for the chiral resolution of *o,p*-DDT in environmental samples. The authors reported achiral and chiral coupling as the most successful process because of the retention time phenomenon, which led to insufficient enantiomeric separation. The drawback of the isomer–enantiomer separation sequence is the relatively high temperature at which the compounds are transferred from the isomeric-selective column into the enantiomer-selective one. This may cause some deterioration in the resolution of the enantiomers, which can be partly compensated by reduction of the upper column temperature. Furthermore, the theory of Jones and Purnell requires isothermal separation, which is not feasible for the separation of complex mixtures of environmental pollutants such as toxaphenes, which show relativity small differences in their retention properties [124]. Therefore, Baycan-Keller and Oehme [125] approximated isothermal conditions using very fast heating ramp, in order to reach the selected isothermal temperature as quickly as possible. These authors also presented calculation procedures that allow the prediction of the best possible isomer–enantiomer separation with the shortest possible tandem system. The latter development is important, because enantiomer separation is usually best at lower temperatures. The procedure suggested by Baycan-Keller and Oehme only requires the measurement of the retention time and of the hold-up time of the single columns. The authors reported that prior calculation of the GC parameters might be useful for the best resolution. Garrison *et al.* [127] reported the chiral resolution of o,p-DDT and DDD by tandem GC using a chiral column of permethyl-trifluoro-acetoxypropyl-γ-CD (Chiraldex G-PT) and an achiral column of XTI-5 95% diphenyl and 95% dimethyl polysiloxane (50 : 50, v/v). The authors also reported poor efficiency of the tandem GC if the order of the columns was reversed.

On the other hand, high resolution gas chromatography (HRGC) has also been used for the successful chiral resolution of environmental pollutants. Many review articles and monographs have been published on the chiral resolution of environmental pollutants by HRGC [20, 27, 28, 86, 87, 107, 128–132]. The chiral resolution of some environmental pollutants by gas chromatography is summarized in Table 6.3. As a specific example, the actual chromatograms of the chiral resolution of α-HCH by gas chromatography are shown in Figure 6.6. The chiral resolution of pollutants in GC is described by the retention time (t), the capacity factor (k), the separation factor (α) and the resolution factor (R_s).

Besides the chiral resolution of pollutants, GC determination of enantiomeric ratios (ERs) can be used for the quantification of the enantiomers in complex matrices, with a known quantity of the pure enantiomer added as an internal standard. With a method of enantiomer labelling, a known quantity of an enantiopure standard is added to the mixture (or an aliquot of it) and the amount of the enantiomer originally present is calculated from the change in the ER after addition of the standard. This method provides precise knowledge of the ERs of the sample and the standards [87]. In addition, the absolute configurations of the pollutants may be determined directly, and free of chiroptical evidence, by GC via co-injection of reference pollutants with known stereochemistry [87].

Table 6.3 The chiral resolution of some environmental pollutants by gas chromatography

Chiral pollutants	Sample matrix	CSPs	Detection	References
α-HCH	–	Heptakis(3-*O*-butyryl-2,6-di-*O*-pentyl)-β-CD	Electrochemical	101
α-HCH	North Sea water	heptakis(3-*O*-butyryl-2,6-di-*O*-pentyl)-β-CD	ECD	102
α-HCH	Water	heptakis(3-*O*-butyryl-2,6-di-*O*-pentyl)-β-CD	ECD	103
γ-PCCH	–	CD CSPs	ECD	104
β-PCCH, α-HCH, β-PCCH and γ-PCCH	–	CD CSP	ECD	105
α-HCH	Marine	50 : 50 mixture of Lipodex C [heptakis (2,3,6-tri-*O*-n-pentyl)-β-CD] and OV-1701CD CSP	ECD	106
α-HCH, β-PCCH, γ-PCCH, *trans*-chlordane *cis*-chlordane, octachlordane MC4, octachlordane MC5, octachlordane MC7 and heptachlor epoxide	Marine	50 % heptakis(2,3,6-tri-*O*-n-pentyl)-β-CD and 50 % OV-1701	ECD	107
α-HCH	Northwater Polynya Arctic marine food web	γ-DEX 120 (20 % nonbonded permethylated β-CD	MSD	108
α-HCH, PCBs, *cis*- and *trans*-chlordane, oxychlordane, and heptachlor epoxide	*Phoca hispida*	β-DEX 120 (20 % nonbonded permethylated β-CD	MS	109

Compound	Sample	CSP/Column	Detection	Ref.
α-HCH	Seawater, air and bulk deposition samples from the North Sea	CD-based CSP	ECD	110
Octa- and nonachlordane, oxychlordane and heptachlor	—	Two β-CD derivatives, dissolved in polysiloxanes	MS	111
Chlordane and its metabolite	Water and air	α- and β-CDs on silicon columns	EID, ECNI and MS	112
cis- and trans-Chlordane	Soil	γ-DEX 120	MS	113
Chlordane	Soil and vegetation	γ-DEX 120	MS	114
Chlordane and its metabolites	Seven species of Arctic sea birds	β-DEX 120	MS	115
PCBs (84, 91, 95, 132, 136 and 149)	—	Chirasil Dex GC	ECD	94
PCBs (45, 95 and 139)	—	Octakis(2,6-di-O-methyl-3-O-n-pentyl)-γ-CD diluted in 1:1 ratio with OV-1701	ECD	95
PCBs (91, 95, 132, 136 and 149)	—	Chirasil-Dex	ECD	95
Nineteen PCBs	Sediment	CD-based CSP	ECD and MS	116
PCBs (132 and 149)	Atlantic Ocean, C. coelolepis	Chirasil-Dex	ECD	96
PCBs (45, 84, 88, 91, 95, 131, 132, 135, 136, 139, 144, 149, 171, 174, 175, 176, 183, 196 and 197)	Dolphin liver	Chirasil-Dex	ECD	117
2,4-DDT and 2,4-DDD	—	tert-Butyldimethylsilyl β-CD 50:50 OV-1701 and	MS	118
Bromocyclen	Water and muscle tissue	heptakis(6-O-tert-butyldimethylsilyl-2,3-di-O-methyl)-β-CD CDs based CSPs	ECD	119

(continued overleaf)

Table 6.3 (*continued*)

Chiral pollutants	Sample matrix	CSPs	Detection	References
HHCB (Galaxolide™) and AHTN (Tonalide™)	Fish and mussels	heptakis(O-(S)-2-hydroxypropyl)-β-CD	MS	120
Eighteen racemic environmental pollutants, i.e. rodenticides (coumachlor and coumafuryl), insecticides (crufomate, bulan, fonofos and mitotane), insect repellent (ethohexadiol), herbicides and fungicides (ancymidol, silvex, napropamide, phenyl mercuric lactate, 2-[3-chlorphenoxy] propionamide and 2-chloropropionic acid) and halocarbons (1,2-dichloropopane, 2-bromo-1-chloropropane, 1,2-dibromo-3-chloropropane, 2,3-dichlorobutane and α-1,2,3,4,5,6-hexacyclohexane)	—	CD-based CSPs	FID	121

Hydroxy acids	—	Z-derivatives of (*R*)-mandelic acid CSPs	67
Hydroxy acids	—	CSP from S-2-hydroxyisopentanoic acid and S-2-hydroxyoctanoic acid by coupling with S-α-phenylethylamine	68
cis-Chlordane	—	Chiraldex G-BP	92

ECD, Electron capture detection; ECNI, electron capture negative ionization; EI, electron ionization; FID, flame ionization detection; MS, mass spectrometry; MSD, mass selective detection.

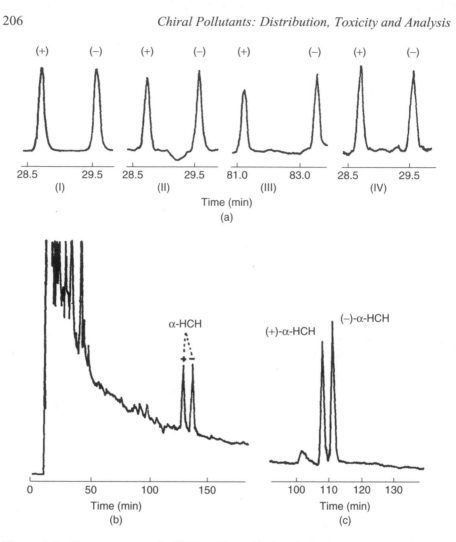

Figure 6.6   Chromatograms of α-HCH on (a) octakis(3-O-butyryl-2,6-di-O-pentyl)-γ-CD in 50 % OV 1701 (I, standard solution; II, sea water; III, bulk deposition extract; IV, air) [110], (b) heptakis(3-O-butyryl-2,6-di-O-pentyl)-β-CD [103] and (c) heptakis(2,3,6-tri-O-pentyl)-β-CD and 50 % OV 1701 [105].

6.4 The Optimization of GC Conditions

Method development on CSPs by GC differs from traditional achiral methods in achieving and optimizing selectivity. The chiral resolution of environmental pollutants in GC is controlled by many parameters. The most important factors are the temperature of the column, the type of CSP, the mobile phase, the amount of injection loading and the sensitivity of the detector. These parameters are discussed herein.

6.4.1 Mobile Phases

In achiral gas chromatography, a carrier gas with an optimal to slightly less than optimal linear velocity is used to control retention. In chiral GC, the highest enantiomeric selectivity is achieved by maximizing the energy differences in the diastereoisomeric complexes formed between each enantiomer and the chiral stationary phase. Therefore, the mobile phase is an important parameter in optimizing the chiral resolution of environmental pollutants by GC. Many individual gases and their mixtures have been used as mobile phases (carrier gases), the most important being helium, hydrogen and nitrogen. To obtain maximum sensitivity and reproducibility, the purity of these gases is very important. The advantages of nitrogen are low cost and the highest efficiency. The disadvantage of nitrogen is its optimal linear velocity of 12 cm s^{-1}; the efficiency is rapidly lost as the linear velocity is increased (the slope of the line for the height equivalent theoretical plate (HETP) versus the linear velocity is steep). Helium is more expensive to use, but the optimal linear velocity is 20 cm s^{-1}, and the slope of the line for the HETP versus the linear velocity is flatter. Hydrogen has an optimal linear velocity of 40 cm s^{-1}, and has both the flattest slope for HETP versus linear velocity and the lowest viscosity. In general, hydrogen forms explosive mixtures with air, and it must be used with care. Helium is a good gas for chiral resolution: it can be used at high linear velocities and it is safe to work with. As a starting point, a 30–40 cm s^{-1} linear velocity is recommended: this will give a good mix of column efficiency and speed of analysis. The flow rates for helium and hydrogen are two to four times those of nitrogen. As the energy differences between the diastereoisomeric complexes need to be exploited through lower temperatures and high linear velocities, the use of hydrogen produces more efficient peaks. For highly volatile pollutants, a low temperature and a low linear velocity are required to achieve sufficient analyte stationary phase interaction for chiral recognition. In these cases, nitrogen is the best carrier gas. The optimal and maximum linear velocities, along with relative viscosities for each carrier gas, are given in Table 6.4.

The flow rate, or the pressure of the mobile phase, is maintained by a trial and error approach. Normally, a low flow rate favours maximal chiral resolution, while a high pressure results in poor chiral resolution – but this does not always mean that a high flow rate is not to be recommended. The flow rate actually depends on the type of pollutant, the CSP used

Table 6.4 The optimum and maximum linear velocities, along with relative viscosities, for each carrier gas and for columns of different lengths [92]

Carrier gas	μ_{opt} (cm s^{-1})	μ_{max} (cm s^{-1})	Relative viscosity (η)
Nitrogen	12	30	1.00
Helium	20	60	1.22
Hydrogen	40	120	0.497

	cm^{-1} s^{-1} psi^{-1}			
Column length (m)	50 °C	100 °C	150 °C	200 °C
20	2.78	2.51	2.29	2.12
30	1.82	1.64	1.50	1.39
50	1.29	1.16	1.06	0.98

and the experimental temperature. It is often desirable to switch between carrier gases. The approximate values given in Table 6.4 can be used to calculate the head pressure of carrier gas needed for a certain linear velocity on a 0.25 mm inner diameter column, or the linear velocity at a certain column head pressure. To measure the average linear velocity, the following equation is used:

$$\mu_{ave} = L/t_0 \qquad (6.1)$$

where L is the length of the column (in cm) and t_0 is the time necessary for an unretained component to pass from the injector to the detector. Good unretained analytes are methane or the hydrocarbon gas from a disposable lighter.

6.4.2 Temperature

The temperature is a crucial parameter in optimization of the chiral resolution of environmental pollutants by GC, because the retention of the pollutant depends on the column temperature. Moreover, the formation of the diastereoisomeric complexes of the pollutants with the CSP also depends on column temperature. Therefore, a very slight variation in temperature may result in poor resolution and a loss of reproducibility. Also, variation of the temperature of the injector and the detector may change the selectivity, sensitivity and the reproducibility. Since the highest enantiomeric selectivity is achieved by maximizing the energy difference in the diastereoisomeric complexes formed between the pollutants and the CSP, these energy differences become smaller with increasing temperature. Therefore, to optimize chiral separation, a lower elution temperature in conjunction with relatively high linear velocities of the carrier gas is the best option. In general, the enantioselectivity decreases with increasing temperature. For separation below 130 °C, a temperature gradient of 1–5 °C min^{-1} is used. Over 130 °C, a gradient of 5–10 °C min^{-1} can be used. To effect faster elution or sharper peaks at temperatures below 130 °C, one should increase the linear gas flow velocity. It is not recommended to heat the CSP column at a rate of more than 15 °C min^{-1}.

While temperature is an important factor in the optimization of chiral resolution by GC, and it is optimized by a trial and error approach, only a few reports are available on the effects of temperature on the chiral resolution of pollutants. Busser and Müller [118] studied the effects of temperature on the chiral resolution of 2,4-DDD, 2,4-DDT, 2,4-DDD and 4,4-DDD. The results are shown in Figure 6.7. The authors used temperatures of 160 °C, with 3 °C min^{-1} variation, 80 °C at 3 °C min^{-1}, 80 °C at 1 °C min^{-1} and 80 °C at 0.5 °C min^{-1}. It may be concluded from Figure 6.7 that the temperature programming of 80 °C at 0.5 °C min^{-1} was found to be the best, since the maximum resolution of these pesticides was achieved. Again, this maximum resolution may be explained on the basis of the stability of the diastereoisomeric complexes (between DDT and DDD and the CSP) and their different retentions on the reported CSP. König *et al.* [68] also investigated the effects of temperature on the chiral resolution of hydroxy acids, using CSPs based on S-α-phenylethylamide (Figure 6.4). The effects of temperature on the chiral resolution of hydroxy acids are summarized in Table 6.5. It may be concluded from this table that a lower temperature normally resulted in the maximum chiral resolution.

At this point, it is also important to mention that all chiral resolution in GC is enthalpy controlled. Therefore, the temperature of the GC machine should be decreased to an acceptable level in order to increase the separation factor (α). The enantioseparation in GC is defined by the free enthalpy (Gibbs energy) $-\Delta_{R,S} (\Delta G)$ of the diastereoisomeric complexation, or between the chiral selector 'CS' and the R- and S-enantiomers of the racemic pollutant. For temperature investigations, the Gibbs–Helmholtz equation may be written as follows:

$$-\Delta_{R,S} (\Delta G) = -\Delta_{R,S} (\Delta G) + T \Delta_{R,S} (\Delta S) = RT \ln K_R/K_S \quad (6.2)$$

where K_R and K_S are the formation constants of the diastereoisomeric complexes between chiral selector 'CS' and the R- and S-enantiomers but, arbitrarily, $K_R > K_S$ [133]. Therefore, Equation (6.2) may be written

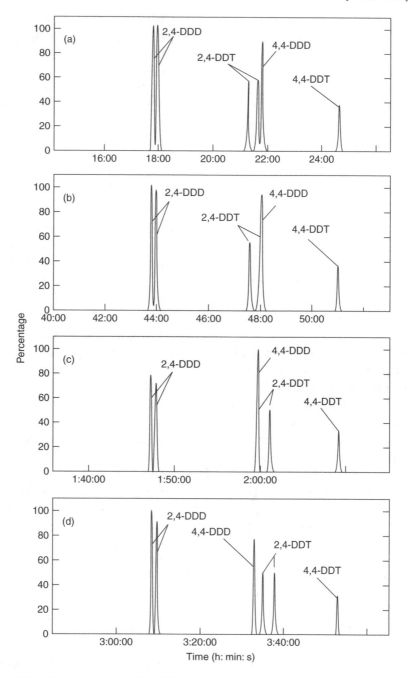

Figure 6.7 Chromatograms of 2,4-DDD, 4,4-DDD, 2,4-DDT and 4,4-DDT pesticides at different temperatures: (a) 160 °C at 3 °C min^{-1}, (b) 80 °C at 3 °C min^{-1}, (c) 80 °C at 1 °C min^{-1} and (d) 80 °C at 0.5 °C min^{-1} [118].

Table 6.5 The effect of temperature on the chiral resolution of hydroxy acids on CSPs based on S-α-phenylethylamide [68]

Hydroxy acid	CSP I		CSP II		CSP III	
	α	T (°C)	α	T (°C)	α	T (°C)
2-Hydroxybutyric acid	1.006	62	1.008	71	1.011	45
3-Hydroxybutyric acid	–	–	–	–	1.007	55
2-Hydroxyisopentanoic acid	1.007	62	1.010	71	1.011	45
2-Hydroxypentanoic acid	1.008	67	1.009	71	1.012	45
2-Hydroxyisohexanoic acid	1.011	62	1.012	71	1.013	45
2-Hydroxyhexanoic acid	1.011	62	1.012	71	1.014	45
2-Hydroxyoctanoic acid	1.006	81	1.015	71	1.011	92
Mandelic acid	1.008	100	1.026	81	1.019	92
Malic acid	–	–	1.011	91	1.012	92
3-Phenyllactic acid	–	–	1.013	81	1.009	100

α, Separation factor; T (°C); column temperature and CSPs.

as follows:

$$-\Delta_{R,S}(\Delta G)/T = -\Delta_{R,S}(\Delta H)/T + \Delta_{R,S}(\Delta S) = RT \ln K_R/K_S \tag{6.3}$$

As expected for an error complexation process, $-\Delta_{R,S}(\Delta H)$ and $\Delta_{R,S}(\Delta S)$ compensate each other in determining $-\Delta_{R,S}(\Delta H)$. Therefore, an isoenantioselective temperature T_{iso} exists at which enantiomers cannot be separated due to peak coalescence [133, 134]. T_{iso} can be expressed by the following equation:

$$T_{iso} = \Delta_{R,S}(\Delta H)/\Delta_{R,S}(\Delta S) \qquad \text{for} \ -\Delta_{R,S}(\Delta GH) = 0 \tag{6.4}$$

It is significant to mention here that below T_{iso} enantioseparation is enthalpy controlled, and the R-enantiomer is eluted first followed by the S-enantiomer, while above T_{iso} enantioseparation is entropy controlled and the reverse order of elution is observed. Therefore, the reverse order of elution occurs due to enthalpy and entropy compensation. Only when the van't Hoff plot is linear over a wide range of temperatures, including T_{iso}, is the thermodynamic principle of entropy/enthalpy compensation established. In principle, temperature-dependent multimodal recognition mechanisms that oppose each other may also account for the observation of an isoenantioselective temperature [87, 135, 136].

6.4.3 The Column Dimensions

The size of the chiral column – that is, the length and diameter – is also important in optimizing the chiral resolution of environmental pollutants.

The maximum chiral resolution can be achieved by increasing and decreasing the column length and diameter, respectively. The typical dimensions of the GC column are 10–30 m × 0.25 mm id, while the film thickness of the chiral stationary phase is usually 0.25 μm. Column miniaturization may have some important merit in terms of a shorter analysis time [137] and the improvement of detectability, and for unified enantioselective capillary chromatography [138, 139]. The latter term implies that one individual column can be used for enantiomer analysis by capillary GC, supercritical fluid chromatography and open tubular electrochromatography. Due to the binary nature of enantioseparation, the whole elution window required for multi-component mixtures need not be exploited unless enantiomers are detected in complex matrices. With shorter columns, the elution temperature can be decreased, whereby the chiral separation factor (α) is increased in the common enthalpy controlled region of enantioselectivity [140, 141]. The loss of efficiency is compensated by the gain in selectivity, leading to a comparable resolution factor (R_s). The shorter analysis time increases the narrowness of the peaks and hence the detectability of the enantiomers. Furthermore, minimization by reduction of the internal diameter of the column to 0.1 or 0.05 mm requires smaller films of the stationary phase in order to keep the phase ratio β constant. The reduced amount of stationary phase decreases the sample capacity. The reduced signal-to-noise ratio may become critical with regard to the precision of the enantiomeric ratio determination [87].

6.4.4 Structures and Types of Chiral Selectors

The efficiency and ease of the chiral resolution of environmental pollutants depend on the type and structure of the chiral selector used in the GC column. Therefore, to achieve the best chiral separation it is important to select a suitable chiral selector. Basically, the chiral resolution of an environmental pollutant occurs due to the formation of a diastereoisomeric complex, and hence the selection of a suitable chiral selector that can form a stable diastereoisomeric complex is essential. As discussed earlier, various chiral selectors have been used for the enantiomeric resolution of environmental pollutants by GC and, among these, CDs and their derivatives have frequently been used. Various CDs and their derivatives used for chiral resolution are given in Table 6.2. Knowledge of the basic chemistry is helpful in selecting a suitable chiral selector for a specific pollutant. The chiral selector should be selected in a way that can provide the maximum number of bonding sites between the chiral selector and the environmental pollutant. However, the selection of the CSP is carried out by a trial and error method.

In spite of the importance of the selection of a suitable chiral selector, there are only few reports on comparative studies of the chiral resolution of environmental pollutants using different chiral selectors. Busser and Müller [112] studied the comparative chiral resolution of chiral chlordane pollutants on permethylated α-CD (PMCD), *tert*-butyldimethylsilyl-β-CD (BSCD) and perethylated-α-CD (PECD) chiral selectors. Chiral columns were prepared by dissolving PMCD in PS086, BSCD in PS086 polysiloxane and PECD in OV1701 stationary phases, respectively. The results of this study are given in Table 6.6. It may be concluded from this table that different patterns of chiral resolution were observed on different CSPs. However, the best chiral resolution was obtained on BSCD CSP. In another study, Wong and Garrison [116] studied the chiral resolution of atropisomers of PCBs on different CD-based chiral GC columns. The authors used Chirasil-Dex, Cyclosil-B, Chiraldex-B-PA and Chiraldex-B-DM chiral columns, and the chromatograms of PCBs obtained on these columns are shown in Figure 6.8. It is clear from this figure that the chiral resolution of PCBs is different on different columns, the best results being obtained on the Chirasil-Dex column. These differing chiral resolutions are due to different bondings between the CSP and PCB. This aspect of different bondings between the CSP and the pollutants will be discussed in detail later, in Section 6.8.

Table 6.6 The chiral resolution factors (R_s) of chlordane pollutants on different CD CSPs [112]

Chlordane	Chiral column		
	PMCD	BSCD	PECD
Heptachlor	0.0	0.0	1.2
trans-Chlordane	1.0	2.1	0.0
cis-Chlordane	0.8	1.0	0.9
MC5	1.3	4.8	0.0
MC6	0.0	0.5	0.0
MC7	2.4	nd	0.0
U82	0.0	0.5	0.0
HEP	0.0	1.2	1.2
endo-HEP	0.0	1.2	4.3
OXY	0.0	1.2	4.3
photo-Heptachlor	0.2	3.6	0.7
photo-cis-Chlordane	0.0	0.0	0.0

nd, Not detected.

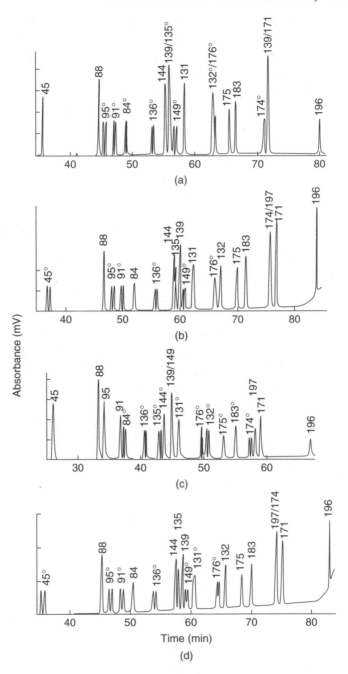

Figure 6.8 Chromatograms of PCBs on different CSPs: (a) Chirasil-Dex, (b) Cyclosil-B, (c) Chiraldex-B-PA and (d) Chiraldex-B-DM [116]. The atropisomers marked with degree signs (°) can be separated on the respective columns.

6.4.5 The Structures of Chiral Pollutants

Optimization of the chiral resolution of environmental pollutants may also be achieved by considering the structures of the pollutants. First of all, the chemical structures of the pollutants should be determined. Then, the maximum number of bonding sites should be calculated and, accordingly, a suitable chiral selector should be chosen. Again, basic knowledge of the chemistry may be useful in finding out the maximum possible number of binding sites between the pollutants and the CSP. Moreover, calculations based on simulations of molecular dynamics can be used to determine the suitable chiral selector. Using these approaches, costly chemicals and time can be saved. A very good example of this may be found in the work of Wong and Garrison [116], who tried to resolve PCB 91 on Chirasil-Dex and Cyclosil-B chiral columns. The chromatograms of chiral resolution of PCB 91 on these columns are given in Figure 6.9, which clearly indicates that the best resolution was achieved on Cyclosil-B. Moreover, the elution order of the atropisomers of PCB 91 is reversed on the two columns. Briefly, the better resolution on Cyclosil-B than on the Chirasil-Dex column may be due to the stronger bonding between PCB 91 and Cyclosil-B in comparison to the Chirasil-Dex column. The reverse order of chiral resolution has already been discussed in Section 6.4.2. Similarly, the chiral resolution of various pollutants may be correlated with the structures of different CSPs.

6.4.6 Detection

In general, pollutants exist in the environment at very low concentrations and, therefore, the detection of these pollutants becomes poor. The selective detection of pollutants in environmental samples is important, since many

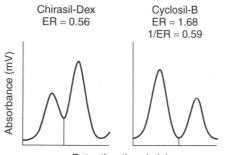

Figure 6.9 The chromatograms of PCB 91 on Chirasil-Dex and Cyclosil-B, showing the reverse order of elution at two different CSPs [116].

other achiral pollutants are often present in the fractions analysed, despite the sophisticated clean-up methods used. The co-elution of chiral pollutants, which interfere, is a problem in chiral GC, particularly since the number of chiral pollutants (enantiomers/isomers) that have to be considered increases but the GC parameters in general are the same as in achiral GC. Therefore, the possibility of interference is increased, and selective detection is of particular importance in chiral GC. Moreover, all of the capillary columns have extremely low capacities, which again makes detection poor. Therefore, the optimal resolution of chiral pollutants can be achieved by using the best detection techniques. Many detection devices, such as electron capture negative ionization (ECNI), electron ionization (EI), flame ionization detection (FID), mass spectrometry (MS), mass-selective detection (MSD), electron capture detection (ECD), nitrogen phosphorus detection (NPD), alkali flame ionization detection (AFID), atomic emission detection (AED), atomic absorption spectrometry (AAS), Fourier transform infrared detection (FT–IR), flame photometric detection (FPD), electron capture detection/nitrogen phosphorus detection (ECD/NPD) and nitrogen phosphorus detection/ion trape detection (NPD/ITD), have been used for the detection of chiral pollutants in chiral GC. It is important to mention here that most of the organochlorine pollutants, including organochlorine, are UV transparent, and hence UV-based detectors are not good for their detection. This makes GC the best technique for chiral resolution of this class of environmental pollutant. ECD, EI, ECNI and MS detectors are extensively used for the detection of chlorinated environmental pollutants. This is due to their potential for increased sensitivity and selectivity towards organochlorine pollutants and their virtual transparency to many of the interfering contaminants. A discussion of the working principles of these detectors is not possible here, and readers should consult other publications on this issue. However, it is essential to mention here that these detectors can be used to detect amounts of environmental pollutants ranging from ppm to ppb levels.

6.4.7 Other Parameters

Besides the parameters discussed above, some other factors are important for the optimization of chiral resolution of environmental pollutants. Sometimes, sample overloading may result in poor and partial resolution. Under such circumstances, the maximum chiral resolution can be achieved by reducing the amount injected into the GC column. Also, some authors have described the coupling of an achiral column with a chiral column in series as the best approach for optimizing the chiral resolution. It is important

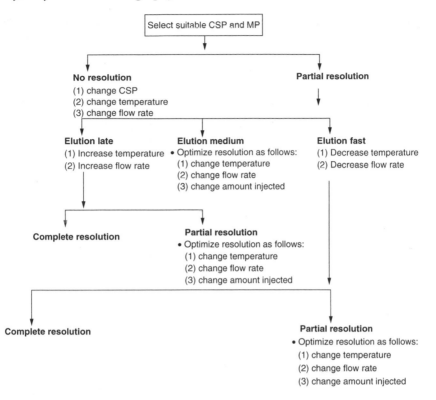

Scheme 6.1 A protocol for the development and optimization of the chiral GC conditions for the enantiomeric resolution of environmental pollutants. Note that this is only a brief outline of the procedure that should be followed.

to mention here that the coupling of chiral with achiral columns, or vice versa, may affect the chiral resolution. In addition, the various achiral pollutants present in the sample may affect the chiral resolution. Therefore, the maximum chiral resolution may be achieved by using an advance clean-up method, prior to GC analysis. Briefly, the optimal chiral resolution of environmental pollutants by GC may be achieved by considering all of the factors discussed above. On the basis of our experience and the available literature, a scheme has been developed to optimize the chiral resolution of the environmental pollutants by GC. This is presented in Scheme 6.1.

6.5 The Reverse Elution Order

One of the most interesting and useful phenomena observed with CD-based CSPs is the ability to reverse the chiral resolution in GC. The significance of this reversal is reliable quantitation of the trace enantiomers. It is always desirable to have the trace component elute first, to avoid interference from the tailing off of the larger component. The reversal has been accomplished in four ways: changing from one CD to another, changing from one phase to another, change of the CD derivatives and operating below ambient temperatures [92]. It has been observed that molecules with limited inclusion complexing strength can reverse their elution order by any one of the first three methods. Aromatic structures have a fixed position in the CD cavity and, as a result, have predictable elution orders for a given CD derivative. Saturated and heterocyclic rings, as well as linear molecules, can occur by restricting the CD cavity (e.g. γ- to β-CD or β- to α-CD). The reversal in sub-ambient conditions is only possible for molecules eluted below $100\,^{\circ}\mathrm{C}$ under normal operating conditions. There are several examples in the literature [116, 142–148] showing the reverse order of elution for chiral pollutants by changing one CD for another. One of these examples is shown in Figure 6.9 [116], where the order of elution of PCB 91 is reversed by replacing Chirasil-Dex by a Cyclosil-B column. The temperature-dependent reason for the reversed order of the elution is described in Section 6.4.2.

6.6 Errors and Problems in Enantioresolution

Despite of great success of GC in the resolution of chiral environmental pollutants, the risk of error should also be considered. Schurig [149, 150] has summarized the possible errors in chiral GC as occurring due to: (i) the decomposition of the chiral pollutants in GC at a high temperature; (ii) the co-elution of impurities, which spuriously increases peak areas; (iii) the racemization of configurationally labile enantiomers, causing peak distortions due to inversion of the configuration during the enantiomeric resolution; (iv) peak distortion caused by inadequate instrumentation; (v) adsorption of the enantiomers on the column wall; and (vi) nonlinear detector response [87].

The first type of error may be minimized via reducing the analysis time by decreasing the column length, using a high gas pressure, keeping within an ambient temperature range, using a CSP that has a small separation factor (α) and, sometimes, by pre-derivatization of the chiral pollutants. The second type of error may be avoided by changing some of the GC parameters, such as the temperature and the mobile phase, and by using pre-columns with differing polarities in multidimensional GC. If the error is due to racemization, it may be overcome by changing the column temperature, by derivatization of the chiral pollutant and, sometimes, by increasing the analysis time. The fourth type of error may be reduced by using high quality instrumentation, with the purest gas possible and the most sensitive detector. The problem due to nonlinearity of the detector response can be avoided by using a good quality detector and running the GC machine overnight or for a long time.

6.7 The Derivatization of Chiral Environmental Pollutants

GC can be used for the analysis of environmental pollutants that are volatile at the normal working temperature. Some pollutants are not volatile and some of them decompose or racemize before the temperature at which they become volatile. Moreover, some of them do not resolve, or only partially resolve, under the varied GC conditions. Therefore, such types of chiral pollutant are converted into their volatile derivatives by reacting them with some suitable achiral reagents. Sometimes, derivatization may also enhance the stereoselective interactions, such as hydrogen bondings, steric effects and so on, with the CSP. The derivatization may also change the volatility, polarity and so on of the pollutants. Furthermore, it may lower the analysis temperature and maximize the energy differences in the diasteroisomeric complexes, which results in a faster analysis time, a greater throughput and better enantioselectivity [151].

A derivatizing reagent should be pure and capable of forming derivatives with a wide range of pollutants. Also, the derivatizing reaction should not involve complex reactions or paraphernalia. The derivatizing reagents are selected depending on the type of chiral pollutant. In general, acylation or alkylation of the chiral pollutants is carried out for this purpose. The first consideration when contemplating derivatization is that the derivatization reaction must not cause racemization. The second criterion is that the derivatization reaction must not produce by-products that interfere with the chiral resolution. Greater care must be taken when the analyte has more than one functional group that is capable of interacting with the derivatizing agent [92].

6.8 Mechanisms of Chiral Resolution

As discussed above, most chiral resolutions of environmental pollutants have been carried out on CDs and their derivative-based CSPs in GC.

Therefore, in this section, the mechanisms of chiral resolution of environmental pollutants on CD-based CSP will be discussed. The mechanisms of chiral resolution in gas and liquid chromatography are more or less similar. There are very few reports dealing with chiral recognition mechanisms in GC; however, some attempts have been made to describe chiral recognition mechanisms in liquid chromatography. Therefore, the approaches used for liquid chromatography may be utilized to explain chiral recognition mechanisms in GC.

The primary C-6 hydroxyl groups, which are free to rotate, can partially block the CD cavity at one end. The mouth of the opposite end of the CD cavity is enriched due to the presence of C-2 and C-3 secondary hydroxyl groups. This arrangement of the CD favours complex formation with a variety of pollutants. The presence of stereospecific glucopyronose units and the restricted conformational freedom and orientation of the secondary hydroxyl groups are considered to be responsible for the chiral recognition capacities of the CDs [152]. Therefore, different diastereoisomeric inclusion complexes are formed by the two enantiomers (Figure 6.10). This has been proven by various data obtained from crystallography, NMR and mass spectrometric studies [13, 54, 153–158]. Basically, in the chiral recognition process, the enantiomers enter into the cavity of the CD molecules and diastereoisomeric complexes are formed; hence these are called inclusion complexes. Therefore, even a slight variation in the chromatographic conditions, such as a change in temperature, derivatization of the CDs or flow of the mobile phase, affects the chiral resolution greatly. Because

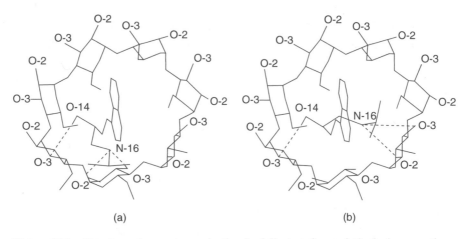

(a)   (b)

Figure 6.10 Structures (computer projections) of diastereoisomeric inclusion complexes of (a) d-propranolol and (b) *l*-propranolol in β-CD. - - - , hydrogen bondings [12].

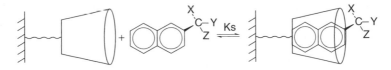

Figure 6.11 A schematic representation of solute inclusion in a cyclodextrin cavity.

of this, many research groups have synthesized various derivatives and utilized them for the chiral resolution of different molecules. Therefore, the selection of the CD is important for stable complex formation with a specific racemic pollutant.

Lipkowitz and Stoehr [153] carried out an NMR study of methyl mandelate on β-CD. The authors conducted molecular dynamics simulations and predicted the guest (molecule) – host (CD) interactions. It was observed that short range dispersion forces rather than long range Coulomb forces were responsible for both complexation and enantiodiscrimination. To explain chiral resolution on CDs, several models of and approaches to guest–host formation have been proposed and discussed [156, 159–161]. The stabilities of the inclusion complexes are controlled by a number of interactions between enantiomers and CSPs, the most important of which are hydrogen bonding, dipole-induced dipole interactions and $\pi-\pi$ forces [153, 162]. Steric effects also play a crucial role in chiral resolution, as complexation is governed by the presence of several groups on the enantiomers [60, 163]. However, some other forces, such as van der Waals, ionic interactions and solvation effects, are also responsible for chiral resolution. These different types of interactions occur between the hydroxyl groups of the CDs and various groups of the enantiomers. These interactions may be enhanced by introducing electronegative or phenyl groups into the CDs (derivatization) and hence CD derivatives are better CSPs than the CSPs obtained from native CDs. Briefly, the enantiomers fit stereogenically in different fashions into the chiral cavity of CD, and are stabilized at different magnitudes by the various interactions discussed above. This arrangement results in chiral separation of the enantiomers. To make the concept clearer, a schematic representation of chiral resolution on these phases is shown in Figure 6.11.

6.9 Conclusions

In spite of the development of more advanced techniques such as HPLC and CE, GC still maintains its status in the chiral resolution of environmental pollutants due to its various advantages. GC is considered to be the best technique for the qualitative and quantitative resolution of chiral environmental pollutants, as most of these pollutants are volatile at the GC working temperature. The use of highly selective and sensitive detectors, as mentioned above, makes GC the technique of choice for the enantiomeric resolution of environmental pollutants. The advanced form of GC – high performance gas chromatography (HRGC) – has resulted in high selectivity, sensitivity and reproducibility. Moreover, multidimensional GC is another development towards the advancement of GC in the chiral resolution of pollutants. To achieve the maximum chiral resolution, the above-mentioned GC parameters should be optimized, and the extraction, clean-up and concentration of the sample should be carried out carefully. Chiral recognition mechanisms in GC are not fully developed and, therefore, more work should be carried out towards their determination, so that GC can be used properly and exactly in the chiral resolution of environmental pollutants. Attempts should also be made to use different chiral selectors in GC columns.

References

1. E. Gil-Av, B. Feibush and R. Charles-Sigler, in A. B. Littlewood, ed., *Gas Chromatography*, Institute of Petroleum, London, 1996, p. 227.
2. J. J. Radial, T. F. Bidleman, B. R. Kerman, M. E. Fox and W. M. J. Strachan, *Environ. Sci. Technol.* **31**, 1940 (1997).
3. T. Koscielski, D. Sybilska and J. Jurczak, *J. Chromatogr.* **280**, 131 (1983).
4. A. Villiers, *C. R. Acad. Sci. Paris* **112**, 536 (1891).
5. F. Schardinger, *Zentr. Bacteriol. Parasitenk. Abt. II*, **29**, 188 (1911).
6. M. Gröger, E. Katharina and A. Woyke, Cyclodextrins, Science Forum an der Universität Siegen, Siegen, 2001.
7. S. M. Han and D. W. Armstrong, in A. M. Krstulovic, ed., *Chiral Separations by HPLC*, Ellis Horwood, Chichester, 1989, p. 208.
8. J. Michon and A. Rassat, *J. Am. Chem. Soc.* **101**, 4337 (1979).
9. S. M. Han, W. M. Atkinson and N. Purdie, *Anal. Chem.* **56**, 2827 (1984).
10. D. A. Lightner, J. K. Gawronski and J. Gawronska, *J. Am. Chem. Soc.* **107**, 2456 (1985).
11. Y. Ihara, E. Nakashini, K. Mamuro and J. Koga, *Bull. Chem. Soc. Jpn* **59**, 1901 (1986).
12. D. W. Armstrong, T. J. Ward, R. D. Armstrong and T. E. Beesley, *Science* **222**, 1132 (1986).
13. J. A. Hamilton and L. Chen, *J. Am. Chem. Soc.* **110**, 5833 (1988).
14. A. M. Stalcup, in G. Subramanian, ed., *A Practical Approach to Chiral Separations by Liquid Chromatography*, VCH, Weinheim, 1994, p. 95.

15. D. W. Armstrong, F. Nome, L. A. Spino and T. D. Golden, *J. Am. Chem. Soc.* **108**, 1418 (1986).

16. M. L. Bender and M. Komiyama, *Cyclodextrin*, Springer, Berlin, 1978.

17. J. Szejtli, *Cyclodextrins and their Inclusion Complexes*, Akademia Kiado, Budapest, 1982.

18. W. L. Hinze, in C. Van Oss, ed., *Separations and Purification Methods*, vol. 10, Dekker, New York, 1981, p. 159.

19. T. Cserhati and E. Forgacs, *Cyclodextrins in Chromatography*, Royal Society of Chemistry, London, 2003.

20. V. Schurig and H. P. Nowotny, *Angew. Chem. Int. Ed. Engl.* **29**, 939 (1990).

21. W. A. Keim, W. Kohnes, W. Meltzow and H. Romer, *J. High Res. Chromatogr.* **14**, 507 (1991).

22. A. Harada, M. Furue and S. Nozakura, *J. Polym. Sci.* **16**, 187 (1978).

23. B. Zsadon, M. Szilasi, K. H. Otta, F. Tudos, E. Fenyvesi and J. Szejtli, *Acta. Chim. Acad. Sci. Hung.* **100**, 265 (1979).

24. B. Zsadon, M. Szilasi, F. Tudos, E. Fenyvesi and J. Szejtli, *Stark* **31**, 11 (1979).

25. B. Zsadon, F. Szilasi, F. Tudos and J. Szejtli, *J. Chromatogr.* **208**, 109 (1981).

26. B. Zsadon, L. Decsei, M. Szilasi and F. Tudos, *J. Chromatogr.* **270**, 127 (1983).

27. V. Schurig, *J. Chromatogr. A* **666**, 111 (1994).

28. W. A. König, *Gas Chromatographic Enantiomer Separation with Modified Cyclodextrins*, Hüthig Buch Verlag, Heidelberg, 1992.

29. Z. Juvancz, G. Alexander and J. Szejtli, *J. High Res. Chromatogr.* **10**, 105 (1987).

30. G. Alexander, Z. Juvancz and J. Szejtli, *J. High Res. Chromatogr.* **11**, 110 (1988).

31. A. Venema and P. J. A. Tolsma, *J. High Res. Chromatogr.* **12**, 32 (1989).

32. V. Schurig and V. Nowotny, *J. Chromatogr.* **441**, 155 (1988).

33. W. A. König, S. Lutz and G. Wenz, *Angew. Chem. Int. Ed. Engl.* **27**, 979 (1988).

34. K. Fujimura T. Ueda, and T. Ando, *Anal. Chem.* **55**, 446 (1983).

35. Y. Kawaguchi, M. Tanaka, M. Nakae, K. Funazo and T. Shono, *Anal. Chem.* **55**, 1852 (1983).

36. M. Tanaka, Y. Kawaguchi, T. Shono, M. Uebori and Y. Kuge, *J. Chromatogr.* **301**, 345 (1984).

37. K. G. Feitsma, B. F. H. Drenth and R. A. De Zeeuw, *J. High Res. Chromatogr.* **7**, 147 (1984).

38. K. G. Feitsma, J. Bosmann, B. F. H. Drenth and R. A. De Zeeuw, *J. Chromatogr.* **333**, 59 (1985).

39. K. Fujimura, M. Kitagawa, H. Takayanagi and T. Ando, *J. Liq. Chromatogr.* **9**, 607 (1986).

40. C. M. Fisher, *Chromatogr. Int.* **5**, 10 (1984).

41. US Pat. 4539399 (1985), D. W. Armstrong.
42. D. W. Armstrong, A. Alak, W. DeMond, W. L. Hinze and T. E. Riehl, *J. Liq. Chromatogr.* **8**, 261 (1985).
43. C. M. Fisher, *Chromatogr. Int.* **8**, 38 (1985).
44. T. E. Beesley, *Am. Lab.* 78 (1985).
45. C. A. Chang, H. Abdel, N. Melchoir, K. Wu, K. H. Pennell and D. W. Armstrong, *J. Chromatogr.* **347**, 51 (1985).
46. T. J. Ward and D. W. Armstrong, *J. Liq. Chromatogr.* **9**, 407 (1986).
47. H. J. Isasaq, *J. Liq. Chromatogr.* **9**, 229 (1986).
48. D. E. Weaver and R. B. L. van Lier, *Anal. Biochem.* **154**, 590 (1986).
49. H. J. Issaq, D. Weiss, C. Ridlon, S. D. Fox and G. M. Muschik, *J. Liq. Chromatogr.* **9**, 1791 (1986).
50. H. J. Issaq, J. H. McConnell, D. E. Weiss, D. G. Williams and J. E. Saavedra, *J. Liq. Chromatogr.* **9**, 1783 (1986).
51. H. J. Issaq, M. Glennon, D. E. Weiss, G. N. Chmurny and J. E. Saavedra, *J. Liq. Chromatogr.* **9**, 2763 (1986).
52. S. L. Abidi, *J. Chromatogr.* **362**, 33 (1986).
53. D. W. Armstrong and W. Li, *Chromatography* **2**, 43 (1987).
54. K. W. Street Jr, *J. Liq. Chromatogr.* **10**, 655 (1987).
55. J. A. Connelly and D. L. Siehl, *Methods Enzym.* **142**, 422 (1987).
56. A. M. Stalcup, S. Chang and D. W. Armstrong, *J. Chromatogr. A* **513**, 181 (1990).
57. D. W. Armstrong, A. M. Stalcup, M. L. Hilton, J. D. Duncan, J. H. Faulkner and S. C. Chang, *Anal. Chem.* **62**, 1610 (1990).
58. A. M. Stalcup, S. C. Chang and D. W. Armstrong, *J. Chromatogr. A* **540**, 113 (1991).
59. A. M. Stalcup and K. H. Gahm, *Anal. Chem.* **68**, 1369 (1996).
60. D. W. Armstrong, S. C. Chang, Y. Wang, H. Ibrahim, G. R. Reid III and T. E. Beesley, *J. Liq. Chromatogr. Rel. Technol.* **20**, 3279 (1997).
61. B. Chankvetadze, E. Yashima and Y. Okamoto, *Chirality* **8**, 402 (1996).
62. F. Blondel, V. Peulon, Y. Prigent, R. Duval and Y. Combret, Abstracts, 8th International Symposium on Chiral Discrimination, Edinburgh, Scotland, 30 June–3 July, 1996, p. 1.
63. C. C. Elfakir, M. Dreux, E. Bourgeaux, H. Leveque and R. Duval, Abstracts, 214th ACS National Meeting, Las Vegas, NV, USA, 7–11 September 1997.
64. I. Caron, C. Elfakir and M. Dreux, Abstracts, 9th International Symposium on Cyclodextrins, Santiago de Compostela, 31 May–3 June 1998.
65. T. E. Bessley and R. P. W. Scott, *Chiral Chromatography*, Wiley, New York, 1998, p. 265.
66. W. A. König, S. Sievers and U. Schulze, *Angew. Chem. Int. Ed. Engl.* **19**, 910 (1980).
67. W. A. König and S. Sievers, *J. Chromatogr.* **200**, 189 (1980).
68. W. A. König, I. Benecke and S. Sievers, *J. Chromatogr.* **217**, 71 (1981).

69. N. Oi, T. Doi, H. Kitahara and Y. Inda, *J. Chromatogr.* **208**, 404 (1981).
70. N. Oi, T. Doi, H. Kitahara and Y. Inda, *J. Chromatogr.* **239**, 493 (1982).
71. N. Oi, H. Kitahara and T. Doi, *J. Chromatogr.* **254**, 282 (1983).
72. H. Frank, G. J. Nicholson and E. Bayer, *J. Chromatogr. Sci.* **15**, 174 (1977).
73. H. Frank, G. J. Nicholson and E. Bayer, *Angew. Chem. Int. Ed. Engl.* **17**, 363 (1978).
74. E. Bayer, *Z. Naturforsch.* **38B**, 1281 (1983).
75. V. Schurig, *Angew. Chem. Int. Ed. Engl.* **23**, 747 (1984).
76. B. Koppenhoefer and E. Bayer, *J. Chromatogr. Libr.* **32**, 1 (1985).
77. T. Saeed, P. Sandra and M. Verzele, *J. Chromatogr.* **186**, 611 (1979).
78. T. Saeed, P. Sandra and M. Verzele, *J. High Res. Chromatogr.* **3**, 35 (1980).
79. W. A. König, *The Practice of Enantiomer Separation by Gas Chromatography*, Hüthig, Heidelberg, 1987, p. 168.
80. W. A. König and I. Benecke, *J. Chromatogr.* **209**, 91 (1981).
81. W. A. König, *J. High Res. Chromatogr.* **5**, 588 (1982).
82. W. A. König, in P. Schreier, ed., *Analysis of Volatiles*, de Gruyter, Berlin, 1984, p. 77.
83. G. Schomburg, I. Benecke and G. Severin, *J. High Res. Chromatogr.* **8**, 391 (1985).
84. V. Schurig, *Angew. Chem. Int. Ed. Engl.* **16**, 1 (1977).
85. V. Schurig and W. Bürkle, *J. Am. Chem. Soc.* **104**, 7573 (1982).
86. P. Thamm, *Methoden der Organischen Chemie*, Houben-Wiley, Vol. 15/1, Thieme, Stuttgart, 1974, p. 869.
87. W. Vetter and V. Schurig, *J. Chromatogr. A* **774**, 143 (1997).
88. D. W. Armstrong, W. Y. Li and J. Pitha, *Anal. Chem.* **62**, 217 (1990).
89. D. W. Armstrong and H. L. Jin, *J. Chromatogr. A* **502**, 154 (1990).
90. D. W. Armstrong, W. Y. Li, C. D. Chang and J. Pitha, *Anal. Chem.* **62**, 914 (1990).
91. W. Y. Li, H. L. Lin and D. W. Armstrong, *J. Chromatogr. A* **509**, 303 (1990).
92. A Guide to the Use of Cyclodextrin Bonded Phases for Chiral Separations by Capillary Gas Chromatography, Advanced Separation Technologies, Inc., Whippany, NJ, 1997.
93. *Chiral Cyclodextrin Capillary GC Columns*, Supelco, Bellefonte, PA, 1988.
94. V. Schurig and A. Glausch, *Naturwiss.* **80**, 468 (1993).
95. A. Glausch, J. Nicholson, M. Fluck and V. Schurig, *J. High Res. Chromatogr.* **17**, 347 (1994).
96. G. P. Blanch, A. Glausch, V. Schurig, R. Serrano and M. J. Gonzalez, *J. High Res. Chromatogr.* **19**, 392 (1996).
97. E. Gil-Av, B. Feibush and R. Charles-Sigler, *Tetra. Lett.*, 1009 (1966).
98. H. Hühnerfuss and R. Kallenborn, *J. Chromatogr. A* **580**, 191 (1992).
99. R. Gatermann, H. Hühnerfuss, G. Rimkus, A. Attar and A. Kettrup, *Chemosphere* **36**, 2535 (1989).

100. R. Kallenborn and H. Hühnerfuss, *Chiral Environmental Pollutants, Trace Analysis and Toxicology*, Springer, Berlin, 2001.
101. R. Kallenborn, H. Hühnerfuss and W. A. König, *Angew. Chem. Int. Ed. Engl.* **103**, 328 (1991).
102. J. Faller, H. Hühnerfuss, W. A. König, R. Krebber and P. Ludwig, *Environ. Sci. Technol.* **25**, 676 (1991).
103. J. Faller, H. Hühnerfuss, W. A. König and P. Ludwig, *Mar. Pollut. Bull.* **22**, 82 (1991).
104. P. Ludwig, H. Hühnerfuss, W. A. König and W. Gunkel, *Mar. Chem.* **38**, 13 (1992).
105. H. Hühnerfuss, J. Faller, W. A. Köning and P. Ludwig, *Environ. Sci. Technol.* **26**, 2127 (1992).
106. M. Möller and H. Hühnerfuss, *J. High Res. Chromatogr.* **16**, 672 (1993).
107. H. Hühnerfuss, J. Faller, R. Kallenborn, W. A. König, P. Ludwig, B. Pfaffenberger, M. Oehme and G. Rimkus, *Chirality* **5**, 393 (1993).
108. J. Moisey, A. T. Fisk, K. A. Hobson and R. J. Norstrom, *Environ. Sci. Technol.* **35**, 1920 (2001).
109. T. A. Fisk, M. Holst, K. A. Hobson, J. Duffe, J. Moisey and R. J. Norstrom, *Arch. Environ. Contam. Toxicol.* **42**, 118 (2002).
110. B. Bethan, W. Dannecker, H. Gerwig, H. Hühnerfuss and M. Schulz, *Chemosphere* **44**, 591 (2001).
111. H. R. Busser and M. D. Müller, *Anal. Chem.* **64**, 3168 (1992).
112. H. R. Busser and M. D. Müller, *Environ. Sci. Technol.* **27**, 1211 (1993).
113. B. D. Eitzer, M. I. Mattina and W. Iannucci-Berger, *Environ. Sci. Technol.* **20**, 2198 (2001).
114. J. C. White, M. I. Mattina, B. D. Eitzer and W. Iannucci-Berger, *Chemosphere* **47**, 639 (2002).
115. A. T. Fisk, J. Moisey, K. A. Hobson, N. J. Karnovsky and R. J. Norstrom, *Environ. Pollut.* **113**, 225 (2001).
116. C. S. Wong and A. W. Garrison, *J. Chromatogr. A* **866**, 213 (2000).
117. L. Ramos, L. M. Hernandez and M. J. Gonzalez, *Anal. Chem.* **71**, 70 (1999).
118. H. R. Busser and M. D. Müller, *Anal. Chem.* **67**, 2691 (1995).
119. B. Bethan, K. Bester, H. Hühnerfuss and G. Rimkus, *Chemosphere* **34**, 2271 (1997).
120. S. Franke, C. Meyer, N. Heinzel, R. Gatermann, H. Hühnerfuss, G. Rimkus, W. A. König and W. Francke, *Chirality* **11**, 795 (1999).
121. D. W. Armstrong, G. L. Reid III, M. L. Hilton and C. D. Chang, *Environ. Pollut.* **79**, 51 (1993).
122. A. Glausch, J. Hahn and V. Schurig, *Chemosphere* **30**, 2079 (1995).
123. S. Reich, L. Jimenez, Marsili, L. M. Hernandez, V. Schurig and M. J. Gonzalez, *Organohal. Compds* **35**, 335 (1998).
124. J. R. Jones and J. H. Purnell, *Anal. Chem.* **62**, 2300 (1990).
125. R. Baycan-Keller and M. Oehme, *J. Chromatogr. A* **837**, 201 (1999).

126. M. Oehme, R. Kallenborn, K. Wiberg and C. Rappe, *J. High Res. Chromatogr.* **17**, 83 (1994).
127. A. W. Garrison, V. A. Nzengung, J. K. Avants, J. J. Ellington and N. L. Wolf, *Organohal. Compds* **31**, 256 (1997).
128. R. H. Liu and W. W. Ku, *J. Chromatogr.* **271**, 309 (1983).
129. H. Hühnerfuss, *GIT Fachz. Lab.* **36**, 489 (1992).
130. H. Hühnerfuss, *Organohal. Compds* **35**, 319 (1998).
131. H. Hühnerfuss, *Chemosphere* **40**, 913 (2000).
132. G. Cancellier, L. D'Acquarica, F. Gasparrini, D. Misiti and C. Vilani, *Chim. Ind. (Milan)* **81**, 475 (1999).
133. B. Beitler and U. Feibush, *J. Chromatogr.* **123**, 149 (1976).
134. B. Koppenhoefer and B. Lin, *J. Chromatogr.* **481**, 17 (1989).
135. A. Dietrich, B. Maas and A Mosandl, *J. High Res. Chromatogr.* **18**, 152 (1995).
136. B. Maas, A. Dietrich, T. Beck, S. Börner and A. Mosandl, *J. Microcol. Sepn* **7**, 65 (1995).
137. M. Lindström, *J. High Res. Chromatogr.* **14**, 765 (1991).
138. V. Schurig, M. Jung, S. Mayer, S. Negura, M. Fluck and H. Jakubetz, *Angew. Chem. Int. Ed. Engl.* **33**, 2222 (1994).
139. V. Schurig, M. Jung, S. Mayer, M. Fluck, S. Negura and H. Jakubetz, *J. Chromatogr. A* **694**, 119 (1995).
140. K. Watabe, R. Charles and E. Dil-Av, *Angew. Chem. Int. Ed. Engl.* **28**, 192 (1989).
141. V. Schurig, J. Ossig and R. Link, *Angew. Chem. Int. Ed. Engl.* **28**, 194 (1989).
142. H. R. Busser and M. D. Müller, *Environ. Sci. Technol.* **28**, 119 (1994).
143. V. Mani and C. Woolley, *LC–GC* **14**, 734 (1995).
144. R. L. Falconer, T. F. Bidleman, D. J. Gregor, R. Semkin and C. Teixeira, *Environ. Sci. Technol.* **29**, 1297 (1995).
145. V. Mani and C. Woolley, *Food & Food Ingred. J. Jpn* **163**, 94 (1995).
146. M. D. Müller, H. R. Busser and C. Rappe, *Chemosphere* **34**, 2407 (1997).
147. L. M. M. Jantunen and T. F. Bidleman, *Arch. Environ. Contam. Toxicol.* **35**, 218 (1998).
148. E. M. Ulrich and R. A. Hites, *Environ. Sci. Technol.* **32**, 1870 (1998).
149. V. Schurig, *J. Chromatogr.* **441**, 135 (1988).
150. V. Schurig, *J. Chromatogr. A* **965**, 315 (2002).
151. L. Heng, L. Jin and D. W. Armstrong, *J. Chromatogr. A* **509**, 303 (1990).
152. K. B. Lipkowitz, K. Green and J. A. Yang, *Chirality* **4**, 205 (1992).
153. K. B. Lipkowitz and C. M. Stoehr, *Chirality* **8**, 341 (1996).
154. R. E. Boeham and D. E. Martire, *Anal. Chem.* **60**, 522 (1988).
155. K. Linnemayr, A. Rizzi and A. Günther, *J. Chromatogr. A* **791**, 299 (1997).
156. N. Morin, Y. C. Guillaumine and J. C. Rouland, *Chromatographia* **48**, 388 (1998).
157. J. Redondo, M. A. Blazquez and A. Torrens, *Chirality* **11**, 694 (1999).

158. K. B. Lipkowitz, *J. Chromatogr. A* **906**, 417 (2001).
159. D. W. Armstrong, C. D. Chang and S. H. Lee, *J. Chromatogr. A* **539**, 83 (1991).
160. D. W. Armstrong, X. Yang, S. M. Han and R. A. Menges, *Anal. Chem.* **59**, 316 (1987).
161. A. Berthod, S. C. Chang and D. W. Armstrong, *Anal. Chem.* **64**, 395 (1992).
162. M. A. Torr, G. Nelson, G. Patonay and I. M. Warner, *Anal. Lett.* **21**, 843 (1988).
163. J. I. Seeman, H. V. Secor, D. W. Armstrong, K. D. Timmons and T. J. Ward, *Anal. Chem.* **60**, 2120 (1988).

Chapter 7

The Analysis of Chiral Pollutants by High Performance Liquid Chromatography

7.1 Introduction

At the present time, high performance liquid chromatography (HPLC) is the most popular and most highly applicable technology in the field of chiral analysis of a variety of racemates. In high performance liquid chromatographic methods, two approaches are used; indirect and direct. The indirect chromatographic separation of racemic mixtures can be achieved by derivatization of the racemic pollutant with a suitable chiral derivatizing agent (CDA), resulting in the formation of diastereoisomeric complexes/salts which, since they have different physical and chemical properties, can be separated from each other by an achiral liquid chromatographic method. The indirect chromatographic approach is not useful and popular, due to certain drawbacks as discussed in Chapter 1. On the other hand, the direct chromatographic approach involves the use of chiral selectors either in the mobile phase, where they are called chiral mobile phase additives (CMPAs), or in the stationary phase, where they are called chiral stationary phases (CSPs). Over the course of time, various types of liquid chromatographic approaches have been developed and used in this application, but HPLC remains the most suitable modality, due to its various advantages

Chiral Pollutants.   I. Ali and H. Y. Aboul-Enein
© 2004 John Wiley & Sons, Ltd   ISBN: 0-470-86780-9

in comparison to other techniques (Chapter 1). Its high speed, sensitivity and reproducibility make HPLC the method of choice in all of the world's analytical laboratories. The availability of chiral selectors in the form of HPLC columns also makes it the technique of choice for the analysis of chiral pollutants. A variety of mobile phases including, normal, reversed and new polar organic phases, are used in HPLC. About 80 % chiral resolution of pharmaceuticals, drugs, agrochemicals and other compounds has been carried out using HPLC [1–6]. HPLC has also been used for the chiral separation of environmental pollutants [7, 8]. Due to the importance and the future scope of HPLC in the analysis of chiral pollutants, attempts have been made in the present chapter to describe the chiral resolution of environmental pollutants by HPLC. Efforts have also been made to explain the optimization and the mechanisms of analysis of chiral environmental pollutants by HPLC.

7.2 Chiral Selectors

It is well known that chiral selectors are required for the enantiomeric resolution of racemic compounds in liquid chromatography. Many chiral compounds have been used as chiral selectors in HPLC. The most important classes of chiral selectors are polysaccharides, cyclodextrins, macrocyclic glycopeptide antibiotics, proteins, crown ethers, ligand exchangers, Pirkle-type I CSPs (see below) and many other compounds. To make the chapter concise, a discussion on the structures and the properties of these chiral selectors has not been included. However, the structures and properties of the cyclodextrins and their derivatives have been discussed in Chapter 6. These chiral selectors are available in the form of columns and capillaries and hence are called chiral stationary phases (CSPs). The development of these CSPs makes HPLC a superb class of liquid chromatographic techniques in the chiral resolution of a wide range of racemic compounds. A list of the most important and useful CSP columns supplied by various commercial manufacturers is given in Table 7.1. These CSPs have been used frequently and successfully for the chiral resolution of many drugs, pharmaceuticals and other compounds. Therefore, they may also be used for the enantiomeric separation of chiral environmental pollutants, and some reports on the chiral separation of environmental pollutants using the above-mentioned CSPs are to be found in the literature. For details regarding the structures and properties of these CSPs, readers should consult the book by Aboul-Enein and Ali [1].

Table 7.1 Various CSPs used in HPLC and supplied by different manufacturers

Commercialized CSPs	Company
Polysaccharide-based CSPs	
Chiralcel OB, Chiralcel OB-H, Chiralcel OJ, Chiralcel OJ-R,Chiralcel CMB, Chiralcel OC, Chiralcel OD, Chiralcel OD-H, Chiralcel OD-R, Chiralcel OD-RH, Chiralcel OF, Chiralcel OG, Chiralcel OA, Chiralcel CTA, Chiralcel OK, Chiralpak AD, Chiralpak AD-R, Chiralpak AD-RH, Chiralpak AR and Chiralpak AS	Daicel Chemical Industries, Tokyo, Japan
Cyclodextrin-based CSPs	
Cyclobond I, II and III; Cyclobond AC, RN, SN; ApHpera ACD and BCD	Advanced Separation Technologies, Inc., Whippany, NJ, USA
Nucleodex β-OH and Nucleodex β-PM	Macherey-Nagel, Düren, Germany
ORpak CD-HQ	
ORpak CDB-453 HQ, ORpak CDBS-453, ORpak CDA-453 HQ and ORpak CDC-453 HQ	Showa Denko, Kanagawa, Japan
Keystone β-OH and Keystone β-PM	Thermo Hypersil, Bellefonte, PA, USA
YMC Chiral cyclodextrin BR, YMC Chiral NEA (R) and YMC Chiral NEA (S)	YMC, Kyoto, Japan
β-Cyclose-6-OH	Chiral Separations, La Frenaye, France
Macrocyclic glycopeptide antibiotic based CSPs	
Chirobiotic R, Chirobiotic T, Chirobiotic V, Chirobiotic TAG and Chirobiotic modified V	Advanced Separation Technologies, Inc., Whippany, NJ, USA
Protein-based CSPs	
Chiral AGP, Chiral HSA and Chiral CBH	Advance Separation Technologies, Inc., Whippany, NJ, USA
Chiral AGP, Chiral CBH and Chiral HSA	Chrom Tech, Ltd, Cheshire, UK
Resolvosil BSA-7 and Resolvosil BSA-7PX	Macherey-Nagel, Düren, Germany
Chiral AGP, Chiral CBH and Chiral HSA	Regis Technologies, Morton Grove, IL, USA

(*continued overleaf*)

Table 7.1 (*continued*)

Commercialized CSPs	Company
AFpak ABA-894	Showa Denko, Kanagawa, Japan
Keystone HSA and Keystone BAS	Thermo Hypersil, Bellefonte, PA, USA
TSKgel Enantio L1 and TSK gel Enantio-OVM	Tosoh, Tokyo, Japan
EnantioPac	LKB Pharmacia, Bromma, Sweden
Bioptic AV-1	GL Sciences, Tokyo, Japan
Ultron ES-BSA, Ultron ES-OVM Column, Ultron ES-OGP Column and Ultron ES-Pepsin	Shinwa Chemical Industries, Kyoto, Japan
Crown ether based CSPs	
Crownpak CR	Chiral Technologies, Inc., Exton, PA, USA
Crownpak CR	Separations Kasunigaseki-Chrome, Tokyo, Japan
Crownpak CR	Daicel Chemical Industries, Tokyo, Japan
Opticrown RCA	USmac Corporation, Winnetka, Glenview, IL, USA
Chiralhyun-CR-1	K-MAC (Korea Materials & Analysis Corp.), South Korea
Chirosil CH RCA	Restech Corporation, Daedeok, Daejon, South Korea
Ligand exchange based CSPs	
Chirosolve	JPS Chemie, Switzerland
Chiralpak WH, Chiralpak WM, Chiralpak WE and Chiralpak MA	Separations Kasunigaseki-Chrome, Tokyo, Japan
Chiralpak WH and Chiralpak WM	Daicel Chemical Industries, Tokyo, Japan
Nucleosil Chiral-1	Macherey-Nagel, Düren, Germany
Phenylglycine and leucine types	Regis Technologies, Morton Grove, IL, USA
Chirex types	Phenomenex, Torrance, USA

Table 7.1 (*continued*)

Commercialized CSPs	Company
Orpak CRX-853	Showa Denko, Kanagawa, Japan
Pirkle-type CSPs	
Opticrown Chiralhyun-Leu-1 and Opticrown Chiralhyun-PG-1	Usmac Corporation, Glenview, IL, USA
Whelk-O 1, Whelk-O 2, Leucine, Phenylglycine, β-Gem 1, α-Burke 1, α-Burke 2, Pirkle 1-J, Naphthylleucine, Ulmo and Dach	Regis Technologies, Morton Grove, IL, USA
Nucleosil Chiral-2	Macherey-Nagel, Düren, Germany
Sumichiral OA	Sumika Chemical Analysis Service, Konohana-ku Osaka, Japan
Kromasil Chiral TBB and Kromasil Chiral DMB	Eka Chemicals Separation Products, Bohus, Sweden
Chirex Type I	Phenomenex, Torrance, CA, USA
Chiris series	IRIS Technologies, Lawrence, KS, USA

7.3 Applications

In spite of the variety of CSPs available for HPLC, it has not been used very frequently for the analysis of chiral environmental pollutants. This is due to the fact that some organochlorine pollutants are transparent to UV radiation and, hence, the very popular UV detector cannot be used in HPLC for the purpose of detecting such xenobiotics. However, HPLC can be coupled with MS, polarimetry and other optical detection techniques for the chiral resolution of such types of pollutant. Apart from these points, some reports are available on the chiral resolution of some UV-absorbing chiral environmental pollutants by HPLC.

Cyclodextrin-based CSPs have been used for the chiral resolution of some pollutants. Schurig *et al.* [9, 10], König *et al.* [11] and Hardt *et al.* [12] used CD-based CSPs for the chiral separation of PCBs. Mannscreck *et al.* [13] resolved PCBs on triacetylcellulose CSP. Ludwig *et al.* [14] used the Chiral AGP column for the enantiomeric analysis of 2-(2,4-dichlorophenoxy)propionic acid (dichlorprop) and its degradation products in a marine microbial community. The authors used a

water–2-propanol–phosphate buffer (10 mM, pH 4.85) (94 : 4 : 2, v/v/v)
as the mobile phase, with detection at 240 nm. Veter *et al.* [15,
16] described the separation of the enantiomers of toxaphene on
tert-butyldimethylsilylated-β-CD-based CSP using HPLC. Blessington
and Crabb [17] used the Chiral AGP column for the enantiomeric
resolution of aryloxypropionate herbicides. The authors used a phosphate
buffer (10 mM, pH 6) – 2-propanol (94 : 4, v/v) as the mobile phase,
with detection at 240 nm. The structures of these herbicides are
given in Figure 7.1. Armstrong *et al.* [18] analysed the enantiomers of
warfarin, coumachlor, coumafuryl, bulan, crufomate, fonofos, anacymidol,
napropamide and 2-(3-chlorophenoxy)-propionamide pollutants by HPLC.
The authors used Chiral AGP and other cyclodextrin-based CSPs. Chu
and Wainer [19] separated the enantiomers of warfarin in serum using
chiral HPLC. Weber *et al.* [20] used permethylated β-CD CSP for the
chiral analysis of phenoxypropionates. Shou and Yang [21] reported the
separation of the enantiomers of 1-hydroxy-3-hydroxymethylcholanthrene
(1-OH-3-OHMC), 3-methylcholanthrene (3MC), *trans-* and *cis*-1,2-
diols and 1-hydroxy-3-methylcholanthrene (1-OH-3MC) using a (*R*)-
N-(3,5-dinitrobenzoyl)phenylglycine (Pirkle-type) CSP. Schneiderheinze
et al. [22] investigated the chiral resolution of phenoxyalkanoic acid
herbicides – that is, 2-(2,4-dichlorophenoxy)propionic acid (2,4-DP) and
2-(4-chloro-2-methylphenoxy)propionic acid (MCPP) – in plant and soil
samples. The authors used a Chirobiotic T column as CSP with
methanol – 1 % triethylammonium acetate, pH 4.1 (60 : 40, v/v) as the

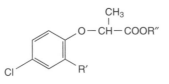

Compound	R′	R″
CMPP	CH$_3$	H
2,4-DP	Cl	H
CMPP-BOE ester	CH$_3$	—CH$_2$CH$_2$OCH$_2$CH$_2$CH$_2$CH$_3$
CMPP-2-EH ester	CH$_3$	—CH$_2$CHCH$_2$CH$_2$CH$_2$CH$_3$ $\quad$ C$_2$H$_5$

Figure 7.1 The chemical structures of aryloxypropionate herbicides [17].

mobile phase. The detection was carried out by chiroptical detector. Caccamese and Principato [23] separated the enantiomers of vincamine alkaloids using Chiralpak AD as CSP. Various mixtures of hexane–2-propanol and hexane–ethanol were used as mobile phases, separately. Möller *et al*. [24] analysed the enantiomers of α-HCH (in the brains of seals) using the Chiraldex column. The mobile phase used was methanol–water (75 : 25, v/v). Ellington *et al*. [25] described the successful separations of the enantiomers of various organophosphorous pesticides using the Chiralpak AD, Chiralpak AS, Chiralcel OD, Chiralcel OJ and Chiralcel OG chiral columns. The authors used different mixtures of heptane and ethanol as the mobile phases. The different organophosphorous pesticides resolved were crotoxyphos, dialifor, fonofos, fenamiphos, fensulfothion, isofenphos, malathion, methamidophos, profenofos, crufomate, prothiophos and trichloronat. Dondi *et al*. [26] resolved chrysanthemic acid [2,2-dimethyl-3-(2-methylpropenyl)-cyclopropanecarboxylic acid] and its halogen-substituted analogues using a terguride-based chiral stationary phase in combination with a UV diode array and chiroptical detectors. Isomers with (1*R*) configuration always eluted before those with (1*S*) configuration. The elution sequence of *cis*- and *trans*-isomers was strongly affected by the mobile phase pH, whereas the enantioselectivity remained the same. This method was used to monitor the hydrolytic degradation products of Cyfluthrin (Baythroid) in soil under laboratory conditions. Müller *et al*. [27] resolved two of the four metolachlor isomers, from racemic metolachlor in enantio- (ee > 98 %) and diastereomerically pure forms, using achiral and chiral high performance liquid chromatographic approaches. The two isomers were identified as the SS- and the RR-isomers by polarimetric measurements, with reference to previous data. Ali and Aboul-Enein [28] determined the chiral resolution of *o,p*-DDT and *o,p*-DDD. The enantiomeric resolution of *o,p*-DDT and *o,p*-DDD has been achieved on the Chiralpak AD-RH, Chiralcel OD-RH and Chiralcel OJ-R chiral stationary phases. The mobile phases used were acetonitrile–water (50 : 50, v/v) and acetonitrile–2-propanol (50 : 50, v/v) at a 1.0 ml min^{-1} flow rate. The detection was carried out at 220 nm for both pesticides. Li *et al*. [29] developed a fast and precise HPLC method for the chiral resolution of phenthoate in soil samples. The authors used the Chiralcel OD column for chiral resolution with hexane–2-propanol (100 : 0.8, v/v) as the mobile phase. Guillaume *et al*. [30] resolved the enantiomers of phenoxy acid herbicides [2-phenoxypropionic acid (2-PPA), 2-(2-chlorophenony)propionic acid (2-CPPA), 2-(3-chlorophenony)propionic acid (3-CPPA), and 2-(4-chlorophenony)propionic acid (4-CPPA)] on

Table 7.2 The chiral resolution of some environmental pollutants by HPLC

Chiral pollutants	Sample matrix	CSPs	Detection	References
PCBs	–	CD-based CSPs	–	9–12
	–	Triacetylcellulose CSP	–	13
2-(2,4-Dichloro-phenoxy)propionic acid (dichlorprop)	Marine microbial community	Chiral AGP	UV, 240 nm	14
Toxaphene	–	*tert*-Butyldi-methylsilylated – β-CD based		15, 16
Aryloxypropionate herbicides	–	Chiral AGP	UV, 240 nm	17
Warfarin, coumachlor, coumafuryl, bulan, crufomate, fonofos, anacymidol, napropamide and 2-(3-chlorophenoxy)-propionamide	–	Chiral AGP and other CD-based CSPs	–	18
Warfarin	Serum	–	–	19
Phenoxypropionates	–	Permethylated β-CD CSP	–	20
1-Hydroxy-3-hydroxy-methylcholanthrene (1-OH-3-OHMC), 3-methylcholanthrene (3MC), *trans*- and *cis*-1,2-diols and 1-hydroxy-3-methylcholanthrene (1-OH-3MC)	–	(*R*)-*N*-(3,5-dinitrobenzoyl)-phenylglycine (Pirkle-type) CSP	–	21
Phenoxyalkanoic acid herbicides, i.e. 2-(2,4-dichlorophenoxy)-propionic acid (2,4-DP) and 2-(4-chloro-2-methylphenoxy)-propionic acid (MCPP)	Plant and soil samples	Chirobiotic T	Chiroptical detector	22
Vincamine alkaloids	–	Chiralpak AD	–	23
α-HCH	Brain of seals	Chiraldex	–	24

Table 7.2 (*continued*)

Chiral pollutants	Sample matrix	CSPs	Detection	References
Organophosphorous pesticides, i.e. crotoxyphos, dialifor, fonofos, fenamiphos, fensulfothion, isofenphos, malathion, methamidophos, profenfos, crufomate, prothiophos and trichloronat	–	Chiralpak AD, Chiralpak AS, Chiralcel OD, Chiralcel OJ and Chiralcel OG	UV, 235 nm, and polarimetry	25
Chrysanthemic acid [2,2-dimethyl-3-(2-methylpropenyl)-cyclopropane-carboxylic acid] and its halogen-substituted analogues	–	Terguride-based CSP	UV diode array and chiroptical detectors	26
Metolachlor	–	–	Polarimeter	27
o,*p*-DDT and *o*,*p*-DDD	–	Chiralpak AD-RH, Chiralcel OD-RH and Chiralcel OJ-R	UV, 220 nm	28
Phenthoate	Soil samples	Chiralcel OD	UV, 230 nm	29
Phenoxy acid herbicides, i.e. [2-phenoxy-propionic acid (2-PPA), 2-(2-chlorophenony) propionic acid (2-CPPA), 2-(3-chlorophenony)-propionic acid (3-CPPA), and 2-(4-chlorophenony)-propionic acid (4-CPPA)]	–	Chirobiotic T	Diode array	30

Chirobiotic T. The authors achieved chiral separation of these pesticides using methanol as the major constituent of the mobile phase, with detection by a diode array detector. The chiral resolution of some environmental pollutants by HLPC is given in Table 7.2. As a specific example, the chromatograms of the resolved enantiomers of crufomate and α-HCH are shown in Figure 7.2. As usual, the enantiomeric resolution is expressed by

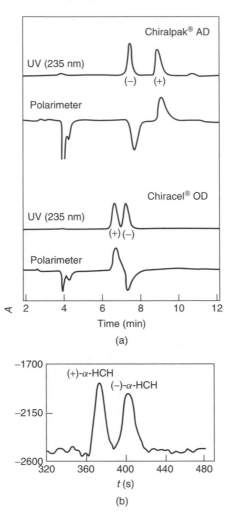

(a)

(b)

Figure 7.2 Chromatograms of chiral resolution of (a) crufomate on Chiralpak AD and Chiralcel OD CSPs [25] and (b) α-HCH on a Chiraldex column [24].

calculating the retention time (t), the capacity factor (k), the separation factor (α) and the resolution factor (R_s).

7.4 The Optimization of HPLC Conditions

As in the case of GC, the optimization of chiral analysis in HPLC is the key issue in obtaining maximum chiral separation. First of all, the selection of the CSP is very important and then the other parameters should be optimized. The different controlling factors for chiral resolution include the composition of the mobile phase, its pH, the temperature, the amount injected on the HPLC machine, detection and so on. The art of the optimization of chiral resolution varies from one CSP to another: therefore, each type of CSP is briefly discussed separately below. The above-mentioned CSPs have frequently been used in HPLC for the chiral resolution of a wide variety of drugs, pharmaceuticals and other compounds [1–7]. Therefore, the following discussion on optimization strategies, which is based on knowledge about their prior use in these areas of application may be used for optimization of the chiral separation of environmental pollutants.

7.4.1 Polysaccharide CSPs

Chiral resolution is sensitive on polysaccharide-based CSPs and, therefore, the optimization of the HPLC conditions on these phases is an important issue, since they are widely used. The selection of the mobile phase is the key aspect in chiral resolution. The mobile phase is selected according to the solubility and structure of the pollutants to be resolved. In the normal phase mode, the use of pure ethanol or 2-propanol is recommended. To decrease the polarity of the mobile phase and increase the retention times of the enantiomers, hexane, cyclohexane, pentane and heptane are used as one of the main constituents of the mobile phase. However, other alcohols are also used in the mobile phase. Normally, if pure ethanol or 2-propanol are not suitable mobile phases, hexane, 2-propanol or ethanol in the ratio of 80 : 20 are used as the mobile phase, and the change in mobile phase composition is carried out on the basis of observations. Finally, the optimization of the chiral resolution is carried out by adding small amounts of amines or acids (0.1–1.0 %). According to Wainer *et al.* [31], the decreasing effect of the retention factor upon increasing the polar modifier content indicates that the competition for the binding sites on the CSP is a saturable process, and that a maximum retention factor effect will be reached at a certain polar modifier concentration. It is interesting to note that only a small change in the value of the retention factor occurs when the polar modifier content is varied from 5 % to 25 %. Ellington *et al.* [25] studied the effect of the concentration of ethanol on the chiral resolution of organophosphorous pesticides. The authors reported poor separation of the enantiomers of fenamiphos, fensulfothion, isofenphos, profenofos, crufomate and trichloronat pesticides at higher concentrations of ethanol. The effect of the concentration of ethanol on the chiral resolution of trichloronat on Chiralcel OJ is shown in Figure 7.3. It may be concluded that 0 % concentration of ethanol is the best, as it gives the maximum

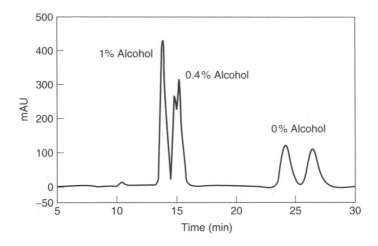

Figure 7.3 The effect of alcohol on the chiral resolution of trichloronate on a Chiralcel OJ column [25].

resolution of trichloronat on the Chiralcel OJ column. Furthermore, the authors reported the reverse order of elution of these pesticides on different CSPs. As an example, the reverse order of crufomate was observed on Chiralpak AD and Chiralcel OD CSPs (Figure 2(a)). Gaffney [32] reported the reverse order of elution of 2-phenoxypropanoic acid on a Chiralcel OB CSP when using different alcohols.

Chiral resolution on polysaccharide-based CSPs in reversed phase mode is carried out using aqueous mobile phases. Again, the selection of the mobile phase depends on the solubility and the properties of the pollutants to be analysed. The choice of the mobile phase in the reversed phase mode is very limited. Water is used as the main constituent of the mobile phases. The modifiers used are acetonitrile, methanol and ethanol. Optimization of the chiral resolution is carried out by adding small percentages of amines or acids (0.1–1.0 %). Some of the resolutions are pH dependent and require a constant pH of the mobile phase. Under such conditions, in general, the resolution is not reproducible when using mobile phases such as water–acetonitrile or water–methanol and, therefore, buffers with some organic modifiers (acetonitrile, methanol etc.) have been used as the mobile phase. The optimization of the resolution is carried out by adjusting the pH values of the buffers and the amounts of the organic modifiers. The most commonly used buffers are perchlorate, acetate and phosphate. The correlation of the separation conditions of neutral, acidic and basic drugs on polysaccharide-based CSPs is presented in Table 7.3. From this table, it

Table 7.3 Correlation of the separation conditions for neutral, acidic and basic compounds/pollutants

Compound	System	
	Normal phase	Reversed phase[a]
Neutral	MP = IPA–hexane, pH has no effect on the resolution	MP = water–ACN, pH has no effect on the resolution
Acidic	MP = IPA–hexane–TFA, pH near 2.0	MP = pH 2.0 perchlorate, acid–ACN
Basic	MP = IPA–hexane–DEA, IPA–hexane–TFA with pH near 2.0, ion-pair separation	MP = pH < 7 buffer–ACN, typical buffer is 0.5 M NaClO₄

[a]Columns are normally not run under basic conditions.
MP, mobile phase; IPA, isopropanol; ACN, acetonitrile; TFA, trifluoroacetic acid; DEA, diethylamine.

may be concluded that a simple mixture of water and an organic modifier is sufficient to obtain a chiral separation of neutral pollutants, because there is no competitive ionic interaction to deal with. For acidic pollutants, it is essential to use an acidic mobile phase, to suppress the dissociation of the analyte and minimize the ionic interactions. For basic pollutants, a mobile phase with a pH greater than 7 is suitable, as the chance of cation formation with the hydronium cation decreased at this pH value and ionic interaction does not occur. The resolution was affected by changing the polarity of the mobile phases. It is interesting to note that the change in resolution with respect to the mobile phase compositions varied from compound to compound. The resolution in reversed phase mode has been improved by adding cations and anions, the most commonly used ions being Na^+, Li^+, K^+, NH_4^+, $N(C_2H_5)_4^+$, ClO_4^-, SCN^-, I^-, NO_3^-, Br^-, Cl^- and AcO^-.

Many researchers have documented the effect of the mobile phase on the enantioselectivity of different racemates on polysaccharide-based CSPs. However, up to the present time, no comprehensive study aimed at identifying the association between the structural features present on solute and appropriate mobile phase conditions has yet been proposed. Piras *et al.* [33] have studied the characteristic features of about 2363 racemic molecules separated on a Chiralcel OD CSP. The mobile phase used for these racemates was compared with their structures, which are available from CHIRBASE (http://www.chirbase.u-3mrs.fr/chirbase/). The data set-up was submitted to data-mining programs for molecular pattern recognition and mobile phase predictions for new cases. Some of the substructural characteristics of the solutes were related to the efficient use of some specific mobile phases. For example, the application of acetonitrile/salt buffer at pH 6–7 was found to be convenient for the reversed phase separation of compounds bearing a tertiary amine functional group. Furthermore, a cluster analysis allowed the arrangement of the mobile phase according to similarities found in molecular patterns of solutes.

The pH value also controls the chiral resolution of different racemic pollutants on polysaccharide-based CSPs, but no reports have been published on this issue considering the pollutants. However, Aboul-Enein and Ali [34] have observed that chiral resolution on polysaccharide-based CSPs is pH dependent under the normal phase mode. It was observed that only partial resolution of certain antifungal agents was achieved at lower pH, while the resolution was improved by increasing the pH using triethylamine on amylose and cellulose chiral columns. The chiral resolution of tetralone derivatives was studied over a range of pH values on a Chiralpak AD-RH column by Aboul-Enein and Ali [35]. The mobile phase that contained triethylamine was more effective in the resolution of these derivatives than the mobile phase that contained acetic acid. The effect of pH on the retention factors of flurbiprofen has been investigated on polysaccharide-based CSPs in the reversed phase mode [36]. Aboul-Enein and Ali [37] have studied the effect of pH on the chiral resolution of flurbiprofen on Chiralpak AD-RH. In the reversed phase mode, chiral resolution on polysaccharide CSPs is pH dependent; therefore, buffers have been used to achieve the best resolution, and it follows that the enantiomeric resolution of the pollutants may be optimized by adjusting the pH.

Temperature also contributes to the chiral resolution of racemic drugs on polysaccharide CSPs. Even then, only a few studies are available that deal with the influence of temperature on chiral resolution. The capacity and the separation factors may be related to temperature using the following equations:

$$\ln k = -\Delta H^\circ / RT + \Delta S^\circ / R + \ln \Phi \qquad (7.1)$$

$$\ln \alpha = \ln(k_2 / k_1) = -\delta H^\circ / RT + \delta S^\circ / R \qquad (7.2)$$

where R is the ideal gas constant, T is the absolute temperature and Φ is the phase ratio [38]. ΔH° and ΔS° represent the enthalpy and entropy terms for each enantiomer, and δH° and δS° represent their differences. k_2 and k_1 are the retention factors of the two resolved enantiomers. According to these equations, both the capacity and the separation factors are controlled by an enthalpic contribution, which decreases with the elevation of temperature, and an entropic contribution, which is temperature-dependent. The selectivity is a compromise between differences in the enantiomeric binding enthalpy and disruptive entropic effects [39]. Ellington *et al.* [25] studied the effect of temperature on the chiral resolution of organophosphorous pesticides and the authors reported the maximum chiral resolution at low temperature. The effect of temperature (from 20 to 60 °C) on the chiral resolution of fensufothion on Chiralcel OJ is shown in Figure 7.4. It may be concluded from this figure that chiral resolution is improved at low temperature and becomes a maximum at 20 °C.

Chiral resolution on polysaccharide CSPs occurs due to the different types of bondings between racemates and CSP, which will be discussed later in this chapter. Therefore, different racemate structures provide different types of bondings and, therefore, the different patterns of chiral recognition observed on different CSPs. Recently, Ali and Aboul-Enein [28] studied the effect of various CSPs on the chiral resolution of *o,p*-DDT and *o,p*-DDD on Chiralpak AD-R, Chiralcel OD-R and Chiralcel OJ-R. The results of this research are given in Table 7.4, which shows that the best resolution of these pesticides was obtained on Chiralpak AD-R under the normal phase mode. The chiral resolution can also be controlled by the flow rate on polysaccharide CSPs. However, there are only few studies that deal with the optimization of chiral resolution by adjusting flow rates. Yashima *et al.* [40] recently studied the effect of the pore size of silica gel, the

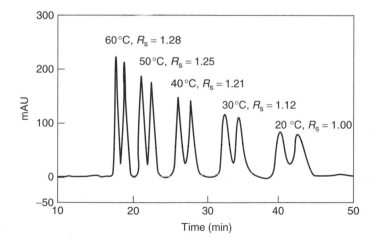

Figure 7.4 The effect of temperature on the chiral resolution of fensulfothion on a Chiralcel OJ column [25].

Table 7.4 The retention factor (k), separation factor (α) and resolution factor (R_s) for the enantiomeric resolution of o,p-DDT and o,p-DDD on polysaccharide-based CSPs under reversed phase [28]

	k_1	k_2	α	R_s
Acetonitrile–water (50 : 50, v/v)				
Chiralpak AD-RH				
o,p-DDT	15.41	19.77	1.24	2.47
o,p-DDD	nr			
Chiralcel OD-RH				
o,p-DDT	4.54	10.28	2.27	2.03
o,p-DDD	nr			
Chiralcel OJ-R				
o,p-DDT	3.49	8.80	2.52	0.80
o,p-DDD	nr			
Acetonitrile–2-propanol (50 : 50, v/v)				
Chiralpak AD-RH				
o,p-DDT	4.74	8.00	1.69	1.00
o,p-DDD	3.26	4.11	1.26	0.60
Chiralcel OD-RH				
o,p-DDT	nr			
o,p-DDD	nr			
Chiralcel OJ-R				
o,p-DDT	nr			
o,p-DDD	nr			

nr, Not resolved.

coating amount and the coating solvent on the chiral discrimination of some aromatic racemates. They concluded that a CSP with a silica gel that had a large pore size and a small surface area showed higher chiral recognition. CSP coated with acetone as the coating solvent was found to show a good chiral resolution capacity. In addition to these parameters, the optimization of chiral resolution on these CSPs can be achieved by varying other HPLC conditions, such as the particle size of the CSP, the dimensions of the column, the concentrations of racemic compounds and the choice of a suitable detector. The protocol for the chiral resolution of pollutants

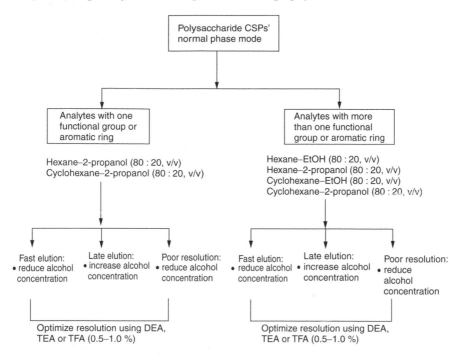

Scheme 7.1 A protocol for the development and optimization of mobile phases on polysaccharide CSPs under normal phase mode. Note that this is only a brief outline of the procedure that should be followed in developing chiral resolution on polysaccharide CSPs under normal phase mode. Other mobile phases may also be used.

on polysaccharide-based CSPs is presented in Schemes 7.1 and 7.2, under normal and reversed phase modes respectively.

7.4.2 Cyclodextrin CSPs

All three CDs, α-, β- and γ-, have different selectivities and enantiorecognition capacities due to their different cavity diameters and lengths. In general, the α-, β- and γ-CDs are suitable for chiral resolution of small, medium and large molecules, respectively. Moreover, with the development of the different derivatives of the CDs, their chiral recognition capacities vary greatly. In view of this, some reports have been published on the chiral resolution on different CD-based CSPs, and the results have been compared and discussed. The formation of CD inclusion complexes is favoured in an aqueous medium, and hence most of the chiral resolutions have been carried out in reversed phase mode. However, given the development of the various derivatives of the CDs, the use of normal and polar organic mobile phases

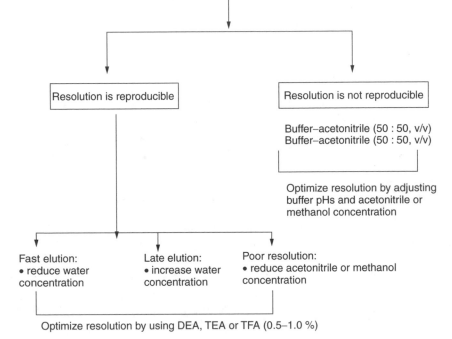

Scheme 7.2 A protocol for the development and optimization of mobile phases on polysaccharide CSPs under reversed phase mode. Note that this is only a brief outline of the procedure that should be followed in developing chiral resolution on polysaccharide CSPs under reversed phase mode. Other mobile phases may also be used.

may be possible. The selectivities of these phases increase in the order of normal > polar organic > reversed phase modes. This is due to certain advantages associated with normal and polar organic phase modes. Due to the possibility of using all three modes of mobile phases, the optimization on these phases can be achieved by varying a number of parameters, such as the composition of the mobile phase, the pH of mobile phase, the use of organic modifiers, the flow rate and the temperature. The art of the optimization of chiral resolution on these CSPs is presented herein.

Most chiral resolutions on CD-based CSPs have been carried out using aqueous mobile phases. Buffers of different concentrations and pH values have been developed and used for this purpose. Triethylammonium acetate (TEAA), phosphate, citrate and acetate are among the most commonly used buffers [41–44]. Also, phosphate buffers, such as sodium, potassium and ammonium phosphate, are commonly used. The stability constant of the complexes decreases due to the addition of organic solvents and hence the organic modifiers are used to optimize the chiral resolution. The most commonly used organic solvents are methanol and acetonitrile. Acetonitrile is a stronger organic modifier than methanol. Some other organic modifiers such as ethanol, 2-propanol, 1-propanol, n-butanol, tetrahydrofuran, triethylamine and dimethylformamide have also been used for the optimization of chiral resolution on CD-based CSPs [41, 42, 45, 46]. The effect of the type and concentration of these organic modifiers varies from one analyte to another, and hence it is very difficult to predict the best strategy for their use as organic modifiers. Sometimes, the use of highly concentrated buffers under the reversed phase mode decreases the life time and efficiency of the column. Therefore, the use of an alternative mobile phase – that is, a normal phase – is an advantage in chiral resolution on these phases. The most commonly used solvents in normal phase mode are hexane, cyclohexane and heptane. However, some other solvents, such as dichloromethane, acetone, propanol, ethyl acetate, ethanol and chloroform, have also been used as components of mobile phases. The concentration of the buffer is a very important aspect of chiral resolution on these phases under the reversed phase mode. The addition of salts into the reversed mobile phase has been found to improve the chiral resolution [44]. In some cases, as the buffer concentration increases, the retention times decrease and the peaks sharpen. Therefore, some studies have been conducted to evaluate the effect of buffer concentration on the chiral resolution on these phases. Haynes *et al.* [47] optimized the chiral resolution of phenoxypropionic acid herbicides by adjusting the concentration of the buffer. Many chiral resolutions have been carried out in the reversed phase mode. Therefore, the pH value of the mobile phase is very important, as it affects the degree of complexation of the analytes with the CD. In 1985, Feitsma *et al.* [43] studied the chiral resolution of aromatic acids. The authors reported that the separation of mandelic acid observed at pH 4.2 disappeared at pH 6.5. It was observed that cyclohexylphenylglycolic acid and cyclohexylphenylacetic acid were highly retained at pH 4.2 and pH 6.5, respectively. Hayens *et al.* [47] reported that the pH value had a major effect on the chiral resolution of phenoxypropionic acid herbicides.

Recently, it has been observed [48] that it is possible to override inclusion complexation in favour of interacting directly with the secondary hydroxyl groups across the larger opening of the cyclodextrin toroid or the appendant carbamate, acetate or hydroxypropyl functional groups. To accomplish this, a polar organic mode, comprising a polar organic solvent such as methanol or acetonitrile, has been developed which produces very efficient separations that have not previously been reported on these phases. Compounds such as warfarin and other analytes have been separated using this polar organic phase mode. This mode is also suitable for molecules that do not contain an aromatic group. Moreover, chiral separations using this polar organic phase mode are more efficient, reproducible and effective. The ratio of acid to base is also crucial and the separation is optimized by the acid and base concentrations. If the analyte is eluting too fast, the concentration of the acid/base is reduced. Conversely, if the analyte is too well retained, the acid/base concentration is increased. The parameters for the concentrations are between 1 % and 0.001 %.

The kinetics of inclusion complex formation between enantiomers and CDs are temperature controlled and, therefore, temperature may be used to optimize chiral resolution on these phases. The degree to which temperature affects the resolution is dependent on the analyte and the CD. In general, a low temperature results in an improvement in chiral resolution. The enantioseparations on these phases depend on the placement of the enantiomers in the cavities of the CDs. Therefore, the flow rate may be used as one of the optimizing parameters by displacing the enantiomers from the cavities of CDs. In this way, optimization of the chiral resolution can be controlled by the flow rate in all three mobile phase modes. Some other factors also contribute to chiral resolution on these phases, including the injection amount, the particle size of the silica gel, the spacer between the CD and the silica gel and the dimensions of the column. Hargitai and Okamoto [46] studied the influence of the pore diameter of the silica gel on the chiral resolution of some compounds. Pore diameters of 60, 100 and 300 Å were used for this purpose. The capacity of the enantioseparations of silica-based CSPs varied from one compound to another. In addition to this, the enantiorecognition on these CSPs may be optimized using columns of differing dimensions. The optimization of the chiral resolution of environmental pollutants on CD-based CSPs is presented in

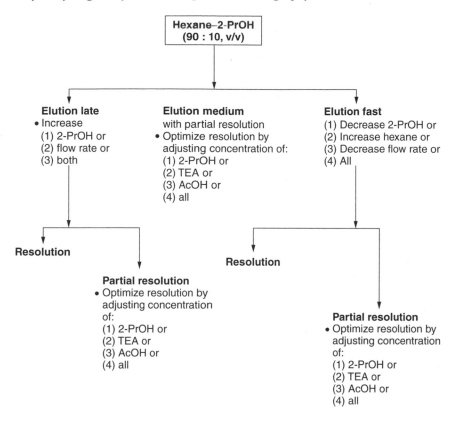

Scheme 7.3 A protocol for the development and optimization of mobile phases on CD-based CSPs under normal phase mode. Note that this is only a brief outline of the procedure that should be followed in developing chiral resolution on CD-based CSPs.

Schemes 7.3, 7.4 and 7.5, under normal, reversed and polar organic phase modes respectively.

7.4.3 Macrocyclic Glycopeptide Antibiotic CSPs

The macrocyclic glycopeptide antibiotics, which are used as chiral selectors in HPLC, are vancomycin and vancomycin aglycon, teicoplanin and teicoplanin aglycon, Ristocetin A, thiostrepton, rifamycin, kanamycin, streptomycin and fradiomycin, and avoparcin. They have differing structures, with different atoms, groups and bridges, along with basket-type moieties. Therefore, a specific antibiotic cannot be used for the chiral resolution of a wide range of racemic compounds: the optimization of

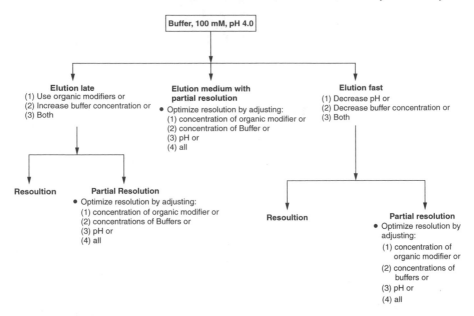

Scheme 7.4 A protocol for the development and optimization of reversed mobile phases on CD-based CSPs under reversed phase mode. Note that this is only a brief outline of the procedure that should be followed in developing a resolution on CD-based CSPs.

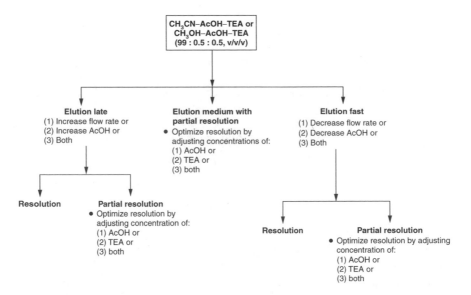

Scheme 7.5 A protocol for the development and optimization of mobile phases on CD-based CSPs under polar organic phase mode. Note that this is only a brief outline of the procedure that should be followed in developing a resolution on CD-based CSPs.

chiral resolution on these CSPs must be achieved by selecting the appropriate antibiotic for a particular racemate. Recently, Aboul-Enein and Ali reviewed chiral resolution on antibiotic CSPs by HPLC [49], and it was observed that it is governed by various HPLC parameters. These antibiotic CSPs may be used in the normal, reversed and new polar ionic phase modes.

Due to the complex structure of these antibiotics, most of them function equally well in reversed, normal and modified polar ionic phases. All three solvent modes generally show different selectivities with different analytes. Sometimes, equivalent separations are obtained in both the normal and the reversed phase mode. This ability to operate in two different solvent modes is an advantage in determining the best preparative methodology where sample solubility is a key issue. In normal phase chromatography, the most commonly used solvents are typically hexane, ethanol, methanol and so on. The optimization of chiral resolution is achieved by adding some other organic solvent, such as acetic acid, tetrahydrofuran (THF), diethylamine (DEA) or triethylamine (TEA) [50, 51].

In a reversed phase system, buffers are mostly used as the mobile phases, with small amounts of organic modifiers. The use of buffers as mobile phases has increased the efficiency of the resolution. Ammonium nitrate, triethylammonium acetate (TEAA) and sodium citrate buffers have been used very successfully. A variety of organic modifiers have been used to alter selectivity [52–54]: acetonitrile, methanol, ethanol, 2-propanol and tetrahydrofuran (THF) have shown good selectivities for various analytes. In the reversed phase mode, the amounts of organic modifiers are typically low, usually of the order of 10–20%. The typical starting composition of the mobile phase is organic modifier – buffer (pH 4.0–7.0) (10:90). The use of alcohols as organic modifiers generally requires higher starting concentrations; for example, 20% for comparable retention when using acetonitrile or tetrahydrofuran in a starting concentration of 10%. The effect of organic solvents on the enantioselectivities also depends on the type of antibiotic. In fact, better recognition is obtained at lower buffer pH values; below or close to the isoelectric point of the antibiotics, especially for vancomycin. Using vancomycin, a low concentration of organic solvents did not significantly influence the separation; while enantioresolution is improved for some compounds with ristocetin A and teicoplanin [55], even at low organic modifier concentrations. The effect of organic modifiers on chiral resolution varies from racemate to racemate [56].

A simplified approach has been proven to be very effective for the resolution of a broad spectrum of racemates. The first consideration in this direction is the structure of the analytes. If the compound has more than one functional group that is capable of interacting with the stationary phase, and at least one of those groups is on or near the stereogenic centre, then the first choice for the mobile phase would be the new polar ionic phase. Due to the strong polar groups present in macrocyclic peptides, it is possible to convert the original mobile phase concept to 100 % methanol, with the acid/base added to effect selectivity. The key factor in obtaining complete resolution is still the ratio of acid to base; the actual concentrations of the acid and base only affect the retention. Therefore, starting with a 1 : 1 ratio, some selectivity is typically observed, and then different ratios of 1 : 2 and 2 : 1 are applied to note the change in resolution indicating the trend. If the analyte is eluting too fast, the acid/base concentration is reduced. Conversely, if the analyte is too well retained, the acid/base concentration is increased. The parameters for the concentrations are between 1 % and 0.001 %. Above 1 % the analyte is too polar and indicates a typical reversed phase system, while below 0.001 % it indicates a normal phase system. Both trifluoroacetic acid (TFA) and acetic acid have been used as the acid component, with ammonium hydroxide and triethyl amine as the base. For an analyte/pollutant that has only one functional group, or for reasons of solubility, typical normal phase solvents (hexane/ethanol) or reversed phase solvents (THF/buffer) are employed. Guillaume *et al.* [30] studied the effect of the methanol concentration on the chiral resolution of phenoxypropionic acid herbicides. The authors reported an enhancement in the separation factors as the methanol concentration increased in the mobile phase.

The pH value is an important controlling factor for enantiomeric resolution in the normal, reversed and new polar ionic phases. In general, buffers are used as the mobile phases to control the pH in HPLC. The pH value of the buffers ranges from 4.0 to 7.0 in a reversed phase system. In general, the analytes interact more favourably at a pH at which they are not ionized. Therefore, the retention and selectivity of molecules that possess ionizable acidic or basic functional groups can be affected by altering the pH. One strategy may be to take advantage of a difference in the pKa values; that is, to keep the analyte of interest neutral and strongly interacting, while keeping other components ionized and poorly retained. Because of the complexities of these interactions, it is necessary to observe the retention and resolution as functions of the pH value, usually testing at pH 4.0 and pH 7.0, or at 0.50 pH unit steps. It has been observed that, most the cases, with an increasing pH value, the values of R_s, k and α decrease. Therefore, the safest and most suitable pH values in reversed phase systems vary from pH 4.0 to pH 7.0 [52, 57]. Enantiomeric resolution by both the normal phase and the new modified polar ionic phase has been achieved below pH

7.0. Guillaume *et al.* [30] studied the effect of pH on the chiral resolution of phenoxypropionic acid herbicides.

Temperature is also an important parameter for controlling the resolution of enantiomers in HPLC. The enthalpy and entropy control of chiral resolution on antibiotic CSPs is similar to the case of polysaccharide-based CSPs. Normally, a low temperature favours an increase in the resolution, which may be due to an increase in the efficiency of the column. It has also been observed that a change in temperature has a greater effect on the retention of solutes in the normal phase compared to the reversed phase. This may be due to the fact that the binding constant of a solute to the macrolide involves several interactive mechanisms that change dramatically with temperature. In the temperature range of 60–80 °C, inclusion complex formation is effectively prevented for most solutes. A lower temperature enhances the weaker bonding forces, and the net result is that chromatographers have an additional powerful means of controlling selectivity and retention. Guillaume *et al.* [30] studied the effect of temperature on the chiral resolution of phenoxypropionic acid herbicides. Apart from the parameters discussed above for the HPLC optimization of chiral resolution on antibiotic CSPs, some other HPLC conditions may be controlled to improve chiral resolution on these CSPs. The effect of the concentrations of antibiotics (on the stationary phase) on enantioresolution varies depending on the type of racemate. The optimization of chiral resolution on macrocyclic glycopeptide antibiotics is shown in Schemes 7.6, 7.7 and 7.8, on vancomycin-, teicoplanin- and restocetin-based CSPs, respectively.

7.4.4 Protein CSPs

The complex structures of proteins, which have different types of groups, loops and bridges, are responsible for the chiral resolution of racemic compounds. Therefore, a small change in the protein molecule results in a drastic change in enantioselectivity. A few reports are available that indicate that different protein structures are responsible for different chiral resolution capacities. One of the advantages of protein-based CSPs is that chiral chromatography is carried out under the reversed phase mode – that is, aqueous mobile phases are frequently used – and, therefore, there is considerable scope for optimizing chiral resolution.

Buffers of differing concentrations and pH values are mostly used for chiral resolution on these CSPs, since these phases are stable under the reversed phase mode. With the exception of a few reports where gradient elution was used, the elution is usually carried out in isocratic mode. The most commonly used buffers were phosphate and borate, which were

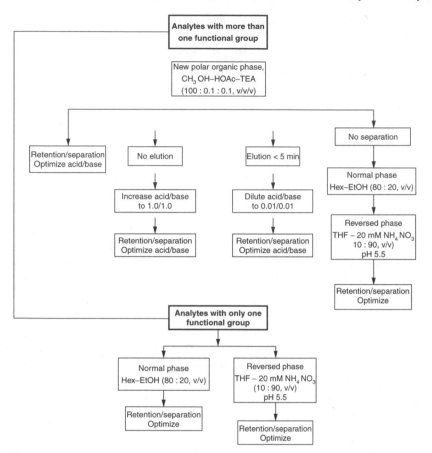

Scheme 7.6 A protocol for the development and optimization of mobile phases on vancomycin CSP. Note that this is only a brief outline of the procedure that should be followed in developing a resolution on vancomycin CSP.

used in a concentration range of 20–100 mM, with a 2.5–8.0 pH range. However, as with all silica-based CSPs, the prolonged use of an alkaline pH buffer (> 8.5) is not suitable. On the other hand, at lower pH, irreversible changes are possible in crosslinked protein phases and, hence, the use of buffers with low pH values for long periods of time is not recommended. Therefore, a buffer that ranges from pH 3 to pH 7 should be chosen. A pH 4.5 ammonium acetate buffer may be useful. For the mobile phase development, any buffer (50 mM, pH 7) can be used and the optimization is carried out by changing the concentration and the pH value. The successful use of organic solvents in the optimization of chiral resolution on these

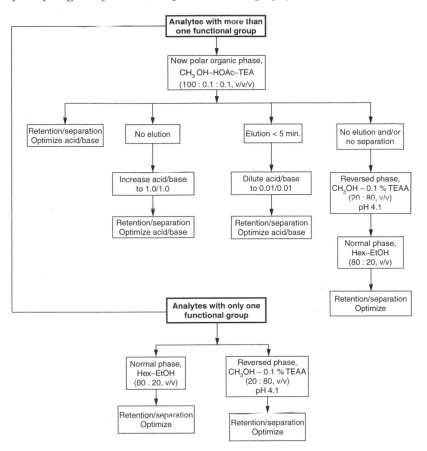

Scheme 7.7 A protocol for the development and optimization of mobile phases on teicoplanin CSP. Note that this is only a brief outline of the procedure that should be followed in developing a resolution on teicoplanin CSP. Other mobile phases may also be used.

CSPs has been reported: the hydrophobic interactions are affected by the use of these solvents, the most important of which are methanol, ethanol, 1-propanol, 2-propanol, acetonitrile and THF. These organic modifiers have been used in the range of 1–10 %. Care must be taken when using these organic modifiers, as they can denature the protein. However, high concentrations of methanol and acetonitrile have been used on some of the crosslinked protein CSPs. The selection of these organic modifiers depends on the structure of the racemic compounds and the CSP used. In some cases, charged modifiers such as hexanoic acid, octanoic acid and

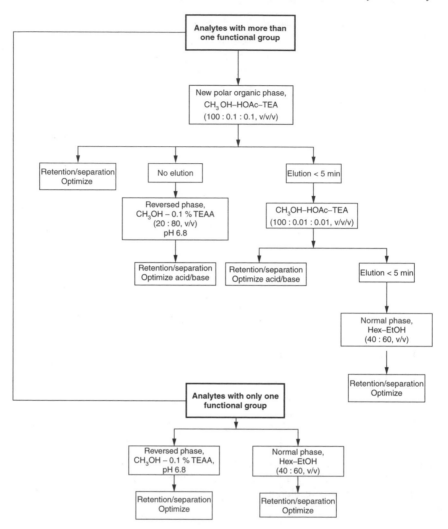

Scheme 7.8 A protocol for the development and optimization of mobile phases on ristocetin CSP under reversed phase mode. Note that this is only a brief outline of the procedure that should be followed in developing a resolution on ristocetin CSP.

quaternary ammonium compounds have also been used for optimum chiral resolution [58, 59]. Oda *et al.* [59] studied the chiral resolution of warfarin on avidin and ovomucoid CSPs using methanol, ethanol, propanol and acetonitrile organic modifiers. In general, the chiral recognition behaviour of these modifiers on avidin and ovomucoid CSPs was in the order methanol > ethanol > propanol > acetonitrile. Briefly, the selection of the mobile phases varies from one solute to another and also depends on the structure of the protein CSP used.

Chiral resolution on protein CSPs is greatly affected by changes in the ionic strength. A decrease in the ionic strength results in an increase in the charge interactions between the analytes and the protein CSP and, accordingly, an increase is observed in the retention of enantiomers. At high ionic strength, hydrophobic interactions are often favoured and this may result in a more complex dependence, yielding a retention minimum at a particular salt (buffer) concentration. Mano *et al.* [60] studied the effect of eluent concentration on the chiral resolution of profens and warfarin. The authors reported 10–50 mM as the best concentration for the buffers. Furthermore, the same authors [59] carried out a comparative chiral resolution using flurbiprofen and warfarin on avidin and ovomucoid columns. The authors claimed that 50–100 mM concentrations of buffers were found to be suitable for good chiral resolution on both CSPs. In the same study, the authors reported the effects of phosphate, citrate, acetate and borate anions on chiral resolution. In general, the chiral resolution capacity was in the order phosphate > citrate > acetate > borate for all racemates and on both CSPs.

The influence of the pH value is more pronounced on charged pollutants than on the uncharged ones. The decrease in the retention of carboxylic acids with an increasing pH value is due to the decreasing number of positive charges on the protein, which leads to fewer charge analytes [58]. Similarly, the retention factors of analytes that contain amino groups interactions with ionic decrease with a reducing pH value due to the decreasing number of negative charges on the protein for charge interactions with the cationic analytes. Therefore, chiral resolution on protein-based CSPs can be controlled by adjusting the pH value of the eluent. The effect of the pH value on the chiral resolution of different racemic compounds on protein-based CSPs has been studied by various workers. Mano *et al.* [60] studied the effect of pH on the chiral resolution of warfarin, using a flavoprotein conjugated silica gel CSP, and the best results were obtained at pH 4.0–4.8.

The effect of temperature on chiral resolution on protein phases is very crucial, but few reports are available that deal with this effect. The loading capacities of protein CSPs may be understood from $\alpha = k_2/k_1$ and $k = k_s + k_n$, where k_s and k_n correspond to selective and nonselective contributions. When the amount of the analyte on the CSP increases, the relative contribution of k_n to the observed k values (k_1 and k_2) also increases. The critical selective sites become saturated, resulting in a decrease in the separation factors, because $k = a_s/a_m$, where a_s and a_m are the amounts of analyte in the stationary and mobile phases, respectively. An increased analyte concentration means a decrease in the observed k values, since the relative amount present in the stationary phase decreases due to the lower affinity of the nonselective sites. Kirkland *et al.* [61] described the loading of $1.5-3.0 \, \text{nM g}^{-1}$ packing producing a $\leq 15\%$ decrease in resolution (a 30% decrease in column plate number) on a $15 \, \text{cm} \times 0.46 \, \text{cm}$ ovomucoid column. According to the authors, samples in the $\mu\text{g l}^{-1}$ range are typical for most analytical applications on a $15 \, \text{cm} \times 0.46 \, \text{cm}$ column of ovomucoid protein.

Besides the parameters discussed above, some other factors, such as the column dimensions, the silica gel particle size and the mobile phase additives, may be used to control chiral resolution on these CSPs. Metal ions are often important structural components in some proteins with regard to the organization of conformation [62], and therefore the effects of metal ions as mobile phase additives may be useful. Oda *et al.* [63] used zinc ions as the mobile phase additive on an avidin CSP to improve the chiral resolution of ibuprofen, and it was found that the chiral resolution had become poor due to the addition of a zinc metal ion in the mobile phase. Recently, Haginaka and Takehira [64] have studied the effect of silica pore size on the chiral resolution of an ovoglycoprotein CSP. The authors have reported the best chiral resolution on a CSP that had a 12 nm silica gel pore size. Furthermore, they have also studied the effect of the loading amount of ovoglycoprotein on the silica gel, and they have found that the best results are obtained on a CSP containing 80 mg of ovoglycoprotein per 1 g of 12 nm pore size silica gel. Fitos *et al.* [65] studied the effect of ibuprofen enantiomers on the stereoselective binding of 3-acyloxy-1,4-benzodiazepines to HSA, using native and sepharose immobilized proteins. This study indicated the different binding natures of the two types of protein in the presence of ibuprofen molecules. Thus, the above-mentioned parameters are also important and are responsible for the differing chiral resolution on these CSPs. The optimization strategy for chiral resolution on protein-based CSPs is given in Scheme 7.9.

7.4.5 Chiral Crown Ether CSPs

The applications of CSPs based on chiral crown ethers (CCEs) are very limited, as these are used only for chiral resolution of pollutants that have amino groups. Aqueous mobile phases containing organic modifiers and acids have been used on these CSPs and, therefore, the composition of the mobile phase is the key parameter in the optimization of chiral resolution. In all applications, aqueous and acidic mobile phases are used, the most

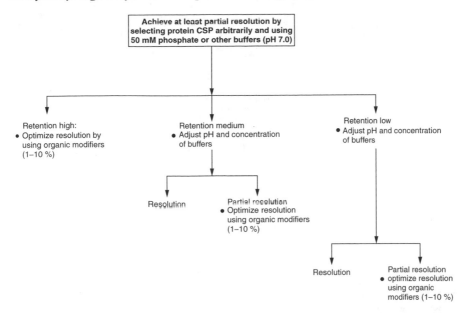

Scheme 7.9 A protocol for the development and optimization of mobile phases on protein CSPs. Note that this is only a brief outline of the procedure that should be followed in developing a resolution on protein CSPs.

commonly used mobile phases being aqueous perchloric acid and aqueous methanol containing sulfuric, trifluoroacetic or perchloric acid separately. Compounds with a higher hydrophobicity generally have longer retention times on CCE-based CSPs and, therefore, organic modifiers are used to optimize the resolution [66]. This optimization is carried out by adjusting the amounts of methanol, sulfuric acid and perchloric acid separately. In general, the separation is increased by an increase in methanol and a decrease in the acid concentrations. The other organic modifiers used are ethanol, acetonitrile and THF, but methanol has been found to be the best one [67–69]. In addition to the composition of the mobile phase, the effects of other parameters, such as the temperature, the flow rate, the pH value and the structure of the analytes, have also been studied, but only a few reports are available in the literature. It has been observed that, in general, lowering of the temperature results in better resolution. The flow rate may be used to optimize the chiral resolution of pollutants on these CSPs. Since all of the mobile phases are acidic in nature, the effect of the pH on chiral resolution is not significant. The optimization of chiral resolution using these CSPs is shown in Scheme 7.10.

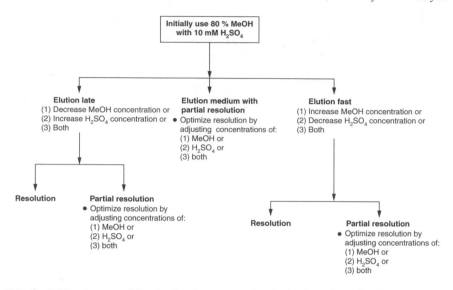

Scheme 7.10   A protocol for the development and optimization of mobile phases on chiral crown ether CSPs. Note that this is only a brief outline of the procedure that should be followed in developing a resolution on chiral crown ether CSPs.

7.4.6   Ligand Exchange CSPs

Chiral resolutions on these CSPs are based on the kinetics of ligand exchangers and enantiomers. The exchange kinetics vary from one chiral selector to another. Therefore, the structures of the chiral selectors are very important from the selectivity point of view. Accordingly, the same racemate may behave in different ways on the different CSPs. Ligand exchange chromatography has been used for the chiral resolution of racemic compounds containing electron-donating atoms and, therefore, its application is confined. In most cases buffers, sometimes containing organic modifiers, have been used as the mobile phase. Therefore, the optimization has been carried out by controlling the compositions of the mobile phases. There are two strategies for the development and use of the mobile phases on these CSPs. With a CSP that has only a chiral ligand, an aqueous mobile phase containing a suitable concentration of a metal ion is used; while in the case of a CSP that contains a metal ion complex as the chiral ligand, a mobile phase without a metal ion is used. In most applications, aqueous solutions of metal ions or buffers have been used as the mobile phases. The most commonly used buffers are ammonium acetate and phosphate. However, the use of a phosphate buffer is avoided if a metal ion is being used as the mobile phase additive, to avoid complex formation between the metal ion and the phosphate, which may block the column. A literature search indicates that these buffers (20–50 mM) have frequently been used for successful chiral resolution, but in some instances organic modifiers have also been used to improve the resolution. In general, acetonitrile has been used as the organic modifier [70]; however, some reports deal with the use of methanol, ethanol and tetrahydrofuran [71–74]. The concentrations of these modifiers vary from 10 % to 30 %. However, some reports have indicated the use of these organic modifiers up to 75 % [73, 74]. In general, the chiral resolution of highly retained solutes is optimized by using organic modifiers. Basically, the organic modifiers reduce the hydrophobic interactions, resulting in an improved resolution.

The pH of the mobile phase has also been recognized as one of the most important controlling factors in chiral resolution on ligand exchange chiral phases [75]. In general, the retention of all racemates increases with an increasing pH value. The selectivity of the separation was only moderately affected by pH changes, while the efficiency of the column showed a different trend, depending on the relative retentions of the racemates. The buffer concentration is also a very important aspect in the optimization of chiral resolution on these CSPs. In general, buffers with the concentration in the range 25–50 mM were used for chiral resolution, with some exceptions. In spite of this, only a few reports are available on the optimization of chiral resolution by varying the ionic strength of the mobile phase. The retention and the selectivity of enantiomeric resolution has also been investigated with respect to metal ion concentrations on these CSPs. Chiral resolution on CSPs containing only chiral ligands has been carried out using different concentrations of metal ions in the mobile phase.

The temperature is also an important parameter in controlling chiral resolution on these CSPs. It has been observed than an increase in temperature results in a decrease in the retention of most racemic compounds. A decrease in the retention times of racemates at high temperature has been observed, which indicates the exothermic nature of the ligand exchange process. In ligand exchange HPLC, the efficiency of chiral resolution may be controlled by adjusting the flow rate of the mobile phase. However, it has been reported that chiral resolution at low flow rates does not follow the well known Van-Deemter plot; this may be due to the slow ligand

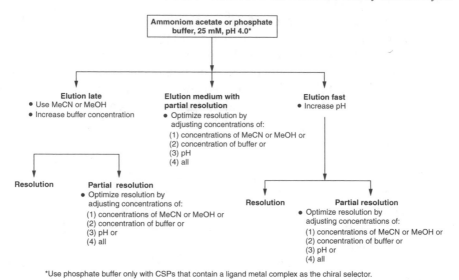

Scheme 7.11 A protocol for the development and optimization of mobile phases on ligand exchange CSPs. Note that this is only a brief outline of the procedure that should be followed in developing a resolution on ligand exchange CSPs.

exchange kinetics. Roumeliotis *et al.* [73] used silica supports composed of different particle sizes, and the authors reported that the best resolution was obtained on a silica gel with a smaller particle size. It has also been reported that the packing of chiral selectors affects chiral resolution: this may again be due to slowness of the ligand exchange kinetics [76–79]. The optimization of chiral pollutants using these ligand exchangers is presented in Scheme 7.11.

7.4.7 Pirkle-type CSPs

Chiral resolution on Pirkle-type CSPs varies from one CSP to another and, therefore, it can be optimized by selecting a suitable chiral selector. Chiral resolutions on π-acidic and π-basic CSPs have been carried out under the normal phase mode. However, some reports are also available dealing with the use of reversed phase eluents, but the prolonged use of a reversed phase mobile phase is not recommended. With the development of more stable new CSPs, the use of the reversed mobile phase mode became possible. Nowadays, both mobile phase modes – that is, normal and reversed – are in use. Therefore, the optimization of chiral resolution on these phases can be achieved by varying the concentration of the mobile phases, including the use of organic modifiers. Due to the development of these new types of CSP (π-acidic–basic types), use of the π-acidic and π-basic types of CSP is no longer common. In view of this, attempts have been made to describe the art of the optimization of chiral resolution using only π-acidic–basic types of CSP.

In general, the normal phase mode has frequently been used for chiral resolution of racemic compounds on Pirkle-type CSPs. Hexane, heptane and cyclohexane are the nonpolar solvents of choice on these phases. Aliphatic alcohols may be considered as hydrogen donors and acceptors, and thus may interact at many points with the aromatic amide groups of CSPs, generating hydrogen bonds. Therefore, the addition of aliphatic alcohols improves chiral resolution, and hence the alcohols are called organic modifiers. The most commonly used alcohol is 2-propanol; however, methanol, ethanol, 1-propanol and n-butanol have also been used. Some reports have also indicated the use of dichloromethane and chloroform as organic modifiers with hexane. In addition to this, acidic and basic additives improve the chromatographic resolution. A small amount of acetic, formic or trifluoroacetic acid improves the peak shape and enantioselectivity for acidic and basic solutes. Sometimes there is a need to combine an acid and an organic amine (e.g. triethylamine) for strong basic racemic compounds. Baeyens *et al.* [80] reported the best resolution of ibuprofen and pirprofen using hexane–2-propanol–acetic acid ($70:30:0.5$, v/v/v) as the mobile phase on a Whelk-O1 CSP.

The effect of the temperature is also important for chiral resolution on these CSPs. Pirkle *et al.* [81] reported inversion of the elution order when the temperature was changed from high to low and vice versa. The temperature effect observed was dependent on the content and polarity of the organic modifier in the mobile phase. The injection amount, the silica gel particle size, the spacer between the chiral selector and silica gel, and the column dimensions may be used as optimization parameters for chiral resolution. Recently, Baeyens *et al.* [80] carried out the chiral resolution of pirprofen on Whelk-O1 columns with internal diameters of 2.1 and 4.6 mm. The authors reported the best resolution on the 4.6 mm column (Figure 7.5). The optimization of chiral resolution on Pirkle-type CSPs is presented in Scheme 7.12.

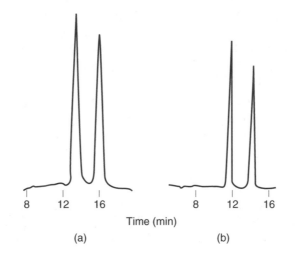

Figure 7.5 A comparison of the chiral resolution of pirprofen on a Whelk-O1 CSP of (a) 2.1 and (b) 4.6 mm internal diameter [80].

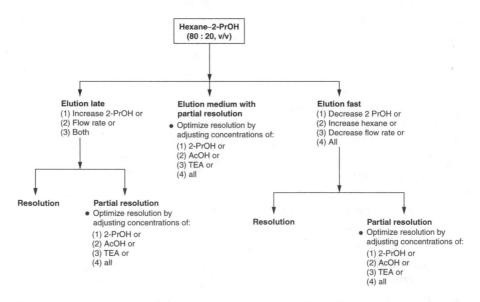

Scheme 7.12 A protocol for the development and optimization of mobile phases on Pirkle-type CSPs under normal phase mode. Note that this is only a brief outline of the procedure that should be followed in developing a resolution on Pirkle-type CSPs. AcOH, acetic acid; TEA, triethylamine.

7.5 Detection

In the chiral analysis of many drugs, pharmaceuticals and agrochemicals by HPLC, detection is mostly achieved using the UV mode [1–7], and hence this detection mode has also been used for the chiral resolution of some environmental pollutants [8]. However, some organochlorine pollutants are transparent to UV radiation; hence, the UV detection mode is not suitable and some other detection devices have to be used in chiral HPLC, the most of which are MS, optical detectors and so on.

7.6 Chiral Recognition Mechanisms

The chiral recognition mechanisms on the above-mentioned CSPs are not fully developed. However, in 1956 Pfeiffer [82] postulated a three-point model for chiral recognition mechanisms; this was subsequently explained in more detail by Pirkle and Pochapsky [83]. According to this model, chiral recognition requires a minimum of three simultaneous interactions between the CSP and at least one of the enantiomers, with at least one of these interactions being stereochemically dependent. This three-point model does not always apply, and Groombridge *et al.* [84] suggested a four-point model for chiral recognition on some protein CSPs. Briefly, it is thought the transitory diastereoisomeric complexes of the enantiomers, which have different physical and chemical properties, are formed on the CSPs, and get resolved during the HPLC process. Chiral recognition mechanisms differ from one CSP to another. It is known that the chiral recognition processes on polysaccharides, CDs, macrocyclic glycopeptide antibiotics, proteins and chiral crown ether based CSPs are more or less similar. The chiral grooves on polysaccharides, the cavities on CDs, the baskets on macrocyclic glycopeptide antibiotics, the bridges and loops on proteins and the cavities on chiral crown ether based CSPs provide a chiral environment for the enantiomers. Two enantiomers get fitted to different extents and hence elute at different rates (retention times). The differing stabilities of the enantiomers on these CSPs are due to their different bondings and interactions, the most important of which are hydrogen bonding, dipole-induced dipole interactions and $\pi-\pi$ complexation, inclusion complexation, anionic and cationic bondings, van der Waals forces and so on [1–7]. Steric effects also play a crucial role in the chiral resolution of racemates. The different binding energies of the diastereoisomeric complexes are due to the various interactions mentioned above.

Pirkle-type CSPs contain a chiral moiety that has an aromatic ring and, therefore, the formation of a $\pi-\pi$ charge transfer diastereoisomeric complex of the enantiomers (with the aromatic group) with a CSP is considered to be essential. In view of this, the π-acidic CSPs are suitable for chiral resolution of π donor solutes, and vice versa. However, the newly developed CSPs that contain both π-acidic and π-basic groups are suitable for the chiral resolution of both types of solutes; that is, π-donor and π-acceptor analytes. Briefly, Pirkle-type CSPs contain a chiral moiety that provides a chiral environment for the enantiomers. Therefore, the enantiomers fit differently on to this chiral moiety (due to their different spatial configurations). Accordingly, the two enantiomers form diastereoisomeric complexes that have different physical and chemical properties, along with different binding energies. Therefore, the two enantiomers elute at different retention times with the flow of the mobile phase, and hence chiral separation occurs.

On ligand exchange CSPs, chiral resolution occurs due to the exchange of chiral ligands and enantiomers on specific metal ions through coordinate bonds. The two enantiomers have different exchange capacities because of the stereospecific nature of the ligand exchange process and hence chiral resolution takes place. Davankov *et al.* [85–87] suggested a theoretical model for the mechanisms of chiral resolution on these CSPs. In this model, the enantiomers are coordinated to the metal ion in different ways, depending on their interactions with the ligands bonded to the stationary phase, which act as chiral selectors. The authors explained that chiral resolution is due to different bondings along with the steric effects that result in the formation of the diastereoisomeric complexes by the two enantiomers. These diastereoisomeric complexes are stabilized at different magnitudes by dipole–dipole interactions, hydrogen bondings, van der Waals forces and steric effects, and elute at different retention times. A general graphical representation of the chiral resolution mechanisms of racemates on the above-mentioned CSPs is shown in Figure 7.6.

7.7 Conclusions

As discussed above, only a few reports are available on the chiral analysis of environmental pollutants and, therefore, HPLC still cannot achieve the status of a routine analytical technique in this field of work. Of course, HPLC has considerable potential in the field of chiral separation of environmental pollutants, due to the great merit of the technique, as already discussed. As discussed above, the different classes of CSPs can be used with a

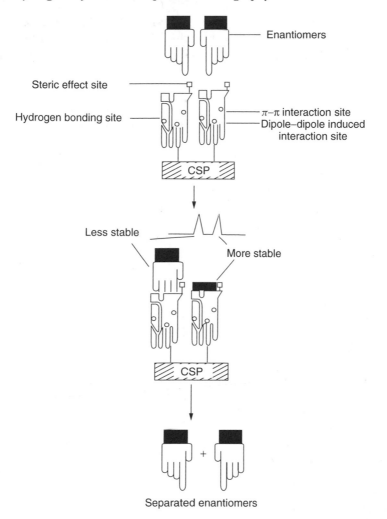

Figure 7.6 A schematic representation of chiral recognition mechanisms.

variety of mobile phases. The wide choice of mobile phases in HPLC also enhances its range of applications. HPLC can be used for the chiral resolution of organochlorine pollutants by coupling it with many of the advanced detection techniques, such as mass spectrometry (MS), optical detection and so on. Due to the rapid development and advancement of HPLC, it continues to replace GC. In addition to this, more experiments should be carried out with the aim of determination of the chiral recognition mechanisms on different CSPs, so that HPLC can be used more precisely and exactly for the chiral separation of environmental pollutants. We very much hope that HPLC will have a bright future and that it will become the techniques of choice for the analysis of chiral environmental pollutants.

References

1. H. Y. Aboul-Enein and I. Ali, *Chiral Separations by Liquid Chromatography and Related Technologies*, Dekker, New York, 2003.
2. T. E. Beesley and R. P. W. Scott, *Chiral Chromatography*, Wiley, New York, 1998.
3. H. Y. Aboul-Enein and I. W. Wainer (eds), *The Impact of Stereochemistry on Drug Development and Use, Chemical Analysis*, vol. 142, Wiley, New York, 1997.
4. G. Subramanian (ed.), *A Practical Approach to Chiral Separations by Liquid Chromatography*, VCH, Weinheim, 1994.
5. S. Allenmark, *Chromatographic Enantioseparation, Methods and Applications*, 2nd edn, Ellis Horwood, New York, 1991.
6. A. M. Krstulovic, *Chiral Separations by HPLC: Applications to Pharmaceutical Compounds*, Ellis Horwood, New York, 1989.
7. H. Hühnerfuss and R. Kallenborn, *J. Chromatogr. A* **580**, 191 (1992).
8. R. Kallenborn and H. Hühnerfuss, *Chiral Environmental Pollutants, Trace Analysis and Toxicology*, Springer-Verlag, Berlin, 2001.
9. V. Schurig and A. Glausch, *Naturwiss.* **80**, 468 (1993).
10. V. Schurig, A. Glausch and M. Fluck, *Tetrahedron Asymm.* **6**, 2161 (1995).
11. W. A. König, B. Gehrcke, T. Runge and C. Wolf, *J. High Res. Chromatogr.* **16**, 376 (1993).
12. I. H. Hardt, C. Wolf, B. Gehrcke, D. H. Hochmuth, B. Pfaffenberger and H. Hühnerfuss, *J. High Res. Chromatogr.* **17**, 859 (1994).
13. A. Mannscreck, N. Puster, L. W. Robertson, F. Oesch and M. Püttman, *Justus. Liebigs. Ann. Chem.*, 2101 (1985).
14. P. Ludwig, W. Gunkel and H. Hühnerfuss, *Chemosphere* **24**, 1423 (1992).
15. W. Vetter, U. Klobes, B. Luckas and G. Hottinger, *Chromatographia* **45**, 255 (1997).
16. W. Vetter, U. Klobes, B. Luckas and G. Hottinger, *Organohal. Compds* **35**, 305 (1998).
17. B. Blessington and N. Crabb, *J. Chromatogr.* **483**, 349 (1989).
18. D. W. Armstrong, G. L. Reid III, M. L. Hilton and C. D. Chang, *Environ. Pollut.* **79**, 51 (1993).
19. Y. Q. Chu and I. W. Wainer, *Pharm. Res.* **5**, 680 (1988).
20. K. Weber, R. Kreuzig and M. Bahadir, *Chemosphere* **35**, 13 (1997).
21. M. Shou and S. Yang, *Chirality* **2**, 141 (1990).
22. J. M. Schneiderheinze, D. W. Armstrong and A. Berthod, *Chirality* **11**, 330 (1999).

23. S. Caccamese and G. Principato, *J. Chromatogr. A* **893**, 47 (2000).
24. W. A. Möller, C. Bretzke, H. Hühnerfuss, R. Kallenborn, J. N. Kinkel, J. Kopf and G. Rimkus, *Angew. Chem. Int. Ed. Engl.* **33**, 882 (1994).
25. J. J. Ellington, J. J. Evans, K. B. Prickett and W. L. Champion Jr, *J. Chromatogr. A* **928**, 145 (2001).
26. M. Dondi, M. Flieger, J. Olsovska, C. M. Polcaro and M. Sinibaldi, *J. Chromatogr. A* **859**, 133 (1999).
27. M. D. Müller, T. Poiger and H. Buser, *J. Agric. Food Chem.* **49**, 42 (2001).
28. I. Ali and H. Y. Aboul-Enein, *Environ. Toxicol.* **17**, 329 (2002).
29. Z. Y. Li, Z. C. Zhang, Q. L. Zhou, R. Y. Gao and Q. S. Wang, *J. Chromatogr. A* **977**, 17 (2002).
30. Y. C. Guillaume, L. Ismaili, T. T. Truong, L. Nicod, J. Millet and M. Thomassin, *Talanta* **58**, 951 (2002).
31. I. W. Wainer, M. C. Alembic and E. Smith, *J. Chromatogr.* **388**, 65 (1987).
32. M. H. Gaffney, *Chromatographia* **27**, 15 (1989).
33. P. Piras, C. Roussel and J. Pierrot-Sanders, *J. Chromatogr. A* **906**, 443 (2001).
34. H. Y. Aboul-Enein and I. Ali, *Fresenius J. Anal. Chem.* **370**, 951 (2001).
35. H. Y. Aboul-Enein and I. Ali, *J. Sepn Sci.* **24**, 831 (2001).
36. K. Tachibana and A. Ohnishi, *J. Chromatogr. A* **906**, 127 (2001).
37. H. Y. Aboul-Enein and I. Ali, *Pharmazie* **57**, 682 (2002).
38. T. O'Brien, L. Crocker, R. Thompson, K. Thompson, P. H. Toma, D. A. Conlon, C. Moeder, G. R. Bicker and N. Grinberg, *Anal. Chem.* **69**, 1999 (1997).
39. R. W. Stringham and J. A. Blackwell, *Anal. Chem.* **69**, 1414 (1997).
40. E. Yashima, P. Sahavattanapong and Y. Okamoto, *Chirality* **8**, 446 (1996).
41. S. M. Han and D. W. Armstrong, in A. M. Krstulovic, ed., *Chiral Separations by HPLC*, Ellis Horwood, Chichester, 1989, p. 208.
42. A. M. Stalcup, in G. Subramanian, ed., *A Practical Approach to Chiral Separations by Liquid Chromatography*, VCH, Weinheim, 1994, p. 95.
43. K. G. Feitsma, J. Bosmann, B. F. H. Drenth and R. A. De Zeeuw, *J. Chromatogr.* **333**, 59 (1985).
44. S. M. Han, *Biomed. Chromatogr.* **11**, 259 (1997).
45. S. Piperaki, A. Kakoulidou-Tsantili and M. Parissi-Poulou, *Chirality* **7**, 257 (1995).
46. T. Hargitai and Y. Okamoto, *J. Liq. Chromatogr.* **16**, 843 (1993).
47. J. L. Haynes III, S. A. Shamsi, F. O'Keefe, R. Darcey and I. M. Warner, *J. Chromatogr. A* **803**, 261 (1998).
48. *A Guide to Using Cyclodextrin Bonded Phases for Liquid Chromatography*, Advanced Separation Technologies, Inc., Whippany, NJ, 1996.
49. H. Y. Aboul-Enein and I. Ali, *Chromatographia* **52**, 679 (2000).
50. D. W. Armstrong, Y. Tang, S. Chen, Y. Zhou, C. Bagwill, R. Chen, *Anal. Chem.* **66**, 1473 (1994).

51. *Chirobiotic Handbook, Guide to Using Macrocyclic Glycopeptide Bonded Phases for Chiral LC Separations*, 2nd edn, Advanced Separation Technologies, Inc., Whippany, NJ, 1999.

52. D. W. Armstrong, K. L. Rundlett and J. R. Chen, *Chirality* **6**, 496 (1994).

53. E. Tesaova, Z. Bosakova and V. Pacakova, *J. Chromatogr. A* **839**, 121 (1999).

54. K. H. Ekborg-Ott, Y. Liu and D. W. Armstrong, *Chirality* **10**, 434 (1998).

55. D. W. Armstrong, Y. Liu and K. Ekborg-Ott, *Chirality* **7**, 474 (1995).

56. A. Peter, G. Torok and D. W. Armstrong, *J. Chromatogr. A* **793**, 283 (1998).

57. H. Y. Aboul-Enein and I. Ali, *IL Farmaco* **57**, 513 (2002).

58. S. Allenmark, in G. Subramanian, ed., *A Practical Approach to Chiral Separations by Liquid Chromatography*, Wiley VCH, Weinheim, 1994, pp. 183–216.

59. Y. Oda, N. Mano, N. Asakawa, Y. Yoshida, T. Sato and T. Nakagawa, *Anal. Sci.* **9**, 221 (1993).

60. N. Mano, Y. Oda, N. Asakawa, Y. Yoshida and T. Sato, *J. Chromatogr. A* **623**, 221 (1992).

61. K. M. Kirkland, K. L. Neilson, D. A. McComb and J. J. DeStefano, *LC–GC* **10**, 322 (1992).

62. J. M. Berg, *J. Biol. Chem.* **265**, 6513 (1990).

63. Y. Oda, N. Mano, N. Asakawa, Y. Yoshida and T. Sato, *J. Liq. Chromatogr.* **17**, 3393 (1994).

64. J. Haginaka and H. Takehira, *J. Chromatogr. A* **773**, 85 (1997).

65. I. Fitos, J. Visy, M. Simonyi and J. Hermansson, *Chirality* **11**, 115 (1999).

66. T. Shinbo, T. Yamaguchi, H. Yanagishita, D. Kitamoto, K. Sakaik and M. Sugiura, *J. Chromatogr. A* **625**, 101 (1992).

67. M. H. Hyun, J. S. Jin and W. Lee, *J. Chromatogr. A* **822**, 155 (1998).

68. M. H. Hyun, J. S. Jin, H. J. Koo and W. Lee, *J. Chromatogr. A* **837**, 75 (1999).

69. M. H. Hyun, S. C. Han, J. S. Jin and W. Lee, *Chromatographia* **52**, 473 (2000).

70. B. Feibush, M. J. Cohen and B. L. Karger, *J. Chromatogr.* **282**, 3 (1983).

71. V. A. Davankov, A. S. Bochkov and Y. P. Belov, *J. Chromatogr.* **218**, 547 (1981).

72. P. Roumeliotis, K. K. Unger, A. A. Kurganov and V. A. Davankov, *J. Chromatogr.* **255**, 51 (1983).

73. P. Roumeliotis, A. A. Kurganov and V. A. Davankov, *J. Chromatogr.* **266**, 439 (1983).

74. C. H. Shieh, B. L. Karger, L. R. Gelber and B. Feibush, *J. Chromatogr.* **406**, 343 (1987).

75. M. Remelli, P. Fornasari, F. Dandi and F. Pulidori, *Chromatographia* **37**, 23 (1993).

76. V. A. Davankov and A. V. Semechkin, *J. Chromatogr.* **141**, 313 (1977).

77. B. Lefebvre, R. Audebert and C. Quivoron, *Israel J. Chem.* **15**, 69 (1977).

78. E. J. Boue, R. Audebert and C. Quivoran, *J. Chromatogr.* **204**, 185 (1981).

79. N. H. C. Cooke, R. L. Viavattene, R. Esksteen, W. S. Wong, G. Davies and B. L. Karger, *J. Chromatogr.* **149**, 391 (1978).
80. W. R. G. Baeyens, G. Van der Weken, H. Y. Aboul-Enein, S. Reygaerts and E. Smet, *Biomed. Chromatogr.* **14**, 58 (2000).
81. W. H. Pirkle and P. G. Murray, *J. High Res. Chromatogr.* **16**, 285 (1993).
82. C. C. Pfeiffer, *Science* **124**, 29 (1956).
83. W. H. Pirkle and T. C. Pochapsky, *Chem. Rev.* **89**, 347 (1989).
84. J. J. Groombridge, C. G. Jones, M. W. Bruford and R. A. Nichols, *Nature* **403**, 615 (2000).
85. German Pat. 1932190 (1969), S. V. Rogozhin and V. A. Davankov.
86. S. V. Rogozhin and V. A. Davankov, *Chem. Lett.*, 490 (1971).
87. V. A. Davankov, *J. Chromatogr. A* **666**, 55 (1994).

Chapter 8

The Analysis of Chiral Pollutants by Micellar Electrokinetic, Capillary Electrochromatographic, Supercritical Fluid and Thin Layer Chromatographic Techniques

8.1 Introduction

High performance liquid chromatography (HPLC) is one of the most popular technologies in analytical science, and many modifications and advancements have been made to this modality from time to time, which have resulted in new types of chromatographic technique. The most important newly developed liquid chromatographic modalities are micellar electrokinetic chromatography (MEKC), capillary electrochromatography (CEC) and supercritical fluid chromatography (SFC). These chromatographic methods have certain advantages over HPLC and are gaining importance day by day. Their high speed, low running cost and good reproducibility are some of the salient features. Recently, some reports have been published on the use of these techniques in chiral separations of drugs and pharmaceuticals [1–8]. Their use in the analysis of chiral pollutants is still limited [9] but is under way, and they will definitely replace other chromatographic

Chiral Pollutants. I. Ali and H. Y. Aboul-Enein
© 2004 John Wiley & Sons, Ltd ISBN: 0-470-86780-9

techniques in chiral separations of environmental pollutants. As in the case of HPLC, two approaches are used in chiral analysis. The first approach is an indirect one, involving the derivatization of the racemic pollutants followed by the separation of the diastereoisomers that have formed. The other approach is to use chiral selectors either as chiral stationary phases (CSPs) or chiral mobile phase additives (CMPAs), depending on the requirements of the technique. The indirect approach is not useful for various reasons, while the direct approach is more popular. The merits and demerits of these direct and indirect approaches have already been discussed in Chapter 7. Due to the growing interest in these newly developed chromatographic methods, attempts have been made in the present chapter to describe the analysis of chiral pollutants using micellar electrokinetic and capillary electro- and supercritical fluid chromatographic techniques. Since chiral separations of environmental pollutants by thin layer chromatography have not been discussed anywhere else in this book, we will also describe the capabilities of TLC in the analysis of chiral environmental pollutant.

8.2 Chiral Selectors

As in the case of HPLC, chiral selectors are required for the chiral analysis of pollutants by the above-mentioned liquid chromatographic modalities. All of the chiral compounds used as chiral selectors in HPLC can be utilized in these techniques. Many chiral compounds have been used as chiral selectors in these modalities, the most important classes being polysaccharides, cyclodextrins, macrocyclic glycopeptide antibiotics, proteins, crown ethers, ligand exchangers and Pirkle-type compounds. The structures and properties of these chiral selectors have not been included herein. However, the structures and properties of the cyclodextrins and their derivatives have been discussed in Chapter 6. These chiral selectors are used as CSPs or CMPAs, depending on the type and nature of the liquid chromatographic technique. Various companies have prepared commercial CSPs for these technologies, which can be used successfully for the chiral analysis of many racemates, including drugs, pharmaceuticals, agrochemicals and other pollutants. For details regarding the structures and properties of these CSPs, readers should consult the book by Aboul-Enein and Ali [10].

8.3 Micellar Electrokinetic Chromatography (MEKC)

Sometimes, a surfactant molecule (at a concentration above its critical micellar concentration) is added for the optimization of chiral resolution in capillary electrophoresis (CE) and the enantioseparation mechanism is shifted towards chromatographic principles. Hence the technique is called micellar electrokinetic chromatography (MEKC). This technique was introduced by Terabe *et al.* [11] in 1984. The improvement in chiral separation occurs due to the formation of a micellar phase. Basically, a surfactant is a molecule that possesses two zones of different polarities, which show special characteristics in solution. Surfactants are divided into three categories; ionic (cationic and anionic), nonionic and zwitterionic. The surfactant molecules form a micelles, and partition of the diastereomeric complexes occurs between these micelle and the mobile phase.

Surfactant molecules containing long alkyl chains (hydrophobic group) and charged or neutral heads (polar group) aggregate in aqueous solutions above their critical micellar concentration (CMC) and form micelles, as shown in Figure 8.1. These types of aggregate are spherical in shape, with polar and hydrophobic groups in the outer and core regions, respectively. The micelle works as a pseudo-stationary phase that possesses a self-mobility that differs from that of the surrounding aqueous phase. Accordingly, the micellar phase acts as a stationary phase, while the aqueous phase acts as the mobile phase in HPLC. The distribution of analytes occurs between the micellar and aqueous phases. The solute and micellar interactions are of three types: (i) the solute is adsorbed on the surface of the micelle by electrostatic or dipole interactions; (ii) the solute behaves as

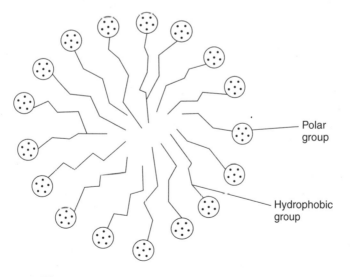

Polar group

Hydrophobic group

Figure 8.1 A schematic representation of a micelle.

a co-surfactant by participating in the formation of the micelle; or (iii) the solute is incorporated into the core of the micelle [12]. The extent of these interactions depends on the nature of the solute and the micelle. Highly polar solutes will mainly be absorbed on the surface of the micelle, while low polar solutes are thought to interact in the core of the micelle. The most important properties of the surfactants are their critical micellar concentration (CMC), the aggregation number and the Kraft point. The CMC depends on the temperature, the salt concentration and buffer additives. The aggregation number indicates the number of surfactant molecules taking part in micelle formation. The Kraft point is the temperature above which the solubility of the surfactant increases steeply due to the formation of micelles.

The best surfactant for MEKC possesses good solubility in a buffer solution, forming a homogeneous micellar solution that is compatible with the detector and has a low viscosity. The most important surfactants are sodium dodecyl sulfate, sodium tetradecyl sulfate, sodium decanesulfonate, sodium *N*-lauryl-*N*-methyllaurate, sodiumpolyoxyethylene dodecyl ether sulfate, sodium *N*-dodecanoyl-L-valinate, sodium cholate, sodium deoxycholate, sodium taurocholate, sodium taurodeoxycholate, potassium perfluoroheptanoate, dodecyltrimethylammonium chloride, dodecyltrimethylammonium bromide, tetradecyltrimethylammonium bromide and cetyltrimethylammonium bromide. Basically, the micelle and the buffer tend to move towards the positive and negative ends, respectively. The movement of the buffer is stronger than that of the micelle, and as a result the micelle and the buffer moves towards the negative end. In this mode of liquid chromatography, the separation depends on the distribution of the analytes between the micellar and aqueous phases. Enantiomers of pollutants form diastereomeric complexes with a chiral selector and these complexes are distributed between the micellar and aqueous phases, as a result of which chiral separation occurs. A schematic of the separation of phenoxy acid herbicides (anionic species) in the presence of a alkylglucoside micellar phase is shown in Figure 8.2. The neutral micelles of the alkylglucoside surfactants migrate at the velocity of the electro-osmotic flow (EOF), the electrophoretic mobility of the herbicides being opposite in direction to the cathodal EOF. Therefore, the effective electrophoretic mobility and the migration time for the herbicides will decrease as the magnitude of their association with the neutral micelle increases. The stronger the interaction between the analyte and the micelle, the higher is the apparent mobility of the herbicides and, consequently, the faster is its migration towards the cathode.

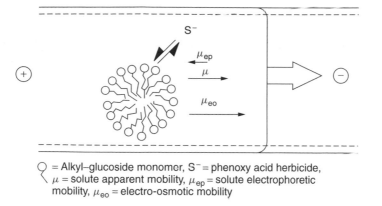

Q = Alkyl–glucoside monomer, S^- = phenoxy acid herbicide,
μ = solute apparent mobility, μ_{ep} = solute electrophoretic
mobility, μ_{eo} = electro-osmotic mobility

Figure 8.2 A schematic representation of the separation of phenoxy acid herbicides (anionic species) in the presence of neutral micelle (alkylglucoside) in MEKC [17].

Many modifications have been made to MEKC from time to time, and hence this modality is gaining importance in the analytical sciences. Szoko *et al.* [13] described an equilibrium binding model of bile salt mediated chiral micellar electrokinetic capillary chromatography in chiral separation science. The optimum concentration of bile salts in chiral separations depends on both the aggregation properties of the surfactant and the stability of the analyte micelle complexes. Pesek *et al.* [14] used an open tubular capillary for chiral separations in electrokinetic chromatography. Peterson and Foley [15] studied the influence of an inorganic counter ion on the chiral separations of many racemates using MEKC. Mazzeo *et al.* [16] developed a novel chiral surfactant, (S)-N-dodecoxycarbonylvaline, for enantiomeric separation by micellar electrokinetic capillary chromatography. Furthermore, the authors reported significantly less background absorbance in the low UV region using this chiral surfactant.

A few reports are available on chiral separations of pollutants using this modality of liquid chromatography. The separated chiral pollutants are 2-(2-chlorophenoxy)propionic acid and 2-(4-chlorophenoxy)propionic acid on n-alkyl-β-D-glucopyranoside [17], ibuprofens on vancomycin [18] and PCBs on γ-cyclodextrin [19]. Marina *et al.* [20] reported chiral separations of polychlorinated biphenyls (PCBs) 45, 84, 88, 91, 95, 132, 136, 139, 149, 171, 183 and 196 by MEKC using cyclodextrin chiral selectors. Mixtures of β- and γ-cyclodextrins were used as chiral modifiers in a 2-(N-cyclohexylamino)ethanesulfonic acid (CHES) buffer containing urea and sodium dodecyl sulfate (SDS) micelles. A mixture of PCBs 45, 88, 91, 95, 136, 139, 149 and 196 was separated into all 16 enantiomers in an analysis time of approximately 35 min. Recently, García-Ruiz *et al.* [21] have used electrokinetic chromatography with cyclodextrin derivatives as the CSPs to achieve rapid enantiomeric separation of chiral PCBs. Thirteen of the 19 chiral PCBs that are stable at room temperature were individually separated into their two enantiomers by using a 2-morpholinoethanesulfonic acid (MES) buffer (pH 6.5) containing carboxymethylated γ-cyclodextrin (CM-β-CD) as pseudo-stationary phase, mixed with β-cyclodextrin (β-CD) or permethylated β-cyclodextrin (PM-β-CD). Copper and Sepaniak [22] reported the analysis of six methyl substituted and three one-position substituted benzopyrene isomers using MEKC. The authors used γ-cyclodextrins as chiral selectors. A comparison of sodium dodecyl sulfate and sodium cholate surfactants was carried out with the aim of separating the isomers, and it was found that sodium dodecyl sulfate is a better surfactant than sodium cholate. The authors used computational calculations to describe the possible separation mechanisms. Chiral analysis by MEKC can be optimized by controlling various experimental factors. The art of the optimization of chiral analysis by CE is discussed in detail in Chapter 9 and, hence, these approaches may be used to control chiral analysis by MEKC. However, the pH of the buffer should be constant, with typical surfactant concentrations ranging from 30 to 100 mM. With a high concentration of cetyltrimethylammonium bromide, the direction of the electro-osmotic flow is reversed and, hence, the polarity of the power supply must also be reversed. The chiral analysis of some pollutants is summarized in Table 8.1. Chromatograms of the chiral analysis of PCBs by MEKC are shown in Figure 8.3.

Table 8.1 The chiral analysis of some environmental pollutants by micellar electrokinetic chromatography (MEKC), capillary electrochromatography (CEC) and thin layer chromatography (TLC)

Chiral pollutants	Sample matrix	CSPs	Detection	References
Micellar electrokinetic chromatography (MEKC)				
2-(2-chlorophenoxy)-propionic acid and 2-(4-chlorophenoxy)-propionic acid	–	n-Alkyl-β-D-glucopyranoside	–	17
Ibuprofen	–	Vancomycin	–	18
PCBs	–	γ-CD	–	19

Table 8.1 (*continued*)

Chiral pollutants	Sample matrix	CSPs	Detection	References
PCBs	–	CDs	–	20
PCBs	–	CDs	–	21
Benzopyrene isomers	–	CDs	–	22
2-Phenoxypropionic acid, warfarin, ibuprofen	–	Proteins		23

Capillary electrochromatography (CEC)

Ibuprofen	–	CDs	–	49
Ibuprofen	–	Avidin		52
Ibuprofen	–	Vancomycin	–	53
Ibuprofen	–	Teicoplanin	–	54
Warfarin	–	HSA	–	51
Warfarin	–	Teicoplanin	–	54
Warfarin	–	Polysaccharide CSPs	–	55–57
Warfarin	–	Polyacrylamide and polysaccharide CSPs	–	58, 59
Warfarin	–	Vancomycin	–	53
Warfarin	–	CDs	–	53
Thalidomide	–	Vancomycin	–	53
Thalidomide	–	Polysaccharide CSPs	–	55–57
Thalidomide	–	Polyacrylamide and polysaccharide CSPs	–	58, 59
2-Phenylpropionic acid and warfarin	–	CDs	–	50
Dichloroprop	–	Vancomycin	–	53

Thin layer chromatography (TLC)

Pyrethrin and pyrethroid	Crops, food and other environment samples	Ligand exchange CSPs	–	75
2-Arylpropionic acid and ibuprofen	–	–	–	76

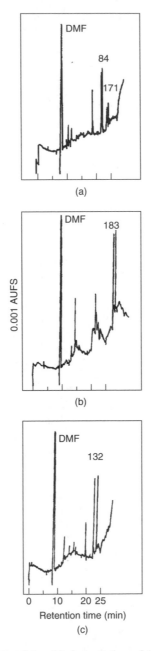

Figure 8.3 Electropherograms of the chiral resolution of (a) PCBs 84 and 171, (b) PCB 183 and (c) PCB 132 by MEKC, using 0.60 M CHES buffer (pH 10.0), 2 M urea, 0.11 M sodium dodecyl sulfate, 0.073 M β-cyclodextrin, 0.022 M γ-cyclodextrin with capillary 65 cm $\times$ 50 μm id, UV 235 nm detection, applied voltage 15 kV, current 41 μA and 45 °C temperature [20].

8.4 Capillary Electrochromatography (CEC)

A hybrid technique between HPLC and CE was developed in 1990, and is known as capillary electrochromatography [24–36]. CEC is expected to combine the high peak efficiency that is characteristic of electrically driven separations with the high separation selectivity of multivariate CSPs available in HPLC. CEC experiments can be carried out on wall coated open tubular capillaries or capillaries packed with particulate or monolithic silica or other inorganic materials, as well as organic polymers. The chromatographic band broadening mechanisms are quite different in the individual modes. The chromatographic and electrophoretic mechanisms work simultaneously in CEC and several combinations are possible. In 1992, Mayer and Schurig [32] used open tubular CEC for enantioseparation. An illustrative quantitative treatment on the effect of capillary diameter and migration mode on the separation results in open tubular CEC was given by Vindevogel and Sandra [37]. The authors pointed out the principal difference between the enantioseparations in capillary electrophoresis and CEC.

The separation occurs on the mobile phase/stationary phase interface and the exchange kinetics between the mobile and stationary phases are important. Again, in this mode of liquid chromatography the separation principle is a mixture of liquid chromatographic and capillary electrophoresis. Enantiomers of pollutants form diastereoisomeric complexes with suitable chiral reagents, and these complexes are distributed between the mobile and stationary phases. The diastereoisomeric complexes possess different physical and chemical properties, and are separated on a solid support (an achiral column/capillary). The formation of diastereoisomeric complexes depends on the type and nature of the chiral mobile phase additives (CMPAs) and the pollutants. In the case of cyclodextrins, inclusion complexes are formed; while in the case of other CMPAs, simple chiral diastereoisomeric complexes are formed. Again, as in the case of CSPs, the formation of the diastereoisomeric complexes is controlled by a number of interactions, such as $\pi-\pi$ complexation, hydrogen bondings, dipole–dipole interactions, ionic bindings and steric effects. These complexes are distributed between the mobile and stationary phases through adsorption and partition phenomena. Also, the different migration times of these diastereomeric complexes depend on their size, charge and interaction with the capillary wall, as a result of which the complexes elute at different time intervals.

During the course of time, many modifications have been reported in CEC and the technique is becoming popular. Gong and Lee [38] used two novel types of substituted cyclam capped β-cyclodextrin bonded silica particles as chiral stationary phases in CEC. These chiral selectors exhibit excellent enantioselectivities in CEC for a wide range of compounds as a result of co-operative functioning of the anchored β-CD and cyclam. After inclusion of the metal ion (Ni^{2+}) from the running buffer into the substituted cyclams and their sidearm ligands, the bonded stationary phases become positively charged and can provide extra electrostatic interactions with ionizable solutes, and thus enhance the dipolar interactions with some polar neutral solutes. This enhances the host–guest interaction with some solutes and improves chiral recognition and enantioselectivity. These new types of stationary phase show great potential for fast chiral separations in CEC. Liu *et al.* [39] developed a chiral stationary phase based on the physical adsorption of avidin on monolithic silica for CEC. The phase ratio of the resulting stationary phase was evaluated using frontal analysis. The method proved to be comparable in phase ratio to the chemical bonding methods used in HPLC. Enantiomer separations were carried out in CEC and capillary liquid chromatography (CLC) modes. Due to its larger phase ratio, the resulting column showed a more powerful separation capability as compared to open tubular CEC (OT-CEC). Tobler *et al.* [40] used low molecular weight chiral cation exchangers – that is, novel chiral stationary phases (3,5-dichlorobenzoyl amino acid and amino phosphonic acid derivatives) – and applied them in the enantioseparation of chiral bases by nonaqueous CEC. Kornysov *et al.* [41] described a continuous bed chiral stationary phase based on the macrocyclic antibiotic vancomycin, for chiral separation in CEC. The CSP was subsequently prepared by immobilizing the vancomycin stationary phase by reductive amination. Preliminary results have indicated that an extremely strong EOF is obtained in both the nonaqueous polar organic (15.2×10^{-5}) cm^2 V^{-1} s^{-1} and the aqueous reversed phase modes of operation (8.5×10^{-5}) cm^2 V^{-1} s^{-1}. Enantioselectivity was obtained for four racemic compounds, the best performance being the case of thalidomide, which was separated in 10 min with a high resolution ($R_s = 2.5$) and efficiency (120 000 plates per metre) values. Similarly, Karlsson *et al.* [42] evaluated the vancomycin chiral stationary phase in CEC using polar organic and reversed phase modes. Desiderio *et al.* [43] used the vancomycin silica stationary phase in packed CEC for chiral separations. Girod *et al.* [44] described enantioseparations in nonaqueous CEC using polysaccharide type chiral stationary phases. Three different polysaccharide derivatives, cellulose tris(3,5-dimethylphenylcarbamate) (Chiralcel OD), amylose tris(3,5-dimethylphenylcarbamate) (Chiralpak AD) and cellulose tris(4-methylbenzoate) (Chiralcel OJ), were used as chiral stationary phases (CSPs). Methanolic or ethanolic ammonium acetate solutions served as a mobile phase. The effects of the type of CSP, the loading of the chiral selector on wide-pore aminopropyl derivatized silica gel, and operational parameters such as the apparent pH, applied voltage on the EOF and chromatographic characteristics (α, N and R_s) were studied.

Wang *et al.* [45] described a novel open tubular capillary electrochromatography (OT-CEC) column coated with 2,6-dibutyl-β-cyclodextrin using a sol-gel technique. In this approach, owing to the three-dimensional network of sol-gel and the strong chemical bond between the stationary phase and the surface of the capillary columns, good chromatographic characteristics and a unique selectivity in separating isomers were shown. The authors achieved high efficiencies of $(5-14) \times 10^4$ plates per metre for the isomeric nitrophenols using sol-gel derived β-CD columns. Lämmerhofer *et al.* [46] used chiral monolithic columns for enantioselective capillary electrochromatography prepared by copolymerization of a monomer with quinidine functionality. The effect of the chromatographic conditions on the performance of chiral monolithic poly(O-[2-(methacryloyloxy)-ethylcarbamoyl]-10,11-dihydroquinidine-co-ethylene dimethacrylate-co-2-hydroxyethyl methacrylate) columns in the capillary electrochromatography of enantiomers has been studied. The flow velocity was found to be proportional to the pore size of the monolith, and to both the pH value and the composition of the mobile phase. The lengths of both the open and the monolithic segments of the capillary column were found to exert a substantial effect on the run times. The use of monoliths as short as 8.5 cm and the short-end injection technique enabled the separations to be achieved in approximately 5 min, despite the high retentivity of the quinidine selector. Very high column efficiencies of close to 250 000 plates per metre and good selectivities were achieved for separations of numerous enantiomers, using chiral monolithic capillaries with optimized chromatographic conditions. Furthermore, the same group [47] used the same chiral selectors in electrochromatography for chiral separations. Otsuka *et al.* [48] described a packed capillary column for enantiomer separations in CEC.

CEC has also been used for the chiral analysis of pollutants, using a variety of chiral mobile phase additives. Cyclodextrin chiral selectors have been used for the chiral resolution of ibuprofen [49], 2-phenylpropionic acid and warfarin [50]. Other chiral selectors that have been resolved include warfarin on human serum albumin [51], ibuprofen on avidin [52], dichloroprop, ibuprofen, warfarin and thalidomide on vancomycin [53], ibuprofen and warfarin on teicoplanin [54], thalidomide and warfarin on

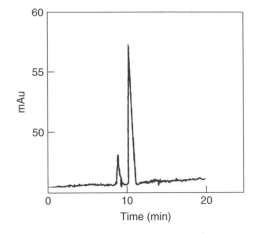

Figure 8.4   Electropherograms of the chiral resolution of thalidomide in CEC on cellulose tris(3,5-dimethylphenylcarbamate) CSP (capillary 32.5 cm × 100 μm id) with applied pressure and voltage 12 bar and 15 kV, respectively, using 10 mM ammonium acetate dissolved in methanol as the mobile phase [59].

polysaccharides [55–57], and warfarin and thalidomide on polyacrylamide and polysaccharide [58, 59]. The chiral resolution of thalidomide using CEC on a cellulose tris(3,5-dimethylphenylcarbamate) CSP is shown in Figure 8.4 [20]. Details of the chiral analysis of environmental pollutants by MEKC, CEC and TLC are given in Table 8.1. Since CEC is a hybrid technique of liquid chromatography and capillary electrophoresis, the optimization of chiral analysis can be achieved by varying the experimental conditions used in liquid chromatography and capillary electrophoresis. Interested readers should consult Chapters 7 and 9, respectively.

8.5 Supercritical Fluid Chromatography (SFC)

Supercritical fluids have been used as mobile phases in liquid chromatography for about 30 years, and this modality is known as supercritical fluid chromatography (SFC). When both the temperature and the pressure of the system exceed the critical values – that is, the critical temperature (T_c) and the critical pressure (P_c) – the fluid is considered to be critical in nature. These fluids have a mixture of the properties of liquids and gases. Supercritical fluids (SFs) are highly compressible like gases, and their density and viscosity can be maintained by changing the pressure and temperature conditions, as in the case of liquids. In chromatographic systems, the solute diffusion coefficients are often of a higher order of magnitude in supercritical fluids than in traditional liquids. On the other hand, the viscosities are lower than those of liquids [60]. At temperatures below T_c and pressures above P_c, the fluid becomes a liquid, while at temperatures above T_c and pressure below P_c, the fluid behaves as a gas. Therefore, supercritical fluids can be used as part of a liquid–gas mixture [61]. The most commonly used supercritical fluids are carbon dioxide, nitrous oxide and trifluoromethane [60–62]. Its compatibility with most detectors, low critical temperature and pressure, low toxicity and environmental burden, and low cost make carbon dioxide the supercritical fluid of choice. The main drawback of supercritical carbon dioxide as a mobile phase is its inability to elute more polar compounds. This can be improved by the addition of organic modifiers to the relatively apolar carbon dioxide. Chiral sub-FC and SFC have been carried out in packed and open tubular columns and capillaries [63].

There has been great progress in the field of separation of chiral compounds by sub- and supercritical fluid chromatography since the first demonstration of a chiral separation by SFC in 1985. Almost all of the chiral selectors used in gas or liquid chromatography have been successfully applied to sub-/supercritical chromatography. The main features of this technique are easier and faster method development, high efficiency, superior and rapid separations of a wide variety of analytes, an extended temperature capability, analytical and preparative scale equipment improvements, and a selection of detection options [63]. Besides, pollutants that cannot be vaporized for GC analysis, but that have no functional groups for sensitive detection with the usual HPLC detectors, can be separated and detected using SFC. The most common detector in SFC is FID, but other HPLC and GC detectors can be used. Compared with GC, the strongly enhanced solvation strength of, for example, supercritical carbon dioxide allows the oven temperature to be reduced significantly. This offers a promising alternative to increasing the enantioselectivity of a chiral stationary phase since, in the usually enthalpy controlled domain of enantiomer separations, the separation factor (α) increases with decreasing separation temperature [64]. The first report on chiral separation by SFC was published in 1985, by Mourier *et al.* [65]. Several papers and reviews have subsequently appeared on the subject [66, 67]. Salvador *et al.* [68] used dimethylated β-cyclodextrin as a mobile phase additive on porous graphite carbon as a solid phase for the chiral analysis of warfarin by SFC. The authors also studied the effects of the concentration of dimethylated β-cyclodextrin, the concentration of the mobile phase, the nature of the polar modifiers, the outlet pressure and the column temperature on chiral resolution. Normally, chiral CSPs made of a variety of chiral compounds are used in SFC and, hence, chiral recognition mechanisms are similar to those used in HPLC. The details of chiral recognition mechanisms using different CSPs are discussed in Chapter 7. Chiral separation can be optimized by using different temperatures and pressures for the critical fluid. Organic modifiers can also be used to obtain good chiral separations. In addition to this, the flow rate of the mobile phase and the amount of chiral pollutant injected are also considerable factors for the optimization of chiral analysis in SFC.

Many modifications and improvements have been reported in SFC for the analysis of racemates. Phinney and Sander [7] described polar additive concentration effects in supercritical fluid chromatography on chiral stationary phases having either a macrocyclic glycopeptide or a derivatized polysaccharide as the chiral selector. Two basic additives, isopropylamine and triethylamine, were added to methanol modifier at various concentrations and the effects on retention, selectivity and resolution were monitored. Many of the analytes failed to elute from the macrocyclic glycopeptide stationary phase in the absence of an additive, and the most noticeable effect of the increasing additive concentration was a significant decrease in retention. However, on the derivatized polysaccharide stationary phase the additives had little effect on retention, but they did foster significant improvements in peak shape and resolution. The combination of the simulated moving bed (SMB) technique with supercritical fluid chromatography (SFC) leads to a process with unique features. Besides the known advantages of the SMB process, the use of supercritical carbon dioxide as the mobile phase offers the advantages of reduction in organic solvents and an easy eluent/solute separation. Because of the low viscosity and high diffusion coefficients of supercritical fluids, a high efficiency is possible. Johannsen *et al.* [69] described preparative chiral separation using simulated moving bed chromatography with supercritical fluids. Liu *et al.* [8] described the chiral recognition capabilities of three macrocyclic glycopeptide chiral selectors – namely, teicoplanin (Chirobiotic T), its aglycone (Chirobiotic TAG) and ristocetin (Chirobiotic R) – with supercritical and subcritical fluid mobile phases. Various amounts of methanol ranging from 7 to 67 % (v/v) were added to carbon dioxide along with small amounts (0.1–0.5 %, v/v) of triethylamine and/or trifluoroacetic acid. The Chirobiotic TAG CSP was the most effective, closely followed by the Chirobiotic T column. Williams *et al.* [70] compared chiral separations on liquid and supercritical fluid chromatographic techniques using commercially available chiral stationary phases (CSPs) bearing three different types of chiral selectors.

Chiral compounds of agricultural interest were used to probe the advantages or limitations of SFC relative to LC for enantiomeric separations. Column equilibrium and other parameter optimization were generally accomplished more rapidly in SFC than in LC. Although improved resolution was often observed in SFC, analysis times were not always lower in SFC than in LC; yet, in some instances, SFC provided separation capabilities that were not readily accessible in LC.

8.6 Thin Layer Chromatography (TLC)

As previously mentioned, the use of thin layer chromatography (TLC) in chiral resolution goes back about 25 years. Most TLC enantioseparations have been carried out in the indirect mode – that is, by preparing diastereoisomers and resolving them on TLC – but this is not feasible in the case of pollutants, as their concentrations are very low. Moreover, many impurities are present with chiral pollutants and, hence, the derivatization of the pollutants in natural samples is very difficult and tedious. The derivatization of racemic mixtures and, subsequently, separation on silica gel or RP TLC plates represents a method of chiral separation. However, only a few reports have appeared on direct enantiomeric separation on chiral TLC plates; that is, using CMPAs or CSPs. Among the direct approaches, the use of CSPs is also very limited. Only ligand exchange based chiral thin layer chromatographic plates are commercially available for the chiral resolution of racemates. However, some reports also deal with the impregnation of TLC plates with suitable chiral selectors for chiral resolution [71]. Several research papers and review articles on chiral resolution by TLC have appeared in the literature [72–74]. In spite of the development of more efficient and rapid chromatographic techniques, TLC still exists as a simple chiral separation method. The main advantages of TLC are its low running cost, easy of handling and co-elution of the racemate and pure enantiomers. The availability of reliable, robust and efficient methods for the resolution of enantiomers by thin layer chromatography is a useful addition to the other techniques of chiral resolution. However, the major weakness of TLC includes the relatively low resolving power, with a high detection limit compared to other liquid chromatographic approaches. Moreover, the wastage of the chiral selector or the chiral thin layer material is another serious drawback.

TLC has been used for the chiral resolution of drugs, pharmaceuticals and agrochemicals [10]. Therefore, it may be used for the chiral resolution of environmental pollutants. For chiral resolution by TLC, the pollutant should be free from other environmental impurities. Chen and Wang [75] resolved the enantiomers of pyrethrin and pyrethroid pesticide residues in crops, foods and other environmental samples. The authors used chiral TLC plates based on the ligand exchange principle. Rosseti *et al.* [76] used chiral high performance thin layer chromatography (HPTLC) for the chiral resolution of 2-arylpropionic acid and ibuprofen. One of the classical approaches to liquid chromatography is paper chromatography, which was also used for chiral resolution about 50 years ago, but at present is virtually obsolete. In paper chromatography, the stationary phase is water-bonded to cellulose (paper material), which is of course chiral in nature and hence provides a chiral surface to the enantiomers. However, some workers have also used chiral mobile phase additives in paper chromatography [77, 78]. Chiral analysis in TLC can be optimized by using different types and concentrations of mobile phases. The pH of the mobile phase is also a very important factor for controlling chiral analysis in this modality of liquid chromatography. The concentrations of the chiral selectors and the amounts of the racemates applied to the TLC plate should also be carefully considered.

In the case of the indirect approach to chiral analysis, the separation of the chiral pollutants in TLC is controlled by adsorption and partition processes. Similarly, in the case of the CMPA mode, diastereoisomeric complexes are formed on TLC plates and are separated due to the adsorption and partition phenomena. When using the direct approach to chiral analysis, the mechanisms of chiral separation are entirely different. As discussed above, only ligand exchange and cellulose based TLC plates are readily available for the direct chiral analysis of racemates. In the case of ligand exchange plates, chiral resolution occurs due to the exchange of chiral ligands and the enantiomers on a specific metal ion through coordinate bonds. The two enantiomers have different exchange capacities because of the stereospecific nature of the ligand exchange process, and hence chiral resolution takes place. Rogozhin and Davankov [30] suggested a theoretical model for the mechanism of chiral resolution on these CSPs. The authors explained that chiral separation is due to different bondings along with the steric effects. Accordingly, diastereoisomeric complexes are formed by the two enantiomers. These diastereoisomeric complexes are stabilized at different magnitudes by dipole–dipole interactions, hydrogen bondings, van der Waals forces and steric effects, and finally get separated. On the other hand, chiral recognition mechanisms are similar to those of the polysaccharide-based CSPs when using TLC plates made of cellulose or other polysaccharide materials. The helical and chiral structures of polysaccharides provide a chiral environment for the enantiomers of the pollutants, and only one enantiomer fits strongly to these structures, in comparison to the other. The stability of the enantiomers on the helical structure is provided by hydrogen bondings, $\pi - \pi$ interactions and dipole-induced dipole attractions. Thus, the strongly bonded enantiomer moves more slowly than the other antipode, and in this way chiral separation occurs.

8.7 LC versus GC

GC has been used widely for the chiral resolution of many chiral environmental pollutants. The use of many of the sensitive detectors in GC makes it an ideal technique for the chiral resolution of pollutants. In spite of this, GC suffers from certain drawbacks, as the efficiency of this technique becomes poor for some pollutants that are volatile only at high temperature; as the extent of the enantiomeric resolution process reduces at high temperature. In addition, some chiral compounds decompose or racemize at the high working temperature of the GC machine. It is also very important to mention here that many chiral CSPs are not available in GC: only cyclodextrin-based GC chiral columns are available, which restricts the application of chiral GC. Sometimes, derivatization of the pollutants is required, which is a time-consuming and tedious job, particularly in environmental samples. In spite of the development of various types of CSPs in LC, it has not been able to achieve a high status in the chiral resolution of pollutants. The reason for this is that most organochlorine pollutants are transparent to UV radiation and, hence, they have to be detected with an ECD detector, which cannot be used in LC. However, many workers have used the HPLC, CEC, MKEC and TLC modalities of liquid chromatography for the chiral resolution of UV-absorbing pollutants. Many other types of detectors, such as MS and optical detectors, have also been used for the detection of chiral pollutants. Therefore, the use of LC is growing rapidly in the field of chiral resolution of environmental pollutants, and it is believed that LC will certainly replace GC. The use of capillaries and capillary columns in CEC is another advancement in liquid chromatography. The greater efficiency of CEC is due to the higher theoretical plate numbers.

8.8 Conclusions

Nowadays, gas chromatography and high performance liquid chromatographic methods are used widely for the chiral analysis of pollutants.

However, certain weaknesses and drawbacks of these modalities, as discussed above, compel scientists to explore new techniques. Various improvements and modifications have been made to GC and HPLC from time to time, which have resulted in the development of the MEKC, CEC and SFC modalities of liquid chromatography for the analysis of chiral racemates. Only a few reports are available on chiral analysis using these modalities, but their high speed, efficiency and reproducibility are attracting increasing attention. Only a few applications of these technologies in chiral analysis have been cited, because these new methods have not yet gained full exposure; in fact, their development is still under way. We hope that these methods will become popular in the chiral analysis of a wide range of racemates in the near future.

References

1. R. Pascoe and J. P. Foley, *Analyst* **127**, 710 (2002).
2. S. J. Thibodeaux, E. Billiot, E. Torres, B. C. Valle and I. M. Warner, *Electrophoresis* **24**, 1077 (2003).
3. K. Otsuka and Terabe, *J. Chromatogr. A* **875**, 163 (2000).
4. K. Kawamura, K. Otsuka and S. Terabe, *J. Chromatogr. A* **924**, 251 (2001).
5. B. Chankvetadz and G. Blaschke, *J. Chromatogr. A* **906**, 309 (2001).
6. G. Gübitz and M. G. Schmid, *Enantiomer* **5**, 5 (2000).
7. K. W. Phinney and L. C. Sander, *Chirality* **15**, 287 (2003).
8. Y. Liu, A. Berthod, C. R. Mitchell, T. L. Xiao, B. Zhang and D. W. Armstrong, *J. Chromatogr. A* **978**, 185 (2002).
9. B. Chankvetadze, *Capillary Electrophoresis in Chiral Analysis*, Wiley, New York, 1997.
10. H. Y. Aboul-Enein and I. Ali, *Chiral Separations by Liquid Chromatography and Related Technologies*, Dekker, New York, 2003.
11. S. Terabe, K. Otsuka, A. Ichikawa and T. Ando, *Anal. Chem.* **56**, 111 (1984).
12. S. Terabe, *J. Pharm. Biomed. Anal.* **10**, 705 (1992).
13. E. Szoko, J. Gyimesi, Z. Szakacs and M. Tarnai, *Electrophoresis* **20**, 2754 (1999).
14. J. J. Pesek, M. T. Matyska and S. Menezes, *J. Chromatogr. A* **20**, 853 (1999).
15. A. G. Peterson and J. P. Foley, *J. Chromatogr. B. Biomed. Sci. Appl.* **18**, 695 (1997).
16. J. R. Mazzeo, E. R. Grover, M. E. Swartz and J. S. Petersen, *J. Chromatogr. A* **30**, 680 (1994).
17. Y. Mechref and Z. El Rassi, *J. Chromatogr. A* **757**, 263 (1997).
18. K. L. Rundlett and D. W. Armstrong, *Anal. Chem.* **67**, 2088 (1995).

19. M. L. Marina, I. Benito, J. C. Diez-Masa and M. J. Gonzalez, *Chromatographia* **42**, 269 (1996).
20. M. L. Marina, I. Benito, J. C. Diez-Masa and M. J. Gonzalez, *J. Chromatogr. A* **752**, 265 (1996).
21. C. Garcia-Ruiz, Y. Martin-Biosca, A. L. Crego and M. L. Marina, *J. Chromatogr. A* **910**, 157 (2001).
22. L. C. Cooper and M. J. Sepaniak, *Anal. Chem.* **66**, 147 (1994).
23. Y. Tanaka, *Chromatography* **23**, 13 (2002).
24. W. R. G. Baeyens, G. Van der Weken, H. Y. Aboul-Enein, S. Reygaerts and E. Smet, *Biomed. Chromatogr.* **14**, 58 (2000).
25. W. H. Pirkle and P. G. Murray, *J. High Res. Chromatogr.* **16**, 285 (1993).
26. C. C. Pfeiffer, *Science* **124**, 29 (1956).
27. W. H. Pirkle and T. C. Pochapsky, *Chem. Rev.* **89**, 347 (1989).
28. J. J. Groombridge, C. G Jones, M. W. Bruford and R. A. Nichols, *Nature* **403**, 615 (2000).
29. German Pat. 1932190 (1969), S. V. Rogozhin and V. A. Davankov.
30. S. V. Rogozhin and V. A. Davankov, *Chem. Lett.* 490 (1971).
31. V. A. Davankov, *J. Chromatogr. A* **666**, 55 (1994).
32. S. Mayer and V. Schurig, *J. High Res. Chromatogr.* **15**, 129 (1992).
33. S. Li and D. Lloyd, *Anal. Chem.* **65**, 3684 (1993).
34. S. Mayer and V. Schurig, *J. Liq. Chromatogr.* **16**, 915 (1993).
35. V. Schurig, M. Jung, S. Mayer, M. Fluck, S. Negura and H. Jakubetz, *J. Chromatogr. A* **694**, 119 (1994).
36. V. Schurig, M. Jung, S. Mayer, S. Negura, M. Fluck and H. Jakubetz, *Angew. Chem.* **106**, 2265 (1994).
37. J. Vindevogl and P. Sandra, *Electrophoresis* **15**, 842 (1994).
38. Y. Gong and H. K. Lee, *Anal. Chem.* **75**, 1348 (2003).
39. Z. Liu, K. Otsuka, S. Terabe, M. Motokawa and N. Tanaka, *Electrophoresis* **23**, 2973 (2002).
40. E. Tobler, M. Lammerhöfer, F. Wuggenig, F. Hammerschmidt and W. Lindner, *Electrophoresis* **23**, 462 (2002).
41. O. Kornysova, P. K. Owens and A. Maruska, *Electrophoresis* **22**, 3335 (2001).
42. C. Karlsson, L. Karlsson, D. W. Armstrong and P. K. Owens, *Anal. Chem.* **72**, 4394 (2000).
43. C. Desiderio, Z. Aturki and S. Fanali, *Electrophoresis* **22**, 535 (2001).
44. M. Girod, B. Chankvetadze and G. Blaschke, *J. Chromatogr. A* **887**, 439 (2000).
45. Y. Wang, Z. Zeng, N. Guan and J. Cheng, *J. Electrophoresis* **22**, 2167 (2001).
46. M. Lämmerhofer, F. Svec and J. M. Frechet, *Anal. Chem.* **72**, 4623 (2000).
47. M. Lämmerhofer, E. C. Peters, C. Yu, F. Svec and J. M. Frechet, *Anal. Chem.* **72**, 4614 (2000).

48. K. Otsuka, C. Mikami and S. Terabe, *J. Chromatogr. A* **887**, 457 (2000).
49. V. Schurig and D. Wistuba, *Electrophoresis* **20**, 2313 (1999).
50. T. Koide and K. Ueno, *J. High Res. Chromatogr.* **23**, 59 (2000).
51. J. Jang and D. Hage, *Anal. Chem.* **66**, 2719 (1994).
52. Z. Liu, K. Otsuka and S. Terabe, *J. Sepn Sci.* **24**, 17 (2001).
53. S. Fanali, P. Catarcini, G. Blaschke and B. Chankvetadze, *Electrophoresis* **22**, 3131 (2001).
54. E. Carlsson, H. Wikström and P. K. Owens, *Chromatographia* **53**, 419 (2001).
55. M. Meyring, B. Chankvetadze and G. Blaschke, *J. Chromatogr. A* **887**, 439 (2000).
56. L. Chankvetadze, I. Kartozia, Y. Yamamoto, B. Chankvetadze, G. Blaschke and Y. Okamoto, *J. Sepn Sci.* **25**, 653 (2002).
57. B. Chankvetadze, I. Kartozia, J. Breitkreutz, Y. Okamoto and G. Blaschke, *Electrophoresis* **22**, 3327 (2001).
58. K. Krause, M. Girod, B. Chankvetadze and G. Blaschke, *J. Chromatogr. A* **837**, 51 (1999).
59. L. Chankvetadze, I. Kartozia, C. Yamamoto, B. Chankvetadze and Y. Okamoto, *Electrophoresis* **23**, 486 (2002).
60. R. C. Weast, ed., *Handbook of Chemistry and Physics*, 54th edn, CRC Press, Cleveland, 1973.
61. T. A. Berger, in R. M. Smith, ed., *Packed Column SFC*, Royal Society of Chemistry, Chromatography Monographs, Cambridge, 1995.
62. P. J. Schoenmakers, in R. M. Smith, ed., *Packed Column SFC*, Royal Society of Chemistry, Chromatography Monographs, Cambridge, 1988.
63. G. Terfloth, *J. Chromatogr. A* **906**, 301 (2001).
64. V. Schurig, A. Ossig and R. Link, *Angew. Chem.* **101**, 197 (1989).
65. P. A. Mourier, E. Eliot, R. H. Caude, R. H. Rosset and A. G. Tambute, *Anal. Chem.* **57**, 2819 (1985).
66. N. Bargmann-Leyder, A. Tambute and M. Caude, *Chirality* **7**, 311 (1995).
67. K. L. Williams, L. C. Sander and S. A. Wise, *J. Chromatogr. A* **746**, 91 (1996).
68. A. Salvador, B. Herbreteau, M. Dreux, A. Karlsson and O. Gyllenhaal, *J. Chromatogr. A* **929**, 101 (2001).
69. M. Johannsen, S. Peper and A. Depta, *J. Biochem. Biophys. Methods* **54**, 85 (2002).
70. K. L. Williams, L. C. Sander and S. A. Wise, *J. Pharm. Biomed. Anal.* **15**, 1789 (1997).
71. R. Bushan and I. Ali, *J. Chromatogr.* **392**, 460 (1987).
72. K. Günther, in J. Sherma and B. Fried, eds, *Handbook of Thin Layer Chromatography*, Dekker, New York, 1991, p. 541.
73. H. Y. Aboul-Enein, M. I. El-Awady, C. M. Heard and P. J. Nicholls, *Biomed. Chromatogr.* **13**, 531 (1999).

74. R. Bhushan and J. Martens, *Biomed. Chromatogr.* **15**, 155 (2001).
75. Z. M. Chen and Y. H. Wang, *J. Chromatogr. A* **754**, 367 (1996).
76. V. Rosseti, A. Lombard and M. Buffa, *J. Pharm. Biomed. Anal.* **4**, 673 (1986).
77. G. B. Bonino and V. Carassiti, *Nature* **167**, 569 (1951).
78. M. Mason and C. P. Berg, *J. Biol. Chem.* **195**, 515 (1952).

Chapter 9

The Analysis of Chiral Environmental Pollutants by Capillary Electrophoresis

9.1 Introduction

As discussed in previous chapters, mostly gas chromatography (GC) and high performance liquid chromatography (HPLC) techniques have been used for the chiral resolution of environmental pollutants. But the high polarity, low vapour pressure and derivatization in some environmental pollutants are factors that complicate GC analysis. On the other hand, due to the inherent limited resolving power of conventional HPLC techniques, the optimization of the chiral resolution of a pollutant often involves complex procedures or numerous experiments, leading to the consumption of large amounts of solvents and sample volumes [1]. Moreover, efficiency in HPLC is limited due to the profile of the laminar flow, mass transfer term and possible additional interactions of enantiomers with the residual silanol groups of the stationary phase. At present, capillary electrophoresis (CE), a versatile technique that offers high speed and sensitivity, is a major trend in analytical science, and some publications on the resolution of chiral environmental pollutants have appeared in recent years. The high efficiency of CE is due to the flat profile originated and the homogeneous partition of the chiral selector in the electrolyte, which in turn minimizes mass transfer. We

Chiral Pollutants.   I. Ali and H. Y. Aboul-Enein
© 2004 John Wiley & Sons, Ltd   ISBN: 0-470-86780-9

have recently reviewed the analysis of chiral pollutants by using capillary electrophoresis [2, 3]. Therefore, attempts have been made to explain the art of the enantiomeric resolution of chiral environmental pollutants by CE. This chapter presents the application of CE for the chiral resolution of environmental pollutants, describing the types of chiral selector, the optimization of the CE parameters and chiral recognition mechanisms.

9.2 Chiral Selectors

As in the case of GC and HPLC, a chiral selector is also required in CE for enantiomeric resolution. In general, chiral compounds are used in background electrolyte (BGE) as additives, and hence they are called chiral selectors or chiral BGE additives. There are only a few publications available that deal with chiral resolution on a capillary coated with a chiral selector in CE [4]. The enantiomeric resolution of chiral pollutants discussed in this chapter has been confined to the use of chiral selectors in the BGE. As a chiral resolution technique, CE has been used widely for the enantiomeric resolution of drugs and pharmaceuticals [5]. Several reviews have also appeared on this issue, and they describe the use of many chiral compounds as chiral BGE additives [6–13]. The most commonly used chiral BGE additives are cyclodextrins, macrocyclic glycopeptide antibiotics, proteins, crown ethers, ligand exchangers, alkaloids [6–16] and so on. The structures and properties of these chiral BGE additives have been described in previous chapters and, therefore, this aspect will not be discussed here again. However, a list of these chiral BGE additives is presented in Table 9.1. α-, β- and γ-Cyclodextrins and their derivatives have been used frequently in chiral CE, because of their water solubility

Table 9.1 The chiral selectors most commonly used in CE

Chiral selectors (chiral BGE additives)	References
Cyclodextrins	5, 9–11
Macrocyclic glycopeptide antibiotics	5
Proteins	5, 12
Crown ethers	5, 15
Alkaloids	5, 14
Polysaccharides	5
Calixarenes	–
Imprinted polymers	16
Ligand exchangers	16

and because they have various functional groups and cavities that are responsible for the formation of diastereomeric inclusion complexes with chiral pollutants. The basic requirements for a suitable chiral selector CE include the following:

- It should be inexpensive in nature, soluble in BGE and capable of forming inclusion complexes with environmental pollutants.

- It should have sufficient groups, atoms, grooves, cavities and so on for complexing with chiral pollutants.

- It should increase the electro-osmotic flow (EOF), or at least it should not affect the EOF.

- The diffusion of the pollutant through the chiral selector, the specific selector–selector interactions and, accordingly, the mass transfer kinetics must be favourable in order to effectively employ the advantages offered by the electrically driven flow.

- It should have suitable chemoselectivity combined with enantioselectivity, as it may be used for the resolution of structurally related pollutants.

- It must be non-UV-absorbing in nature as, generally, detection in CE is carried out using a UV detector.

9.3 Applications

For a number of years, CE has been used for the chiral resolution of pesticides, polynuclear–aromatic hydrocarbons, amines, carbonyl compounds, surfactants, dyes and other toxic compounds. CE has also been utilized to separate the structural isomers of various toxic pollutants, such as phenols, polyaromatic hydrocarbons and so on. Weseloh *et al.* [17] investigated a CE method for the resolution of biphenyls using a phosphate buffer as BGE with cyclodextrin as the chiral additive. Sarac *et al.* [18] resolved the enantiomers of 2-hydrazino-2-methyl-3-(3,4-dihydroxyphenyl) propionic acid using cyclodextrin as the BGE additive. The cyclodextrins used were native, neutral and ionic in nature, with a phosphate buffer as the BGE. Miura *et al.* [19] used CE for the chiral resolution of seven phenoxy acid herbicides, using methylated cyclodextrins as the BGE additives. Furthermore, the same group [20] resolved MCPP, DCPP, 2,4-D, 2,4-CPPA, 2,4,5-T, 2,3-CPPA, 2,2-CPPA, 2-PPA and silvex pesticides, using cyclodextrins with negatively charged sulfonyl groups as chiral BGE additives.

Gomez-Gomar *et al.* [21] investigated the simultaneous enantioselective separation of ($\pm$)-cizolirtine and its impurities: ($\pm$)-N-desmethylcizolirtine, ($\pm$)-cizolirtine-N-oxide and ($\pm$)-5-(-hydroxybenzyl)-1-methylpyrazole by capillary electrophoresis. Otsuka *et al.* [22] coupled capillary electrophoresis with mass spectrometry (CE–MS), and this combination was used for the chiral separation of phenoxy acid herbicide. The authors also described electrospray ionization (ESI) as an ionization method for a CE–MS interface. In general, nonvolatile additives in separation solutions sometimes decrease the MS sensitivity and/or signal intensity. However, heptakis(2,3,6-tri-O-methyl)-β-cyclodextrin (TM-β-CD), which migrated directly into the ESI interface, was used as a chiral selector. Using the negative-ionization mode, along with a methanol–water–formic acid solution as a sheath liquid and nitrogen as a sheath gas, the stereoselective resolution and detection of three phenoxy acid herbicide enantiomers was successfully achieved with a 20 mM TM-β-CD in a 50 mM ammonium acetate buffer (pH 4.6). Zerbinati *et al.* [23] resolved four enantiomers of the mecoprop and dichlorprop herbicides using an ethyl carbonate derivative of β-cyclodextrin (β-CD), with three substituents per molecule of hydroxypropyl-β-CD and native β-CD. The performance of these chiral selectors has been quantified by means of two-level full factorial designs and the inclusion constants were calculated from CE migration time data. The chiral resolution of environmental pollutants by CE is summarized in Table 9.2. The actual chromatograms

Table 9.2 The chiral resolution of some environmental pollutants by capillary electrophoresis and by micellar electrokinetic chromatography

Chiral pollutants	Electrolytes	Detection	References
Capillary electrophoresis			
Fenoprop, mecoprop and dichlorprop	20 mM tributyl-β-CD in 50 mM, ammonium acetate, pH 4.6	MS	22
2-Phenoxypropionic acid, dichloroprop, fenoxaprop, fluaziprop, haloxyfop and diclofop enantiomers	75 mM Britton– Robinson buffer with 6 mM vancomycin	–	24
Imazaquin isomer	50 mM sodium acetate, 10 mM dimethyl-β-CD, pH 4.6	–	25

Table 9.2 (*continued*)

Chiral pollutants	Electrolytes	Detection	References
Phenoxy acid herbicides	200 mM sodium phosphate, pH 6.5, with various concentrations of OG and NG	–	
Diclofop	50 mM sodium acetate, 10 mM trimethyl-β-CD, pH 3.6	–	25
Imazametha-benz isomers	50 mM sodium acetate, 10 mM dimethyl-β-CD, pH 4.6	–	25
2-(2-Methyl-4-chloro-phenoxy) propionic acid	0.05 M lithium acetate containing α-cyclodextrins	UV, 200 nm	26
2-(2-Methyl-4,6-dichlorophenoxy) propionic acid	0.05 M lithium acetate containing β-CD	UV, 200 nm	26
2-(2,4-Dichlorophenoxy) propionic acid	0.05 M lithium acetate containing heptakis-(2,6-di-O-methyl)-β-CD	UV, 200 nm	26
2-(2-Methyl-4-chloro-phenoxy) propionic acid and 2-(2,4-dichlorophenoxy)-propionic acid	0.03 M lithium acetate containing heptakis-(2,6-di-O-methyl)-β-CD	UV, 200 nm	27
2(2-Methyl-chlorophenoxy) propionic acid	0.05 M NaOAc, pH 4.5, α-CD	UV, 230 nm	28
Ethofamesate and napropamide	Buffer, pH 9, with SBE-β-CD	–	29
1,1-Binaphthol, 1,1'-Binaphthyl-2-2'-dicarboxylic acid and 1,1'-binaphthyl-2,2'-dihydrogen phosphate	0.01 M + 0.006 M borate (pH 9.0)	UV, 254 nm	30

(*continued overleaf*)

Table 9.2   (*continued*)

Chiral pollutants	Electrolytes	Detection	References
1,1-Binaphthol, 1,1-binaphthyl-2,2-dicarboxylic acid and 1,1′-binaphthyl-2,2′-dihydrogen phosphate	Various phosphate buffers with different concentrations and pHs, containing α-, β- and γ-CDs separately	UV, 214 nm	31
1,1-Binaphthyl-2,2-dicarboxylic acid and 1,1′-binaphthyl-2,2-dihydrogen phosphate and 2,2′-dihydroxy-1,1′-binaphthyl-3,3′-dicarboxylic acid	0.04 M carbonate buffer (pH 9.0) concentrations; noncyclic oligosaccharides	UV, 215–235 nm	32
Phenoxy acid	0.1 M phosphate buffer, pH 6, with:		
	vancomycin	–	33
	ristocetin	–	33, 34
	teicoplanin	–	33, 35
	OG	–	36
Phenoxy acid derivatives	β-CD and TM-β-CD	–	28, 37
Polychlorinated biphenyls	0.1 M CHES, pH 10, 2 M urea and 0.11 M SDS, containing γ-CD	–	38
Silvex	0.4 M borate, pH 10, containing Deoxy big CHAP	–	39

Micellar electrokinetic chromatography

Phenoxy acid herbicides	0.1 M phosphate and acetate buffer containing OM and CDs	–	40
	0.4 M Na borate (pH 10) with N,N-bis(3D-gluconamidopropyl) deoxycholamide	UV, 240 nm	39

Table 9.2 (*continued*)

Chiral pollutants	Electrolytes	Detection	References
PCBs	0.09 M CHES 0.11 M SDS, 2 M urea (pH 10) with β- and γ-CDs separately	UV, 235 nm	41
	CHES containing, SDS, bile salts and CDs and CDs	–	42
	buffers with CD derivatives	–	43
1,1′-Bi-2-naphthol	0.016 M NaCl + MeOH (pH 8.1–8.3) with sodium deoxycholate and polyoxyethylene ethers separately	UV, 210 nm	44
1,1′-Bi-2-naphthol, 1,1′-binaphthyl dicarboxylic acid and 1,1′-binaphthyl diyl hydrogen phosphate	Sodium cholate, sodium deoxycholate and sodium taurodeoxycholate	UV (laser etched flow cells and modified detector)	45
1,1′-Bi-2-naphthol and 1,1′-binaphthyl 2,2′-diyl hydrogen phosphate	0.025 M–0.20 M sodium borate (pH 8.11), with N,N'-bis-(3D-gluconamidopropyl)-cholamine	UV, 240 nm	39
1,1′-Binaphthyl-2,2′-diamine	Sodium borate + 15 % MeOH with N,N'-bis-(3D-gluconamidopropyl)-cholamine	UV, 240 nm	39
1,1′-Bi-2-naphthol and 1,1′-binaphthyl 2,2′-diyl hydrogen phosphate	0.03 M sodium dihydrogen phosphate + 0.01 M sodium borate (pH 8.0), with n-dodecyl-β-D-glucopyranoside-4,6-hydrogen phosphate Na salt and n-Dodecyl-β-D-glucopyranoside-6-hydrogen sulfate Na salt separately	UV, 214–254 nm	30

(*continued overleaf*)

Table 9.2   (*continued*)

Chiral pollutants	Electrolytes	Detection	References
1,1'-Bi-2-naphthol, 1.1'-binaphthyl-2,2'-dicarboxylic acid and 1,1'-2,2'-diyl hydrogen phosphate	0.025 M phosphate (pH 8.0) with sodium deoxycholate and sodium deoxycholate and α-CD separately	UV, 241 nm	30
Diniconazole and uniconazole	0.1 M SDS, 2 M urea, 0.1 M borate and 5 % 2-Me-2-PrOH	UV, 254 nm	46
1.1'-Bi-2-naphthol, 2, 2, 2'-trifluoro-1-(9-anthryl)ethanol and 1,1'-binaphthyl-2,2'-diyl hydrogen phosphate	0.05 SDS, 0.02 M phosphate (pH 9), Na d-camphor sulfonate with γ-CD and β-CD derivatives separately	UV, 220 nm	47
1.1'-Bi-2-naphthol and 1,1'-binaphthyl-2,2'-diyl hydrogen phosphate	0.025 M borate (pH 9.0) γ-CD, poly-(Na N-undecylenyl D-valinate)	UV, 280 nm	48

of the resolved enantiomers of phenoxy acid herbicides are shown in Figure 9.1. The separation and identification of the pollutants are defined by the migration time (t), the electrophoretic mobility (μ_{ep}) (k), the separation factor (α) and the resolution factor (R_s).

9.4 The Optimization of CE Conditions

In general, chiral resolution by CE is very sensitive and is controlled by a number of parameters. The optimization factors may be categorized into two classes, the independent and dependent. The independent parameters, which are under the direct control of the operator, include the choice of buffer, the pH of the buffer, the ionic strength of the buffer, the type of chiral selector, the applied voltage, the temperature of the capillary, the dimensions of the capillary, the BGE additives and some other parameters. On the other hand, the dependent parameters, which are those that are directly affected by the independent parameters, and are not under the direct control of the operator, are the field strength (V m^{-1}), the EOF, the Joule heating, the BGE viscosity, the sample diffusion, the sample mobility, the sample charge, the sample size and shape, the interaction of the sample with the capillary and the BGE, the molar absorptivity and so on. Therefore,

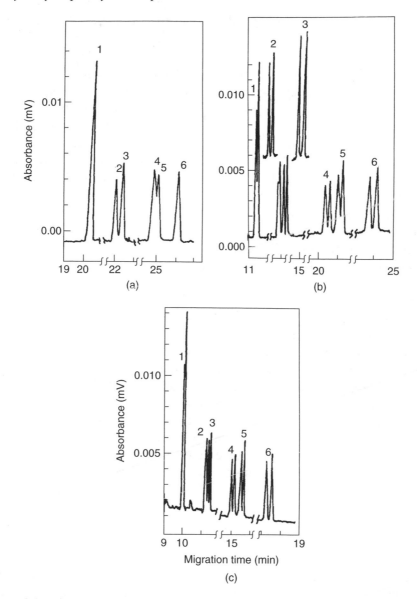

Figure 9.1 Electropherograms of phenoxy acid herbicides, using 175 mM sodium phosphate, pH 6.5, containing (a) 10 mM (b) 60 mM and (c) 100 mM n-octyl-β-D-maltopyranoside (OM), respectively, as BGEs with a fused silica capillary (57 cm × 50 μm id) and 25 kV as the applied voltage. 1 = silvex, 2 = dichlorprop, 3 = mecoprop, 4 = 2, 4-CPPA, 5 = 2, 3-CPPA, 6 = 2, 2-CPPA and 7 = 2-PPA [37].

the optimization of the chiral resolution can be controlled by varying all of these parameters.

9.4.1 The Composition of the Background Electrolyte

In CE, the background electrolyte (BGE) is used to maintain a high voltage gradient across the sample-containing solution in the capillary, and this requires that the conductivity of the electrolyte should be higher than the conductivity of the sample. Therefore, buffers are used as the BGE in most chiral CE applications: the use of buffers is essential to control the pH of the BGE. The most commonly used buffers for chiral resolution are phosphate, acetate, borate, ammonium citrate, tris, 2-(*N*-cyclohexylamino)ethanesulfonic acid (CHES), morpholinoethanesulfonic acid (MES), piperazine-*N*,*N*'-bis(2-ethanesulfonic acid) [PIPES], *N*-2-hydroxyethylpiperazine-*N*'-2-ethanesulfonic acid (HEPES) and so on, which are used at different concentrations and pH values. For the optimum chiral resolution of environmental pollutants, the identity and concentration of the electrolyte must be chosen carefully. The selection of the BGEs depends on their conductivity and the type of the environmental pollutant to be resolved. The relative conductivities of different electrolytes can be estimated from their condosities (defined as the concentration of sodium chloride, which has the same electrical conductance as the substance under study) [49]. A wide variety of electrolytes can be used to prepare buffers for chiral CE. Low-UV absorbing components are required for the preparation of the buffers if a UV detector is used. Also, volatile components are required in the case of MS or ICP detection methods. However, these conditions substantially limit the choice to a moderate number of electrolytes. The pH of the BGE is also another factor that determines the choice of buffer. For low-pH buffers, phosphates and citrates have commonly been used, although the latter absorb strongly at wavelengths < 260 nm. For basic buffers, borate, tris, CAPS and so on are used as suitable BGEs. A list of useful buffers along with their pH values and working wavelengths is given in Table 9.3 [50].

Crego and Marina [1] and Gübitz and Schmid [16] reviewed the chiral resolution of environmental pollutants by CE, and they described phosphate, borate, acetate, CHES and carbonate as suitable BGEs. Welseloh *et al.* [17] used a pH 2.4 30 mM phosphate buffer as the BGE for the chiral resolution of biphenyls. Mechref and El Rassi [40] resolved phenoxy acid herbicides using 175 mM phosphate (pH 6.5) [40] and 200 mM borate (pH 10.0) buffers [37] as BGEs. Furthermore, Tsunoi *et al.* [20] used a mixture of 0.1 M borate and 0.05 M phosphate buffers (pH 9.0) as BGE for the chiral

Table 9.3 Commonly used buffers with suitable pHs and wavelengths for the chiral separation of environmental pollutants in CE [50]

Buffer	pH	Wavelength (nm)
Phosphate	1.14–3.14	195
Citrate	3.06–5.40	260
Acetate	3.76–5.76	220
MES	5.15–7.15	230
PIPES	5.80–7.80	215
Phosphate	6.20–8.20	195
HEPES	6.55–8.55	230
Tricine	7.15–9.15	230
Tris	7.30–9.30	220
Borate	8.14–10.14	180
CHES	9.50–	< 190

CHES, 2-(*N*-cyclohexylamino)ethanesulfonic acid; MES, morpholinoethanesulfonic acid; PIPES, piperazine-*N*,*N'*-bis(2-ethanesulfonic acid); HEPES, *N*-2-hydroxyethylpiperazine-*N'*-2-ethanesulfonic acid.

resolution of phenoxy acid herbicides. Therefore, for chiral resolution, the nature and the type of BGE depend on the racemic pollutants and chiral selectors used.

9.4.2 The pH of the Background Electrolyte

The pH of the BGE is an important factor in optimizing chiral resolution, as it is thought to be responsible for the stability of the diastereomeric complexes formed between the enantiomers and the chiral selector. However, an increase in the buffer pH, from pH 4 to pH 9, may result in an increase in the EOF; therefore, the analysis time may be reduced by increasing the pH. It is also important to note that the pH value of the buffer may be altered in a secondary manner; that is, by other parameters such as temperature, ion depletion and so on. The suitable pH ranges for various buffers are summarized in Table 9.3 [50]. The literature reported herein indicates that a wide range of pHs have been used for the chiral resolution of environmental pollutants. Some reports indicate chiral resolution at acidic pH values, while others indicate basic pH values, which reveals that the pH requirements depend upon the type of the buffer used and other CE conditions. In general, a low pH is used to resolve cationic pollutants, while a high pH is required for the chiral resolution of anionic pollutants.

Crego and Marina Crego [1] reviewed the chiral resolution of environmental pollutants by CE, and reported the use of pH values ranging from

4.5 to 10.5. Weseloh *et al.* [17] studied the chiral resolution of biphenyls, using a pH 2.4 phosphate buffer is various concentrations. Mechref and El Rassi [40] studied the effect of the pH of the phosphate buffer on the chiral resolution of herbicides. The authors varied the pH from 5 to 8, and the results are shown in Figure 9.2: it may be concluded from this figure that values between pH 5 and pH 7 were found to be suitable for chiral resolution. This is due to the ionic nature of the herbicides studied (silvex, dichlorprop, mecoprop, 2,4-CPPA, 2,3-CPPA and 2,2-CPPA), which interact with the n-octyl-β-D-maltopyranoside (OM) surfactant used by the authors: hence the effect of the pH on the chiral resolution was markedly observed. The same authors also resolved these herbicides using a mixture of pH 5.0 phosphate and borate buffers [37]. Recently, Tsunoi *et al.* [20] have resolved phenoxy acid herbicides using pH 9.0 phosphate and borate buffers. Sarac *et al.* [18] reported 2.5 as the best pH for a phosphate buffer for the chiral resolution of propionic acid derivatives. Briefly, the pH value depends on the type of chiral environmental pollutant, the chiral selector and the buffer used.

9.4.3 The Ionic Strength of the Background Electrolyte

An increasing ionic strength decreases the EOF and, consequently, an increase in the separation time occurs in chiral CE. Additionally, an

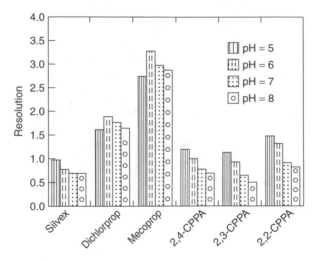

Figure 9.2 The effect of pH on the chiral resolution of phenoxy acid herbicides, using 100 mM sodium phosphate–sodium acetate buffers containing 60 mM OM as BGE. Other conditions as in Figure 9.1 [37].

increasing ionic strength also increases the current at a constant voltage, to the point at which adequate thermostating of the capillary becomes a concern. Furthermore, an increasing ionic strength decreases the formation of diastereomers and wall interactions. Therefore, the selection of the ionic strength depends on several parameters, such as the capillary length and diameter, the applied voltage and the efficiency of the capillary thermostating condition. The ionic strength also provides stability to diastereomers in some cases. It is very important to mention here that at high buffer concentrations excessive Joule heating occurs, which affects chiral resolution. The buffers that are problematic in this context are those that have electrolytes such as chloride, citrate and sulfate. However, the heating problem can be solved by decreasing the applied voltage, increasing the length and decreasing the internal diameter of the capillary. Therefore, optimization of the chiral resolution of environmental pollutants may be achieved by varying the ionic strength of the BGE. A literature search suggests that relatively little work has been carried out on this aspect of the optimization of chiral resolution. However, Mechref and El Rassi [40] studied the effect of ionic strength on the chiral resolution of herbicides, varying the ionic strength from 50 to 200 mM, and their findings are plotted in Figure 9.3.

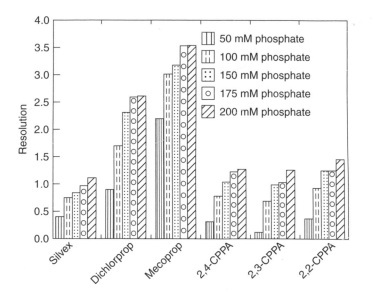

Figure 9.3 The effect of ionic strength on the chiral resolution of phenoxy acid herbicides, using sodium phosphate buffer containing 60 mM OM as BGE. Other conditions as in Figure 9.1 [37].

It may be seen from this figure that, in general, the resolution increased at high ionic strength (200 mM). The authors described the best resolution at high ionic strength as being due to the salting out effect, which is thought to increase the nonpolar interactions with the chiral micelles.

9.4.4 Structures and Types of Chiral Selectors

The enantiomeric resolution of environmental pollutants is mostly carried out using chiral selectors in the BGE. The chiral selectors are thought to interact with the enantiomers of the environmental pollutants: this will be discussed in detail later, in Section 9.7. Therefore, the selection of a suitable chiral selector for a specific chiral environmental pollutant is a key parameter. The selection of the chiral selector is governed by considering the interaction sites between the chiral selector and the environmental pollutants. Knowledge of the basic chemistry is helpful in selecting suitable chiral selectors. The chiral selectors that are most commonly used in CE are listed in Table 9.1. These commonly used cyclodextrins (α-, β-, γ-) form diastereomeric complexes with several environmental chiral pollutants, due to their ability to form inclusion complexes. Also, their good solubility in aqueous buffers and their low cost make these the chiral selectors of choice in CE. Mechref and El Rassi [37] used α-, β- and γ-CDs and their derivatives for the chiral resolution of phenoxy acid herbicides. The chiral resolution of these herbicides is shown in Table 9.4, which indicates 2,3,6-tri-O-methyl-β-CD as the best chiral selector. Furthermore, these authors also studied the effect of the concentration of these CDs on the enantiomeric resolution. The concentrations were varied from 5 to 30 mM and the results are shown in Figure 9.4, which indicates an increase in the resolution at higher concentrations of β-CD and HP-β-CD, while the resolution decreases at higher concentrations of DM-β-CD and TM-β-CD. These different trends of chiral resolution with these CDs at higher concentrations may be caused by some other interactions, such as steric effects.

9.4.5 The Applied Voltage

The applied voltage is one of the most important aspects in the optimization of the chiral resolution of environmental pollutants by CE. In general, a voltage increase results into an increase in the EOF, a shorter migration time, sharper peaks and, sometimes, improved resolution. Therefore, it is advisable to start with a moderate voltage value; that is, between 10 and 15 kV. Increasing the voltage also has a number of disadvantages. If the

Table 9.4 The chiral resolution of phenoxy acid herbicides (underivatized and derivatized with 7-aminonaphthalene-1,3-disulfonic acid) using various cyclodextrins as chiral selectors [37]

Herbicides	α-CD	β-CD	γ-CD	DM-β-CD	HP-β-CD	TM-β-CD
Silvex						
Underivatized	0.00	0.39	0.00	0.00	0.10	3.03
Derivatized	0.00	2.54	2.43	0.00	0.20	0.85
Mecoprop						
Underivatized	0.80	0.15	0.00	1.44	0.57	2.49
Derivatized	0.29	0.00	0.53	0.45	0.00	7.04
Dichlorprop						
Underivatized	0.67	0.00	0.00	0.00	0.75	1.05
Derivatized	0.14	0.00	0.00	0.00	0.00	3.76
2,4-CPPA						
Underivatized	0.43	1.00	0.00	0.00	0.35	1.08
Derivatized	0.00	3.43	0.00	1.28	2.45	0.20
2,3-CPPA						
Underivatized	0.27	1.68	0.00	0.00	1.00	0.00
Derivatized	2.63	3.33	0.00	1.28	2.45	0.20
2,2-CPPA						
Underivatized	0.00	0.00	0.00	0.00	0.00	0.00
Derivatized	0.00	0.00	0.00	0.79	1.29	1.76
2-PPA						
Underivatized	0.00	0.00	0.00	0.00	0.00	0.00
Derivatized	0.00	5.68	0.00	1.89	1.39	0.00

DM-β-CD, 2,6-di-*O*-methyl-β-cyclodextrin; HP-β-CD; hydroxypropyl-β-cyclodextrin; TM-β-CD, 2,3,6-tri-*O*-methyl-β-cyclodextrin; 2,2-CPPA: 2-(2-chlorophenoxy) propionic acid; 2,3-CPPA, 2-(3-chlorophenoxy) propionic acid; 2,4-CPPA, 2-(4-chlorophenoxy) propionic acid; 2-PPA, 2-phenoxypropionic acid.

ionic strength of the sample matrix is greater than the EOF, the increasing production of Joule heat cannot be efficiently dissipated. Additionally, the heating of the capillary results in a decrease in the viscosity of the BGE and hence reproducibility is lost. Nonreproducibility is observed due to an increase in both the ionic mobility and pollutant diffusion. The magnitude of the voltage also depends on the type of buffer used. Nelson *et al.* [51] reported no heating of the capillary up to a voltage of 30 kV when borate buffer was used, while heating of the capillary was observed even at 10 and 12 kV using CAPS and phosphate buffers, respectively. As reported by Crego and Marina [1], 20–30 kV has been used as the applied voltage with an acetate buffer.

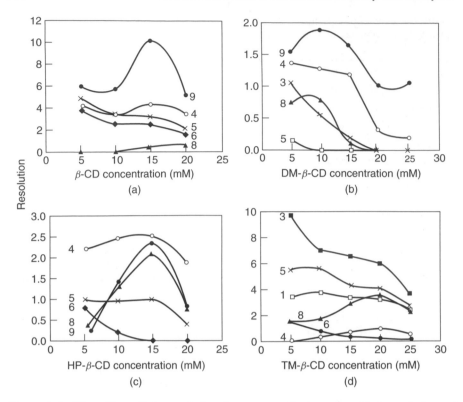

Figure 9.4 The effect of the cyclodextrin concentration on the chiral resolution of phenoxy acid herbicides, using 25 mM sodium phosphate and 600 mM borate buffers (pH 5.0) containing different concentrations of (a) β-CD, (b) DM-β-CD, (c) HP-β-CD and (d) TM-β-CD as BGEs. 1 = dichlorprop, 3 = mecoprop, 4 = 2, 3-CPPA, 5 = 2, 4-CPPA, 6 = silvex, 8 = 2, 2-CPPA, 9 = 2-PPA [34].

9.4.6 Temperature

In general, at high temperature the viscosity of the BGE decreases, which results in a short analysis time and poor resolution. Also, it is important to note that when the sample introduction is hydrostatic, the sample volume increases at a higher temperature, which sometimes results in poor resolution. At high temperature, concurrent changes in the buffer pH and peak broadening also occur. Briefly, temperature variation can be used to optimize chiral resolution by CE, but it has not been used as a routine optimization parameter, because control of the experimental temperature is difficult in the present CE model. However, Mechref and El Rassi [40] studied the effect of temperature on the chiral resolution of herbicides: the

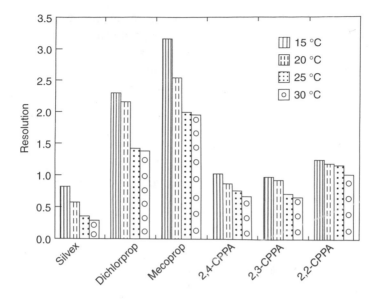

Figure 9.5 The effect of temperature on chiral resolution of phenoxy acid herbicides, using 150 mM sodium phosphate buffer (pH 6.5) containing 60 mM OM as BGE. Other conditions as in Figure 9.1 [37].

results are shown in Figure 9.5, which indicates that the chiral resolution improved at a lower temperature.

9.4.7 The Structures of Chiral Pollutants

The chiral resolution of environmental pollutants by CE depends on the formation of diastereomeric complexes and, therefore, the structures and sizes of the chiral pollutants are responsible for their enantiomeric resolution. To study this aspect, phenoxy acid herbicides (see Table 9.4) may be considered as the best class of chiral pollutant. Mechref and El Rassi [40] studied these herbicides using cyclodextrins as chiral selectors. It has been reported that the chiral resolution of these herbicides was in the order 2-PPA > 2,2-CPPA > 2,3-CPPA. 2-PPA has no chlorine atom on the phenyl ring, while 2,2-CPPA and 2,3-CPPA have chlorine atoms in the *ortho-* and *meta-* positions, respectively. Therefore, it may be concluded that the chlorine atom creates some sort of hindrance in the formation of diastereomeric complexes. Furthermore, it may be observed that the *ortho-* position creates a greater strain in comparison to the *meta-* position in the formation of diastereoisomeric complexes, and hence the above-mentioned order of resolution is observed. Briefly, the steric effect due to the presence of chlorine atoms plays a dominant role in the chiral resolution of these herbicides. Similarly, the differing patterns of chiral resolution of other environmental pollutants may be correlated with their structures. Finally, the optimum chiral resolution of environmental pollutants may be achieved by considering the interactions among these chiral selectors and the environmental pollutants.

9.4.8 Organic Modifiers

Generally, buffers of different concentrations and pH values are used as the background electrolytes (BGEs) for the chiral resolution of environmental pollutants by CE. However, the use of some organic solvents in BGEs, called organic modifiers, may be useful. The addition of these organic modifiers may change the EOF, the formation of the diastereomeric complexes and the interactions of the diastereoisomeric complexes with the capillary wall, and may bring about decreases in the conductivity, thermal diffusion and so on. Therefore, chiral resolution may be optimized by using different types of organic solvent at different concentrations. The most important organic modifiers for this purpose are acetonitrile, methanol, ethanol and 1,4-diaminobutane. Care should be taken in adjusting the concentrations of these organic modifiers, as higher concentrations may precipitate the buffer constituents, which may block the capillary. Mechref and El Rassi [40] studied the effect of the concentration of methanol on the chiral resolution of herbicides. The authors used 5 % and 10 % methanol concentrations in the phosphate and acetate buffer mixture: the results are shown in Figure 9.6, which indicates that 5 % methanol is the best concentration to achieve resolution.

9.4.9 Other Parameters

Besides the parameters discussed above, some other factors can also be used to optimize chiral resolution by CE. These parameters include the reversal of polarity, the volume of sample injected, the use of EOF modifiers and pre-derivatization of the chiral pollutants with a suitable reagent. In the normal CE machine, the anode (+) and cathode (−) are always at the inlet and outlet ends, respectively. In this modality, the EOF always tends to travel towards the cathode (detector). On the other hand, in the reverse mode, the direction of the EOF is away from the detector, and hence only negatively charged diastereomeric complexes with an electrophoretic mobility greater than the EOF will pass the detector. This format is typically used with capillaries that are coated with substances that reverse the net charge of the

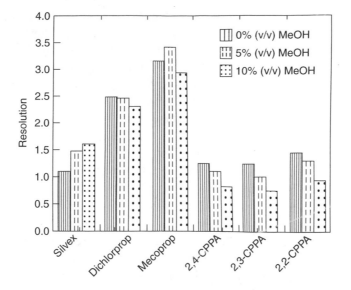

Figure 9.6 The effect of methanol concentration on the chiral resolution of phenoxy acid herbicides, using 200 mM sodium phosphate–sodium acetate buffers (pH 6.5) containing 60 mM OM as BGE. Other conditions as in Figure 9.1 [37].

inner wall (reverse EOF), or when the diastereomeric complexes are all net negatively charged.

Sometimes, partial resolution of chiral pollutants is observed due to sample overloading. Under such circumstances, the maximum resolution can be achieved by reducing the sample volume or decreasing the concentration of the pollutants. Mechref and El Rassi [37] studied the chiral resolution of phenoxy herbicides, both derivatized with 7-aminonaphthalene-1,3-disulfonic acid and underivatized, using cyclodextrin in the BGE. The authors reported better resolution of derivatized herbicides (Table 9.4). Chiral resolution may also be optimized by coating the capillary wall with a suitable coating agent, such as polyacrylamide, alkylhydroxyalkyl cellulose, polyvinyl alcohol or tetra-alkyl ammonium ions. Additionally, varying the length and diameter of the capillary may help to optimize the chiral resolution.

Sometimes, the addition of surfactants (at a concentration above the critical micellar concentration) can optimize the chiral resolution in CE, this improvement occurring due to the formation of a micellar phase. The chiral resolution mechanism changes slightly towards chromatographic behaviour and, therefore, this modality of CE is called micellar electrokinetic chromatography (MEKC). Basically, a surfactant is a molecule possessing two zones of different polarities, which shows special characteristics in solution form. As mentioned previously, surfactants are divided into three categories; ionic (cationic and anionic), nonionic and zwitterionic. The surfactant molecules form a micelle and the partition of the diastereomeric complexes occurs between this micelle and the BGE.

Mechref and El Rassi [40] studied the effect of different concentrations of n-octyl-β-D-maltopyranoside (OM) as the surfactant, using 30–100 mM concentrations (Figure 9.7). The figure indicates that a 60 mM concentration is the best for chiral resolution of the reported herbicides. The authors also reported a decrease in the EOF with the increasing OM concentration: this decrease was more pronounced for a more hydrophobic pollutant such as silvex. According to the authors, OM forms an upper layer of the micelle in which the herbicides are likely to solubilize; hence the ease of diastereomeric complex formation, which results in maximum resolution. Marina *et al.* [41] separated the enantiomers of polychlorinated biphenyls (PCBs) (45, 84, 88, 91, 95, 132, 136, 139, 149, 171, 183 and 196), using mixtures of β- and γ-cyclodextrins, by micellar electrokinetic chromatography. The BGE

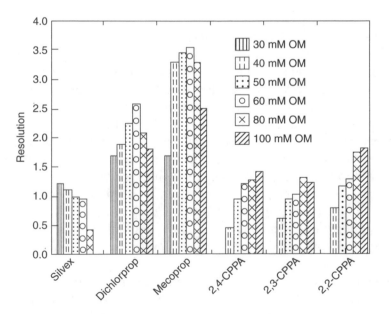

Figure 9.7 The effect of the OM concentration on the chiral resolution of phenoxy acid herbicides, using 175 mM sodium phosphate buffer (pH 6.5) as BGE. Other conditions as in Figure 9.1 [37].

used was 2-(*N*-cyclohexylamino)ethanesulfonic acid (CHES), containing urea and sodium dodecyl sulfate (SDS) micelles. The authors reported the enantiomeric mixture of PCBs 45, 88, 91, 95, 136, 139, 149 and 196 in a single run of approximately 35 min. The same authors [42] resolved the enantiomers of PCBs 84, 95 and 176, using mixtures of bile salts (sodium cholate) in micellar electrokinetic chromatography, using a CHES buffer that contained SDS. However, up to 15 PCBs were resolved by the further addition of γ-cyclodextrin to the above-mentioned CE conditions. The effect of the concentration of γ-cyclodextrin on the chiral resolution of the PCBs was also studied. Similarly, Lin *et al.* [43] investigated the chiral resolution of PCBs using four CD derivatives. 2-Hydroxyl-γ-CD resulted in the best resolution of PCB 84. The authors also investigated the effect of organic modifiers (methanol, acetonitrile and 2-propanol) on chiral resolution; 2-propanol was found to be the best organic modifier.

9.4.10 Optimization by Dependent Variables

As mentioned earlier, the dependent parameters – the field strength ($V\ m^{-1}$), the EOF, Joule heating, the BGE viscosity, the sample diffusion, mobility, charge, size and shape, the interaction of the sample with the capillary and the BGE, the molar absorptivity and so on – also contribute to the chiral resolution of pollutants. It is very interesting to note that these parameters cannot be controlled directly: however, the variation can be carried out through the independent variables. For instance, optimization of the chiral resolution of pollutants may be achieved by changing the temperature of the capillary, which may result in a decrease in the viscosity, an increase in the EOF and so on. On the basis of experience, variation of the independent parameters may be useful in controlling the optimization of the chiral resolution of pollutants via the dependent variables. A method development for the chiral resolution of environmental pollutants is presented in Scheme 9.1. First of all, the CE parameters are selected and set on the CE machine and then the experiment is started: finally, the chiral resolution is optimized by varying the above-mentioned parameters. The chiral resolution is expressed in the form of separation and resolution factors as in chromatography, these being calculated from the electropherograms (peaks).

9.5 Detection

In general, chiral pollutants occur in the environment at low concentrations and, therefore, sensitive detection is essential. The most commonly used

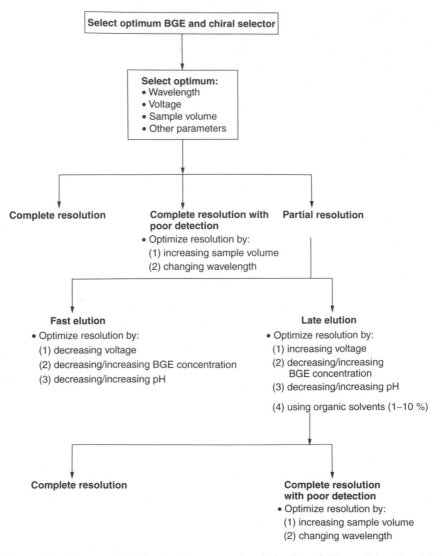

Scheme 9.1 A protocol for the development and optimization of CE conditions for chiral resolution. Note that this is only a brief outline of the procedure that should be followed. Other variations are also possible.

types of detector in chiral CE are the UV, electrochemical and fluorescence detectors and the mass spectrometer. Mostly, the detection of the chiral resolution of drugs and pharmaceuticals in CE has been achieved in UV mode [1, 19, 37, 40] and, therefore, the UV mode may be used for the detection of environmental pollutants. The selection of the UV wavelength depends on the type of buffer, the chiral selector and the specific environmental pollutant. Suitable wavelengths for various buffers are given in Table 9.3. However, the concentration and the sensitivity of UV detection are restricted in so far as the capillary diameter limits the optical path length. It has been observed that some pollutants, especially organochlorine pesticides, are not UV-sensitive and, therefore, for such applications, electrochemical detectors, the mass spectrometer and so on are the best types to use. Some chiral selectors, such as proteins and macrocyclic glycopeptide antibiotics, absorb UV radiation and hence detection becomes poor.

Only few reports are available in the literature that deal with the detection limits for chiral resolution of pollutants by CE, the pollutant concentrations ranging from mg l^{-1} to ng l^{-1}. Tsunoi *et al.* [20] carried out an extensive study on the determination of detection limits for the chiral resolution of herbicides. The authors used a wavelength of 230 nm for detection and the minimum detection limit achieved was 4.7×10^{-3} M for 2,4-dichlorophenoxy acetic acid. On the other hand, Mechref and El Rassi [37] reported better detection limits, for herbicides, in the derivatized mode in comparison to the underivatized mode. For example, the detection limit was enhanced by almost one order of magnitude, from 1×10^{-4} M (10 pmol) to 3×10^{-5} M (0.36 pmol). In the same study, the authors reported 2.5×10^{-6} M and 1×10^{-9} M as the detection limits for herbicides by fluorescence and laser induced fluorescence detectors, respectively.

9.6 Validation of Methods

CE is a fast-growing, efficient and versatile technique for the chiral resolution of drugs, pharmaceuticals and environmental pollutants, but its working conditions still need to be improved to achieve reproducibility. To the best of our knowledge, only a few reports are available in the literature on method validation related to the determination of environmental pollutants by CE [52]. Therefore, at the present time, more emphasis should be placed on method validation. For routine analysis, it is essential to keep the migration time constant in order to allow automatic peak identification by means of commercial data analysis software. Automatic peak identification and quantification are only possible if the relative standard deviation of the migration time is less than 0.5 % [53]. Several reviews have been published that describe improvements in the reproducibility of results [53–59].

Another approach to qualitative and quantitative reproducibilities is the transformation of the total time scale (the x-scale) of the electrophoretic data to the corresponding effective mobility scale (μ-scale) [60, 61]. The conversion leads to a better interpretation of the obtained electropherograms in terms of separation, and enables better direct comparison of the electropherograms and easier peak tracking when trying to identify single components from complex matrices, especially when UV-visible signatures of the components are also available [61]. A quantitative improvement has been achieved on the μ-scale, with significantly better peak area precision, which equates to better precision in quantitative analysis than with the primary time scale integration. However, electrophoretic-based data processing CE software is needed to be able to handle the electrophoretic data directly.

Therefore, it may be assumed that the selectivity of different environmental pollutants by CE is quite good: however, reproducibility is still a problem and many attempts have been made to find a solution. Organic solvents have been added to BGEs to improve reproducibility. It has also been reported that the organic solvents ameliorate the solubility of the hydrophobic complexes, reduce their adsorption on the capillary wall, regulate the distribution of diastereomeric complexes between the aqueous phase and the micellar phase, adjust the viscosity of the separation medium and, accordingly, accomplish the improvement in reproducibility. A constant pH of the electrolyte solution is very important from the selectivity and reproducibility points of view. The pH controls the behaviour of the EOF, the acid/base dissociation equilibria of complexes and the state of existing complexes. Therefore, selectivity and reproducibility can be improved by adjusting the pH of the BGE.

9.7 Mechanisms of Chiral Resolution

It is well known that a chiral environment is essential for enantiomeric resolution in chromatography, and this is also true in the case of chiral CE. In CE, this chiral situation is provided by the chiral compound used in the BGE, which is called a chiral selector or a chiral BGE additive. Basically, the chiral recognition mechanisms in CE are similar to those in chromatography using a chiral mobile phase additive mode, except that in the case of CE the resolution occurs through the different migration velocities of the diastereoisomeric complexes. Chiral resolution takes place due to the formation of diastereomeric complexes between the enantiomers of the pollutants and the chiral selector, and this depends on the type and nature of the chiral selectors used and the pollutants.

In the case of cyclodextrins, inclusion complexes are formed, and the formation of diastereomeric complexes is controlled by a number of interactions, such as $\pi-\pi$ complexation, hydrogen bondings, dipole–dipole interactions, ionic bindings and steric effects. Zerbinati *et al.* [23] used ethylcarbonate-β-CD, hydroxypropyl-β-CD and native α-CD for the chiral resolution of mecoprop and dichlorprop. The authors calculated the performances of these chiral selectors by means of two-level full factorial design, and they calculated the inclusion constants from CE migration time data. Furthermore, they have proposed a possible structure for the inclusion complexes, on the basis of molecular mechanics simulations. Recently, Chankvetadze *et al.* [62] have explained chiral recognition mechanisms in cyclodextrin using UV, NMR and electrospray ionization mass spectrometric methods. Furthermore, the authors determined the structures of the diastereomeric complexes by X-ray crystallographic methods.

Macrocyclic antibiotics are similar to cyclodextrins in some ways. Most of them contain ionizable groups and, consequently, their charge and possibly their three-dimensional conformation, can vary with the pH of the BGE. The complex structures of the antibiotics, which contain different chiral centres, inclusion cavities, aromatic rings, pyranose and furanose sugar moieties, rings and bridges, along with several hydrogen donor and acceptor sites and other groups, are responsible for their surprising chiral selectivities. This allows for an excellent potential to resolve a greater variety of racemates. The possible interactions involved in the formation of diastereomeric complexes are $\pi-\pi$ complexation, hydrogen bonding, inclusion complexation, dipole interactions, steric interactions, and anionic and cationic bindings. Accordingly, diastereomeric complexes that possess different physical and chemical properties become separated on the capillary path (achiral phase). The different migration times of these complexes depend on their sizes, their charges and their interactions with the capillary wall, and as a result these complexes elute at different time intervals.

9.8 CE versus HPLC

As previously mentioned, high performance liquid chromatography (HPLC) is now frequently used as the method of choice for the chiral resolution of a wide range of racemic compounds, including drugs, pharmaceuticals, agrochemicals and environmental pollutants. The wide application of HPLC is due to the development of various chiral stationary phases and its reproducibility. However, HPLC suffers from certain drawbacks, as the chiral selectors are fixed on the stationary phase and hence the concentrations of the chiral selectors cannot be varied. Also, a large amount of costly solvent is consumed to establish the chiral resolution procedure. Furthermore, the poor efficiency of HPLC is due to the laminar flow profile, the mass transfer term and possible additional interactions of enantiomers with the residual silanol groups on the stationary phase.

On the other hand, chiral resolution in CE is achieved using chiral selectors in the BGE: it is very fast, sensitive and it involves the use of inexpensive buffers. Besides, the high efficiency of CE is due to the flat flow profile originated and the homogeneous partition of the chiral selector in the electrolyte, which in turn, minimizes the mass transfer. In general, the theoretical plate number in CE is higher than in HPLC and, therefore, good resolution is achieved in CE. Also, more than one chiral selector can be used simultaneously to obtain the best resolution. However, reproducibility is a major problem in CE and, therefore, the technique is not popular for routine chiral resolution. The other drawbacks of CE include the wastage of the chiral selector used in the BGE. Furthermore, chiroptical detection techniques such as polarimetry and circular dichroism cannot be used as the chiral pollutants elute in the form of diastereoisomers which are achiral in nature. Moreover, some of the well known chiral selectors may not be soluble in the BGE and, thus, the use of a stationary bed of a chiral selector may allow the advantages of chiral stationary beds, which are inherent in HPLC, to be applied to an electrically driven technique such as CE. This will allow CE to be coupled with mass spectrometry, polarimetry, circular dichroism and UV detection without any problems. Briefly, at present CE is not a popular technique as compared to HPLC for chiral resolution due to reproducibility problems but, of course, it has a bright future.

9.9 Conclusions

The chiral resolution of environmental pollutants at trace levels is a very important and challenging issue nowadays. Gas chromatography and high performance liquid chromatography have been used for the chiral resolution of environmental pollutants and, in recent years, capillary electrophoresis has also been used for this purpose. A search of the literature cited in this chapter indicates a few reports on chiral resolution of environmental pollutants by CE, but the technique has not achieved a respectable status in the routine chiral analysis of these pollutants due to its poor reproducibility and detection limits. Many workers have suggested different modifications to make CE the method of choice: to obtain good reproducibility, selection of the capillary wall chemistry, the pH and ionic strength of the BGE, the chiral selectors and the detectors, and optimization of the BGE, have been investigated [63–69].

Apart from the points discussed above, some other aspects should also be addressed so that CE can be used as a routine method in this field. These include the development of new chiral selectors and detector devices. The nonreproducibility of the methods may be due to the heating of the BGE after a long run of the CE machine. Therefore, to keep the temperature constant throughout experiments, a cooling device should be included. There are only a few reports that deal with method validation. To make the developed method more applicable, the validation of the methodology should be determined. Chiral capillaries should be developed and the CE machine should be coupled with a mass spectrometer and with polarimetric and circular dichroism detectors, which may result in good reproducibility and low detection limits. Not all of the capabilities and possibilities of CE as a chiral resolution technique have been explored as yet, and a lot of work remains to be done to advance the use of CE for the chiral resolution of environmental pollutants. CE will definitely prove itself as the best technique within the coming few years, and it will achieve a reputable status as a technique for routine analysis in most environmental laboratories.

References

1. A. L. Crego and M. L. Marina, *J. Liq. Chromatogr. Rel. Technol.* **20**, 1337 (1997).
2. I. Ali and H. Y. Aboul-Enein, in J. Cazes, Ed., *Encyclopedia of Chromatography*, Dekker, New York, 2004, in press.
3. I. Ali, V. K. Gupta and H. Y. Aboul-Enein, *Electrophoresis* **24**, 1360 (2003).
4. M. Jung and S. Mayer, V. Schurig, *LC–GC* **7**, 340 (1994).
5. B. Chankvetadze, *Capillary Electrophoresis in Chiral Analysis*, Wiley, New York, 1997.
6. G. Blaschke and B. Chankvetadze, *J. Chromatogr. A* **875**, 3 (2000).
7. S. Zaugg and W. Thormann, *J. Chromatogr. A* **875**, 27 (2000).
8. H. Wan and L. G. Blomberg, *J. Chromatogr. A* **875**, 43 (2000).
9. S. Fanali, *J. Chromatogr. A* **875**, 89 (2000).
10. M. Fillet and Ph. Hubert, J. Crommen, *J. Chromatogr. A* **875**, 123 (2000).
11. B. Koppenhoefer, X. Zhu, A. Jakob, S. Wuerthner and B. Lin, *J. Chromatogr. A* **875**, 135 (2000).
12. J. Haginaka, *J. Chromatogr. A* **875**, 235 (2000).

13. F. Wang and M. G. Khaledi, *J. Chromatogr. A* **875**, 277 (2000).
14. V. Piette, M. Fillet, W. Lindner and J. Crommen, *J. Chromatogr. A* **875**, 353 (2000).
15. Y. Tanaka, K. Otsuka and S. Terabe, *J. Chromatogr. A* **875**, 323 (2000).
16. G. Gübitz and M. G. Schmid, *J. Chromatogr. A* **792**, 179 (1997).
17. G. Weseloh, C. Wolf and W. A. König, *Chirality* **8**, 441 (1996).
18. S. Sarac, B. Chankvetadze and G. Blaschke, *J. Chromatogr. A* **875**, 379 (2000).
19. M. Miura, Y. Terashita, K. Funazo and M. Tanaka, *J. Chromatogr. A* **846**, 359 (1999).
20. S. Tsunoi, H. Harino, M. Miura, M. Eguchi and M. Tanaka, *Anal. Sci.* **16**, 991 (2000).
21. A. Gomez-Gomar, E. Ortega, C. Calvet, R. Merce and J. Frigola, *J. Chromatogr. A* **950**, 257 (2002).
22. K. Otsuka, J. S. Smith, J. Grainger, J. R. Barr, D. G. Patterson Jr., N. Tanaka and S. Terabe, *J. Chromatogr. A* **817**, 75 (1998).
23. O. Zerbinati, F. Trotta and C. Giovannoli, *J. Chromatogr. A* **875**, 423 (2000).
24. C. Desiderio, C. M. Polcaro, P. Padiglioni and S. Fanali, *J. Chromatogr. A* **781**, 503 (1997).
25. K. V. Penmetsa, R. B. Leidy and D. Shea, *J. Chromatogr. A* **790**, 225 (1997).
26. M. W. F. Nielen, *J. Chromatogr. A* **637**, 81 (1993).
27. M. W. F. Nielen, *Trends Anal. Chem.* **12**, 345 (1993).
28. A. W. Garrison, P. Schmitt and A. Kettrup, *J. Chromatogr. A* **688**, 317 (1994).
29. C. Copper, J. B. Davis and M. J. Sepaniak, *Chirality* **7**, 401 (1995).
30. H. Nishi, *J. High Res. Chromatogr.* **18**, 695 (1995).
31. K. Kano, K. Minami, K. Horiguchi, T. Ishimura and M. Kodera, *J. Chromatogr. A* **694**, 307 (1995).
32. C. Desiderio, C. Palcaro and S. Fanali, *Electrophoresis* **18**, 227 (1997).
33. M. P. Gasper, A. Berthod, U. B. Nair and D. W. Armstrong, *Anal. Chem.* **68**, 2501 (1996).
34. D. W. Armstrong, M. P. Gasper and K. L. Rundlet, *J. Chromatogr. A* **689**, 285 (1995).
35. K. L. Rundlet, M. P. Gasper, E. Y. Zhou and D. W. Armstrong, *Chirality* **8**, 88 (1996).
36. Y. Mechref and Z. El Rassi, *J. Chromatogr. A* **757**, 263 (1997).
37. Y. Mechref and Z. El Rassi, *Anal. Chem.* **68**, 1771 (1996).
38. M. L. Marina, I. Benito, J. C. Diez-Maser and M. J. Gonzalez, *Chromatographia* **42**, 269 (1996).
39. Y. Mechref and Z. El Rassi, *J. Chromatogr. A* **724**, 285 (1996).
40. Y. Mechref and Z. El Rassi, *Chirality* **8**, 518 (1996).
41. M. L. Marina, I. Benito, J. C. Diez-Masa and M. J. Gonzalez, *J. Chromatogr. A* **752**, 265 (1996).
42. A. L. Crego, M. A. Garcia and M. L. Marina, *J. Microcol. Sepn* **12**, 33 (2000).
43. W. C. Lin, C. C. Chang and C. H. Kuei, *J. Microcol. Sepn* **11**, 231 (1999).

44. J. G. Clothier and S. A. Tomellini, *J. Chromatogr. A* **723**, 179 (1996).
45. R. O. Cole, M. J. Sepaniak and W. L. Hinze, *J. High Res. Chromatogr.* **13**, 579 (1990).
46. R. Furut and T. Doi, *J. Chromatogr. A* **676**, 431 (1994).
47. H. Nishi, T. Fukuyarna and S. Terabe, *J. Chromatogr.* **553**, 503 (1991).
48. J. Wang and I. M. Warner, *J. Chromatogr. A* **711**, 297 (1995).
49. A. V. Wolf, G. B. Morden and P. G. Phoebe, in R. C. Weast, M. J. Astle, W. H. Beyer, eds, *CRC Handbook of Chemistry and Physics*, 68th edn, CRC Press, Boca Raton, FL, 1987, p. D219.
50. R. P. Oda and J. P. Landers, in J. P. Landers, ed., *Handbook of Capillary Electrophoresis*, CRC Press, London, 1994.
51. R. J. Nelson, A. Paulus, A. S. Cohen, A. Guttmann and B. L. Karger, *J. Chromatogr.* **480**, 111 (1989).
52. E. Dabek-Zlotorzynska, R. Aranda-Rodriguez and K. Keppel-Jones, *Electrophoresis* **22**, 4262 (2001).
53. G. Raber and H. Greschonig, *J. Chromatogr. A* **890**, 355 (2000).
54. A. Faller and H. Engelhardt, *J. Chromatogr. A* **853**, 83 (1999).
55. E. Dabek-Zlotorzynska, M. Piechowski, M. McGrath and E. P. C. Lai, *J. Chromatogr. A* **910**, 331 (2001).
56. M. Macka, C. Johns, P. Doble and P. R. Haddad, *LC–GC* **19**, 38 (2001).
57. M. Macka, C. Johns, P. Doble and P. R. Haddad, *LC–GC* **19**, 178 (2001).
58. P. Doble and P. R. Haddad, *J. Chromatogr. A* **834**, 189 (1999).
59. E. Dabek-Zlotorzynska and K. Keppel-Jones, *LC–GC* **18**, 950 (2000).
60. N. Ikuta, Y. Yamada, T. Yoshiyama and T. Hirokawa, *J. Chromatogr. A* **894**, 11 (2000).
61. P. Schmitt-Kopplin, A. V. Garmash, A. V. Kudryavtsev, P. Menzinger, I. V. Perminova, N. Hertkorn, D. Freitag, V. S. Petrosyan and A. Kettrup, *Electrophoresis* **22**, 77 (2001).
62. B. Chankvetadze, N. Burjanadze, G. Pintore, D. Bergenthal, K. Bergander, C. Mühlenbrock, J. Breitkreuz and G. Blaschke, *J. Chromatogr. A* **875**, 471 (2000).
63. V. Pacakova, P. Coufal and K. Stulik, *J. Chromatogr. A* **834**, 257 (1999).
64. B. F. Liu, B. L. Liu and J. K. Cheng, *J. Chromatogr. A* **834**, 277 (1999).
65. S. M. Valsecchi and S. Polesello, *J. Chromatogr. A* **834**, 363 (1999).
66. A. R. Timerbaev and W. Buchberger, *J. Chromatogr. A* **834**, 117 (1999).
67. M. Macka and P. R. Haddad, *Electrophoresis* **18**, 2482 (1997).
68. J. Horvath and V. Dolnike, *Electrophoresis* **22**, 644 (2001).
69. B. X. Mayer, *J. Chromatogr. A* **907**, 21 (2001).

Chapter 10

Perspectives on the Analysis of Chiral Pollutants

10.1 Introduction

Most aspects of the analysis of chiral pollutants have been discussed in Chapters 1–9. However, recent reports have raised new concerns that chiral pollutants may include some chemicals that stereoselectively disrupt the normal development, growth and other life activities of various organisms, including human beings [1]. In recent years, several chiral pharmaceuticals and drugs have been reported in various components of the environment; these may be treated as pollutants [2–4] if they enter the human body through the food chain. Hegeman and Laane [5] presented a hypothetical model of enantiomer fractions for chiral compounds in different components of the ecosystem (Figure 10.1), which indicates a wide distribution of chiral pollutants. But it is unfortunate that environmental studies have historically neglected to determine the adverse effects associated with particular enantiomers, and to determine which enantiomers may persist in the environment. Consequently, much of the existing environmental data on chiral pollutants may only represent the presence of relatively innocuous enantiomers [6]. However, enantioselective biological effects associated with chiral chemicals are well recognized in the pharmaceuticals industry, where 50 of the top 100 most widely used drugs are marketed as single enantiomers in order to avoid adverse side effects [6, 7]. Moreover,

Chiral Pollutants. I. Ali and H. Y. Aboul-Enein
© 2004 John Wiley & Sons, Ltd ISBN: 0-470-86780-9

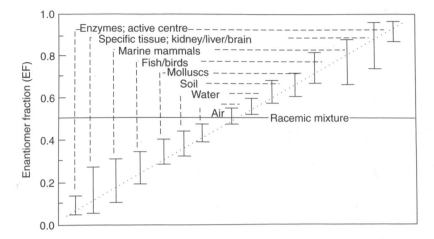

Figure 10.1 The hypothetical distribution of enantiomer fractions (EFs) for a chiral compound in different compartments of the ecosystem. The deviations from the racemic mixture are above and below the line EF = 0.5 [5].

regulations have been formulated for the development and testing of chiral pharmaceuticals and drugs before they are put on the market. A literature search reveals a few reports that deal with the legal implications of chirality [8–11]. In 1993 Witte *et al.* [12], and in 1994 Rauws and Groen [13], reviewed the status of the regulatory aspects of chiral medicinal production, considering the pharmaceuticals industries in the USA, Japan and some European countries. In these countries, distinct authorities are responsible for the control and approval of newly developed drugs (both chiral and nonchiral).

10.2 Regulations on Chirality in the USA

In the United States, new drug applications must be submitted to the Food and Drug Administration (FDA), together with the appropriate chemical, manufacturing and control data, such as methods and specifications, the results of stability tests, proper labelling, details of pharmacological activity, the pharmacokinetic profile, toxicology studies and impurity limits [14]. In 1987, for the first time, the FDA published a set of guidelines on the submission of new chiral drug applications. Each new drug submission should show the molecular structure and the chiral centres of the drug. The FDA also emphasizes the need for toxicological studies for each enantiomer before the drug is launched on to the market. An investigational exemption (IND) for a new drug is made possible as long as the appropriate chemical, manufacturing and control data are included for the individual enantiomers. The possibilities of bridging data can be considered on a case-by-case basis, by consultation between the manufacturers and the authorities. Clinical trials should be carried out under a new IND for each enantiomer. In 1992, the FDA issued another set of guidelines stressing the use of a single enantiomer [11]. However, if a racemate is to be marketed instead of a single enantiomer, justification for this course of action should be provided in detail. According to these guidelines, complete information – with details of all synthetic, analytical and enantiomer activity – should be included. The FDA also urges applicants to provide information on pharmacokinetics, on the occurrence of metabolic chiral inversion *in vivo* (if possible) and on the differences in toxicology between the enantiomers [14]. Additionally, the evaluation should include data regarding the conversion of the studied eutomer into the distomer. DeCamp [15] summarized the FDA statement as having significant implications for chemists who are working on the development and validation of analytical controls for chiral drug substances and products. The testing of the bulk drug, the manufacturing of the finished product, the design of stability testing protocols and the labelling of the drug must all take the chirality of the active ingredient into consideration. It has recently been reported that the FDA might approve a single enantiomer of a racemate as a new drug, which would offer new marketing opportunities for racemic drugs that are about to go off patent [16]. According to Stinson [17], the production and marketing of single enantiomers was likely to grow by more than 30 % by the end of 2000. These products are chiral drugs that have already been approved in the form of racemates, but that are being redeveloped as single enantiomers.

10.3 Regulations on Chirality in Europe

Guidelines on the quality, safety and efficacy of medicinal products for human use have also been formulated in Europe [18]. Witte *et al.* [12] reviewed the guidelines on chirality in the European Community. In general, the European guidelines are similar to those of the FDA, as both are based on the same body of scientific knowledge [13]. These guidelines have been published by the Committee for Proprietary Medicinal Products (CPMP), in two volumes (II and III) of the rules governing medicinal products in the European Community. Volume II, which describes the marketing authorizations for medicinal products in the European Community countries, was published in 1989. This volume also contains the export report on chiral drugs, which summarizes the specific requirements, the intention being that they should consist of a critical evaluation of the quality of the product and details of the investigations carried out in a variety of organisms, including human beings, such that all of the relevant data is made available for evaluation. The stereochemical configuration should be clear prior to marketing the drug. If a racemic mixture of a chiral drug is already on the market, the toxicological properties of the individual enantiomers should be provided. Bridging studies that utilize data for racemic drugs that are already on the market are not allowed. Volume III, which was published by the CPMP in 1990, describes the analytical validation of chiral drugs. This report stresses the development and use of analytical methods for chiral drugs. In 1993, the CPMP issued new guidelines on the chirality of drugs. Landoni *et al.* [11] reported that many of the FDA's common regulations are incorporated in this report, the most important points of which concern the toxicological effects of the enantiomers and complete information on analytical methods and on the activity of the enantiomers.

10.4 Regulations on Chirality in Japan

The registration and approval situation in Japan is different from that of the USA and the European countries. In Japan, no official statements have been issued on the regulation of chiral drugs [12, 19]. However, the Japanese registration authorities recommend adherence to the guidelines adopted in other countries. The Japanese authorities also recommend investigations into the absorption, distribution, metabolism and excretion of each of the enantiomers of a particular drug.

10.5 Regulations on Chirality in Other Countries

Due to the direct effects of chirality on human health, an awareness of chirality is growing worldwide. Landoni *et al.* [11] reviewed the regulations on chirality in various other countries. The authors reported that Australia and Canada are also formulating regulations on chirality in drugs, and that some Nordic countries are also moving in this direction. The authors also predicted that harmonization will take place on this issue in the world in the year to come. Rauws and Groen [13] discussed the current emphasis on the two most important cases: (i) the number of extra requirements that are justified for new racemates; and (ii) how few extra requirements are acceptable in the development of a single enantiomer from an approved racemate. At the moment, the opportunities for early harmonization look favourable and the formulation of an international guidance document seems feasible.

10.6 The Capabilities of Chiral Analysis Technologies

As discussed in Chapter 1, many different technologies have been used to analyse chiral pollutants and drugs, the most important being chromatography, capillary electrophoresis and various sensing techniques. While advances in these methodologies have been reported from time to time, the analysis of chiral pollutants is still a challenging job because of the sample preparation that is required for this type of analysis in biological and other environmental samples, where many impurities are present. Moreover, the detection of some pollutants using conventional detectors is very poor and, therefore, special techniques, such as mass spectrometry and nuclear magnetic resonance, are required for this purpose. Moreover, the coupling of such sophisticated types of detection techniques is very difficult, and reproducibility sometimes becomes poor. Furthermore, a universal chiral selector – which is the main requirement in chiral analysis – is still lacking. To summarize, there is a great need to develop more advanced methods for the analysis of chiral pollutants. The developed technologies should be fast, inexpensive and highly reproducible in nature, would undoubtedly be a boon in the scientific world.

10.7 The Large-scale Production of Pure Enantiomers

It is well known that, in the synthesis of chiral drugs, agrochemicals and other compounds, a racemic mixture is obtained from prochiral or achiral precursors. Therefore, the production of optically active pure enantiomers depends on separation methods. Economic interests are very important in the development of optically pure compounds, and so these separation methods (for racemic compounds) are more economical than asymmetric synthesis methods; hence most pharmaceutical companies prepare single enantiomers by using separation techniques. At the preparative scale, many separation technologies, including preparative HPLC, column chromatography and simulated moving bed chromatography, have been tried. The latter technology has been developed recently and has been shown to be very successful for chiral separation at the preparative scale: sizeable amounts of enantiomers can be separated using this technique. However, the large-scale separation of enantiomers is a costly and time-consuming affair. Separation methods are affordable in the drug industry, where small amounts are required to combat a particular disease. But the scenario is different in the case of agrochemicals, where large quantities are required for agricultural, forestry and various domestic purposes. Therefore, the present methodologies for the large-scale production of pure enantiomers need to be improved.

10.8 The Impact of Chirality on Economic Growth

The economic growth of any country is directly related to health, a disease-free atmosphere and an eco-friendly environment. Therefore, to avoid any outbreaks of disease and other tragedies due to the adverse effects of chirality, it is essential to monitor chiral environmental pollutants. As discussed above, chiral agrochemicals are carcinogenic in nature and hence may create serious problems. Accordingly, it can be seen that chirality indirectly affects the world's economic growth.

Many pharmaceutical companies have started to market optically active pure enantiomers of some drugs [20–24], particularly anti-inflammatory, analgesic, anti-viral, anti-cancer, cardiovascular and other pharmaceutical products, which are used in the treatment of central nervous system disorders and dermatological, gastrointestinal, ophthalmic and respiratory diseases [25]. The racemate versus enantiomer debate has resulted in a new commercial strategy, the so-called racemic switch, which denotes the development of a single enantiomer from a drug that was first approved as a racemate. From a search of the literature, it has been found that in 1997 the worldwide consumption of drugs cost about $US 300 billion [25–27]. It has also been estimated that about 28 % of this consumption involved the marketing of drugs with a single enantiomer [25]. It is very interesting to note that the 21 % market share of enantiomeric pure drugs in 1996 increased to 28 % in 1997 [25], and that it continues to increase, due to population growth and the development of new biological assay of racemic drugs. To summarize, all of the pharmaceutical companies are competing strongly on cost and in the marketing of new drugs (optically active pure forms), and they are also demanding economical methods for the preparation of pure enantiomeric drugs.

At present, the economic impact of the industrial production of chiral drugs is huge and more than 50% of the 500 top-selling drugs are single enantiomers. Sales have continued to increase tremendously, and the worldwide annual sales of enantiomeric drugs exceeded $US 100 billion for the first time in the year 2000, with chiral drugs representing close to one-third of all sales worldwide. While some chiral switches may be of less apparent benefit – or, indeed, detrimental in some cases – encouragement by the regulatory agencies, and the ability to extend the life cycle of a drug that is coming off patent, continue to promote the trend. However, it may turn out that the ability to provide chiral templates, and thereby attack the key targets of selectivity and specificity, will lead to the greatest benefits. Research into new chemical entities that can interact specifically with enzyme families may potentially lead to new therapies for complex disease processes. The approach is designed to create a product that is made to measure, rather than off the peg [28, 29].

10.9 Conclusions

Some advanced and developed countries are aware of the adverse effects of chirality and have formulated certain guidelines for the marketing of chiral drugs. Unfortunately, however, the world's developing and underdeveloped countries still lag far behind on this issue, and are not considering the seriousness of toxicology due to chirality. Moreover, no country has considered the adverse and toxic effects associated with the different enantiomers of agrochemicals and other pollutants. In the USA, almost all of the environmental regulations that are based on toxicological studies treat racemates as single molecules, with the same properties [5, 30, 31]. Similarly, in the European Union, the possible different effects of the enantiomers of chiral pollutants are not taken into consideration.

Nowadays, many types of diseases are being diagnosed in human beings, which may be correlated with the pollution of our environment, including the action of chiral pollutants. Basically, the lethality and toxicity due to chirality are long-term effects and hence people are not considering these on top priority. However, toxicity due to chirality works as a slow poison and results in various health problems. The present reduced life expectancy in some underdeveloped countries may be correlated with the notorious effects of these pollutants. Therefore, it is important to carry out studies on the biodegradation and toxicology of chiral pollutants. Correspondingly, research into enantioselective biodegradation and toxicology is needed in order to investigate the distribution effects and ultimate fate of chiral agrochemicals.

In view of these points, it is essential to consider chirality in drugs and agrochemicals all around the world. A sound and uniform policy should be developed regarding the regulation of chirality at the international level. Since there are no regulations on chiral agrochemicals, there is a very urgent need to develop guidelines before exposing our environment to these notorious chiral compounds. Environmental authorities worldwide should come forward on this issue, both to initiate guidelines and to formulate a harmonized policy on the use of chiral drugs and agrochemicals. To summarize, immediate action is required to intensify studies and increase the level of awareness of the toxicity and carcinogenicity that is due to chirality.

References

1. P. F. Hoekstra, B. K. Burnison, T. Neheli and D. C. G. Muir, *Toxicol. Lett.* **125**, 75 (2001).
2. M. L. Richardson and J. M. Bowron, *J. Pharm. Pharmacol.* **37**, 1 (1985).
3. K. Kümmerer, A. Al-Ahmadi, B. Bertram and M. Wießler, *Chemosphere* **40**, 767 (2000).
4. M. Winkler, J. R. Lawrence and T. R. Neu, *Water Res.* **35**, 3197 (2001).
5. W. J. M. Hegeman and W. P. M. Laane, *Rev. Environ. Contam. Toxicol.* **173**, 85 (2002).
6. D. L. Lewis, A. W. Garrison, K. E. Wommack, A. Whittemore, P. Steudler and J. Melillo, *Nature* **401**, 898 (1999).
7. A. Williams, *Pestic. Sci.* **46**, 3 (1996).
8. J. R. Brown, *Drug Chirality: Impact on Pharmaceutical Regulation*, Legal Studies and Services Limited, Healthcare and Regulatory Affairs Division, London Press Centre, London, 1990.
9. A. J. Hutt, *Chirality* **3**, 161 (1991).
10. E. J. Ariëns, *Trends Pharmacol. Sci.* **14**, 68 (1993).
11. M. F. Landoni, A. I. Soraci, P. Delatour and P. Lees, *J. Vet. Pharmacol. Therap.* **20**, 1 (1997).
12. D. T. Witte, E. Kees, J. P. Franke and R. A. Zeeuw, *Pharmacy World Sci.* **15**, 10 (1993).
13. A. G. Rauws and K. Groen, *Chirality* **6**, 72 (1994).
14. FDA's policy statement for the development of new stereoisomeric drugs, *Chirality* **4**, 338 (1992).
15. W. H. DeCamp, *J. Pharm. Biomed. Anal.* **11**, 1167 (1993).
16. S. C. Stinson, *Chem. Eng. News* **75**, 28 (1997).

17. S. C. Stinson, *Chem. Eng. News* **73**, 44 (1995).
18. *The Rules Governing Medicinal Products in the European Union, Guidelines on the Quality, Safety and Efficacy of Medicinal Products for Human Use,* Clinical Investigation of Chiral Active Substances, 1991.
19. H. Shindo and J. Caldwell, *Chirality* **3**, 91 (1991).
20. A. M. Krstulovic, ed., *Chiral Separations by HPLC: Applications to Pharmaceutical Compounds,* Ellis Horwood, New York, 1989.
21. S. Allenmark, *Chromatographic Enantioseparation: Methods and Applications,* 2nd edn, Ellis Horwood, New York, 1991.
22. G. Subramanian, ed., *A Practical Approach to Chiral Separations by Liquid Chromatography,* VCH, Weinheim, 1994.
23. H. Y. Aboul-Enein and I. W. Wainer, eds, *The Impact of Stereochemistry on Drug Development and Use,* Chemical Analysis, Vol. 142, Wiley, New York, 1997.
24. P. N. Patil and D. D. Miller, in W. J. Lough and I. W. Wainer, eds, *Chirality in Natural and Applied Science,* CRC Press, Boca Raton, FL, 2003, pp. 139–178.
25. S. C. Stinson, *Chem. Eng. News* **76**, 83 (1998).
26. S. C. Stinson, *Chem. Eng. News* **77**, 101 (1999).
27. A. M. Thayer, *Chem. Eng. News* **76**, 25 (1998).
28. D. Burke and D. J. Henderson, *Br. J. Anaesth.* **88**, 563 (2002).
29. I. K. Reddy, T. R. Kommuru, A. A. Zaghloul and M. A. Khan, *Crit. Rev. Ther. Drug Carrier Syst.* **17**, 285 (2000).
30. E. H. P. Kohler, W. Angst, W. Giger, C. Kanz, S. Müller and M. J. F. Suter, *Chimia* **51**, 947 (1997).
31. J. M. Schneiderheinze, D. W. Armstrong and A. Berthod, *Chirality* **11**, 330 (1999).

Index

Chiral Pollutants.   I. Ali and H. Y. Aboul-Enein
© 2004 John Wiley & Sons, Ltd   ISBN: 0-470-86780-9

With kind thanks to Ann Lloyd-Griffiths for compilation of this index.